Fencing for Conservation

Michael J. Somers • Matthew W. Hayward
Editors

Fencing for Conservation

Restriction of Evolutionary Potential or a Riposte to Threatening Processes?

Springer

Editors
Michael J. Somers
Centre for Wildlife Management
Centre for Invasion Biology
Mammal Research Institute
University of Pretoria
Pretoria, South Africa
mjs@up.ac.za

Matthew W. Hayward
Scotia Sanctuary
Australian Wildlife Conservancy
Wentworth, Australia
and
Centre for African Conservation Ecology
Nelson Mandela Metropolitan University
Port Elizabeth, South Africa
and
Mammal Research Institute
Polish Academy of Science
Białowieża, Poland
and
School of Biological
Earth and Environmental Science
University of New South Wales
Sydney, Australia
matt@australianwildlife.org

ISBN 978-1-4614-0901-4 e-ISBN 978-1-4614-0902-1
DOI 10.1007/978-1-4614-0902-1
Springer New York Dordrecht Heidelberg London

Library of Congress Control Number: 2011938462

Printed on acid-free paper

Springer is part of Springer Science+Business Media (www.springer.com)

This book is dedicated to the late Simon Thirgood (right hand side in photo) who inspired us with his good humour, enthusiasm and great contribution to conservation, we all miss him.

Foreword

This book draws together practical experience regarding fences and their ecological effects from around the world.

Fences are such a familiar part of life on our crowded planet that we easily forget that they are socially and evolutionarily a recent phenomenon. The "enclosure" movement in Britain, just a few one hundred years ago, was part of the huge social transformation that led to the industrial era, with all its benefits and problems. Before that, land was mostly held communally, and grazing animals moved freely. The trend to private ownership of land and the associated containment of the animals it supported came later to the colonies. The invention of barbed wire is one of the technologies that underpinned the conquest of the American prairies, the Australian outback and the South African veld.

Changes of that magnitude, both socially and ecologically, have consequences – some beneficial, others less so. Fences are designed to be barriers to animal movement. As such, they reduce the functional scale of an important element of grazing ecosystems. Seasonal migrations become impossible, gene-flow is restricted and dispersal is constrained. Sometimes it is possible to artificially substitute these processes or compensate for their absence. In other cases, there is a realisation that managing at larger scales is both easier and less expensive, and fences have come down.

Within the African continent, South Africa represents one pole of an ongoing debate about how and when wildlife movement should be constrained. East Africa represents the other pole. In South Africa, wildlife and other land uses are strictly spatially separated by fences. The fencing technology to do so has evolved greatly from the four-strand barbed wire of our cattle-ranching forebears. The cable fence for containing elephants, the high multistrand smooth-wire fence for antelopes and the electric fence for carnivores are examples. In East Africa, the preference has always been for an intermingling of wildlife with cattle ranching and crop agriculture. The benefit is protected areas without hard boundaries. A negative consequence is human–wildlife conflict. The two poles are less sharply defined than they once were – the benefits of fewer, more strategic fences and more selectively porous boundaries are better understood in South Africa, and a rapidly filling landscape is forcing East Africa into a stricter separation of incompatible land uses.

Our relationship with fences therefore contains some paradoxes. We recognise the wry wisdom in the saying "Good fences make good neighbours". At the same time, we have to agree with Robert Frost: "Something there is that doesn't love a wall".

Pretoria, South Africa Bob Scholes

Acknowledgements

This book arose from a symposium on "Fencing for Conservation" at the 2008 Society for Conservation Biology meeting in Chattanooga, Tennessee, U.S.A. and would not have been possible without the support and enthusiasm from the presenters (many of whom are chapter authors here).

We sincerely thank the following for excellent and timely reviews: Marco Apollonio, Doug Armstrong, Dave Balfour, Andrew Balmford, J du P. Bothma, Mark Boyce, Tony Cathcart, Markus Gusset, David Cumming, Ken Ferguson, Sam Ferreira, Robert Guldemond, Alison Hester, David Hetherington, Heribert Hofer, Grant Hopcraft, Chris Johnson, Ferran Jori, Marcella Kelly, Kyran Kunkel, Peter Lindsay, Matt Linkie, Hamish McCallum, Gus Mills, Katherine Moseby, Norman Owen-Smith, Daniel Parker, John Read, Phil Seddon, Jeff Short and Murray Williams. Michele Walters, Kristoffer Everatt and Leah Andresen are thanked for helping proof read some of the chapters. Without the efforts of these people, this book would not be as thorough or accurate as we hope it is.

Acknowledgements

Contents

Contributors

Laura Adam Mammal Research Institute, Department of Zoology and Entomology, University of Pretoria, Pretoria, South Africa

Andrew L. Beck Animal, Plant and Environmental Sciences, University of the Witwatersrand, Johannesburg, South Africa

Mark S. Boyce Department of Biological Sciences, University of Alberta, Edmonton, AB, Canada

Joseph Bull Wildlife Conservation Research Unit, Department of Zoology, University of Oxford, Oxford, UK

Bruce Burns School of Biological Sciences, University of Auckland, Auckland, New Zealand

Susan Canney Spatial Ecology and Epidemiology group, Department of Zoology, University of Oxford, Oxford, UK

Sandra M.C. Cavalcanti Instituto Pró-Carnívoros, Av. Horácio Neto 1030, Atibaia, SP, 12945-010, Brazil

Richard M. Cowling Department of Botany, Nelson Mandela Metropolitan University, Port Elizabeth, South Africa

Peter G. Crawshaw CENAP/ICMBio, Av. dos Bandeirantes, s/nº, Balneário MunicipalAtibaia, Brazil

Fredrik Dalerum Centre for Wildlife Management, Mammal Research Institute, Department of Zoology and Entomology, University of Pretoria, Pretoria, South Africa

Tim Day Xcluder Pest Proof Fencing, Rotorua, New Zealand

Chris R. Dickman Institute of Wildlife Research, School of Biological Sciences, University of Sydney, Sydney, Australia

Marion L. East Leibniz Institute for Zoo and Wildlife Research, Berlin, Germany

Ken Ferguson Institute of Biodiversity, Animal Health and Comparative Medicine, University of Glasgow, Glasgow, Scotland

Michelle E. Gadd Division of International Conservation, U.S. Fish and Wildlife Service, Arlington, VA 22203, USA

C. Cormack Gates Faculaty of Environmental Design, University of Calgary, Calgary, AB, Canada

Katja V. Goller Leibniz Institute for Zoo and Wildlife Research, Berlin, Germany

Markus Gusset Wildlife Conservation Research Unit, Department of Zoology, University of Oxford, Oxford, UK

Matthew W. Hayward Scotia Sanctuary, Australian Wildlife Conservancy, Wentworth, Australia

Centre for African Conservation Ecology, Nelson Mandela Metropolitan University, Port Elizabeth, South Africa

Mammal Research Institute, Polish Academy of Science, Białowieża, Poland

School of Biological, Earth and Environmental Science, University of New South Wales, Sydney, Australia

Heribert Hofer Leibniz Institute for Zoo and Wildlife Research, Berlin, Germany

John Innes Landcare Research, Hamilton, New Zealand

Andrew Jakes Faculaty of Environmental Design, University of Calgary, Calgary, AB, Canada

W. Jędrzejewski Mammal Research Institute, Polish Academy of Science, Białowieża, Poland

Paul Jones Alberta Conservation Association, Lethbridge, AB, Canada

Ferran Jori UPR AGIRs, CIRAD, Montpellier, France,
Research Fellow at the Mammal Research Institute, Department of Zoology and Entomology, University of Pretoria, Pretoria, South Africa

Andrew T. Knight Department of Botany, Nelson Mandela Metropolitan University, Port Elizabeth, South Africa

Department of Conservation Ecology and Entomology, Stellenbosch University, Matieland, South Africa

R. Kowalczyk Mammal Research Institute, Polish Academy of Science, Białowieża, Poland

Kyran Kunkel World Wildlife Fund, Northern Great Plains Office, Bozeman, MT, USA

Peter A. Lindsey Mammal Research Institute, University of Pretoria, Pretoria, South Africa

David W. Macdonald Wildlife Conservation Research Unit, Department of Zoology, University of Oxford, Oxford, UK

Nicky Marlow Department of Environment and Conservation and Invasive Animals Cooperative Research Centre, Western Australian Wildlife Research Centre, Wanneroo, Australia

Chap L. Masterson Zvakanaka Wild Vets, Hluhluwe, South Africa

Julius W. Nyahongo School of Natural Sciences and Mathematics, University of Dodoma, Dodoma, Tanzania

Stephanie Romañach African Wildlife Conservation Fund, Doral, FL, USA

Christopher Sandom Wildlife Conservation Research Unit, Department of Zoology, University of Oxford, Oxford, UK

K. Schmidt Mammal Research Institute, Polish Academy of Science, Białowieża, Poland

Bob Scholes CSIR Natural Resources and Environment, Pretoria 0001, South Africa

Rob Slotow Amarula Elephant Research Programme, School of Biological and Conservation Sciences, Westville Campus, University of KwaZulu-Natal, Durban, South Africa

Michael J. Somers Centre for Wildlife Management, Centre for Invasion Biology, Mammal Research Institute, University of Pretoria, Pretoria, South Africa

Michael Suitor Faculaty of Environmental Design, University of Calgary, Calgary, AB, Canada

Paul J. de Tores Department of Environment and Conservation and Invasive Animals Cooperative Research Centre, Western Australian Wildlife Research Centre, Wanneroo, Australia

Fernando R. Tortato Panthera, R. Barão de Melgaço, 2754, 8° andar, sala 803, Cuiabá, MS, Brazil

Kevin Wilson Venture North EcoRec Enterprises Ltd, Edmonton, Canada

Chapter 1
An Introduction to Fencing for Conservation

Matthew W. Hayward and Michael J. Somers

Conservation managers are tasked with the difficult job of conserving biodiversity, often with limited information and poorly defined goals (Hayward 2009). There are numerous techniques available to achieve this objective; however, the critical element is the separation of biodiversity from the processes threatening it.

Fencing is one element of the conservation managers' arsenal that can separate threats from biodiversity. Fencing evolved in the Neolithic Age to solve social conflict by separating resources (often food) from threats to the retention of those resources (often wildlife or pillaging humans) (Kotchemidova 2008). By fencing, we refer to both physical barriers, such as a standard post and wire fence, and metaphorical barriers, such as inhospitable land, gardens of chilli, walls of noise, effective anti-poaching patrols or buffers of poison (Hayward and Kerley 2009). The essential element linking these management strategies is that they act to keep biodiversity separated from the factors that threaten it.

The use of conservation fencing varies throughout the world. Australia has had some staggering conservation successes attributable to conservation fencing. Early marsupial reintroduction programmes largely failed because introduced predators decimated founder populations (Short et al. 1992). This discovery led to the

M.W. Hayward (✉)
Scotia Sanctuary, Australian Wildlife Conservancy, CARE P.O,
Wentworth, NSW 2648, Australia

Centre for African Conservation Ecology, Nelson Mandela Metropolitan University,
Port Elizabeth, South Africa

Mammal Research Institute, Polish Academy of Science, Białowieża, Poland

School of Biological, Earth and Environmental Science, University of New South Wales,
Sydney, Australia
e-mail: hayers111@aol.com

M.J. Somers
Centre for Wildlife Management, Centre for Invasion Biology, Mammal Research Institute,
University of Pretoria, Pretoria, South Africa

M.J. Somers and M.W. Hayward (eds.), *Fencing for Conservation: Restriction of Evolutionary Potential or a Riposte to Threatening Processes?*,
DOI 10.1007/978-1-4614-0902-1_1,

reintroduction in the Australian mainland of numerous highly endangered species that were previously restricted to islands that were free of introduced European red foxes *Vulpes vulpes* and feral cats *Felis catus*. Peninsulas around Western Australia's Shark Bay were fenced and the introduced species eradicated before species such as burrowing bettongs *Bettongia lesueur*, western barred bandicoot *Perameles bougainville* and the greater stick-nest rat *Leporillus conditor* were reintroduced (Short and Turner 2000; Richards et al. 2001; Richards and Short 2003). This may not be the conservation panacea that is hoped for, as there are examples of poorly maintained conservation fences and poorly designed fencing projects that illustrate the complexity of the issue (see de Tores and Marlow 2012).

Mainland feral-free islands have since been created in Australia. Scotia Sanctuary, in far western New South Wales, possesses an 8000 ha fenced area that is completely free from cats, foxes, dogs *Canis familiaris*, rabbits *Oryctolagus cuniculus*, goats *Capra aegagrus* and sheep *Ovis aries*. Reintroduction programmes initiated by Earth Sanctuaries Ltd, and continued and expanded by Australian Wildlife Conservancy, have seen numbats *Myrmecobius fasciatus*, bridled nailtail wallabies *Onychogalea fraenata*, greater stick-nest rats, burrowing bettongs, brush-tailed bettongs *Bettongia penicillata* and bilbies *Macrotis lagotis* reintroduced in the past decade. Arid Recovery is another private organization that has fenced a large area and reintroduced numerous previously extirpated mammals (Moseby and O'Donnell 2003). However, these efforts are hindered to an extent in Australia by the retention of ownership of natural capital by the federal government.

Despite this success, the government organizations tasked with conserving Australia's biodiversity have rarely initiated widespread fencing programmes of national parks. Most national parks are, in fact, bound by a fence – a post-and-barbed-wire stock fence aimed at excluding the domestic livestock that surround these isolated conservation areas, while foxes, cats, cane toads *Bufo marinus* and rabbits enter at will to wreak devastation on native biodiversity. There are some notable exceptions such as the 1 km^2 area within Dryandra Nature Reserve (Western Australian Department of Environment and Conservation), Venus Bay Conservation Park (South Australian Department of Environment and Heritage) and Mulligan's Flat Nature Reserve (ACT Department of Territory and Municipal Services) to name a few.

New Zealand conservation managers (and more importantly private groups) have embraced conservation fencing even more so than Australians. Again, it is the private sector leading the way with 43 of 64 fenced, feral-free reserves funded by private donors and community organizations (Burns et al. 2012). For example, the Karori Sanctuary, in Wellington, is surrounded by a 2-m high wire-mesh fence that excludes a range of invasive species from brushtail possums *Trichosurus vulpecula* to house mice *Mus domesticus* (KRWSSC 1994). Such a large tract of feral-free land in the middle of the nation's capital city is a triumph for conservation, as the reserve appears to act as a source population for numerous, previously rare, native birds to repopulate the rest of the city.

India has had a long history of conservation action, but managers have only recently begun considering fencing as a conservation tool. Again, the fencing in

India is designed to separate threats from biodiversity; however, in this case the conflict between humans and wildlife often comes from within conservation areas as villages frequently exist within the boundary of such places (Ravi Chellam, Wildlife Conservation Society – India, *pers comm.*).

Re-wilding projects are also likely to require fencing to ensure exotic and/or dangerous species do not spread from the release sites. This is critical in the Pleistocene period, re-wilding projects of North America where ecologically equivalent species are proposed to be introduced (Donlan et al. 2005, 2006), and to a lesser extent in European re-wilding, where ecological substitution is not planned (see, Zimov 2005; Hetherington 2006; Marris 2009; Sandom et al. 2012).

Management strategies in some parts of Africa also include conservation fencing. Unlike in Australia where the federal government retains ownership of natural capital, in southern Africa the private sector has led the conservation field through innovative strategies (Slotow and Hunter 2009). By law, all reserves housing dangerous wildlife in South Africa are required to erect game-proof fencing. Yet, despite this requirement, the ranges of numerous species of such dangerous game are expanding in South Africa (Gusset et al. 2006; Hayward et al. 2007a, b; Hunter 2007), and appropriate metapopulation management plans are in place to ensure that loss of genetic diversity does not eventuate (Davies-Mostert et al. 2009).

In East Africa, fencing is rarely used, other than on some private game reserves (e.g. Lewa Wildlife Conservancy, Kenya). The vast Serengeti ecosystem is largely free of fencing and it is its size that lessens the need for fencing (see East et al. 2012). Yet East Africa is faced with an unprecedented decline in wildlife numbers (Caro and Scholte 2007; Stoner et al. 2007).

The success of fencing as a conservation tool can be simplistically assessed by comparing large predator distribution in East and southern Africa. In East Africa, large predator distributional ranges are increasingly being restricted to conservation areas and populations are declining due to human persecution inside and outside conservation areas (Maddox 2003; Bauer and Van der Merwe 2004). Conversely, in southern Africa, fencing separates predators from persecuting humans, which also isolates populations; however, predator numbers are stable in reserves and distributions are increasing due to numerous reintroduction programmes (Hofmeyr 1997; Gusset et al. 2006; Hayward et al. 2007a, b; Hunter 2007).

Fencing can also cause conservation problems (Hayward and Kerley 2009). In Yellowstone National Park, USA, migratory ungulates are funnelled through narrow necks of conservation land that could easily be cut off by the construction of livestock fences (Berger 2004). Ungulate migrations elsewhere in the USA are also restricted by barriers, including livestock fences, roads, railways and human development (see Berger et al. 2008; Harris et al. 2009; Gadd 2012; Gates et al. 2012, Simon Thirgood, pers. comm. SCB meeting 2008 Chattanooga).

Poorly aligned fences have caused mass mortality events in Botswana and Australia when migrations were cut off from critical water sources (see Owens and Owens 1985; Mbaiwa and Mbaiwa 2006; Hayward and Kerley 2009; Gadd 2012). Once an area is bound by a fence, it becomes isolated. The 1,600 km^2 Bialowieza Forest in eastern Europe is the home of the European bison *Bison bonasus;* however, it is cut in two by the border fence of the European Union, effectively artificially separating the bison population in two (see Kowalczyk et al. 2012). This fence was not designed with conservation in mind, and as such population isolation may lead to numerous problems, such as loss of genetic diversity; management intervention is, therefore, likely to be required on all but the largest fenced conservation areas.

There are also more subtle impacts of fencing, such as fences facilitating predation. For example, some raptors are known to use fences to hunt from, with an increase in local hunting success reducing the recruitment of wading birds nesting close to the fences (Johnson and Oring 2002).

This book investigates fencing for conservation from a range of disparate viewpoints. As ever, our perspective is derived from our experiences, and the authors of the following chapters have experienced issues associated with fencing and conservation in Africa, Asia, Australia, Europe and the Americas. The authors discuss their experiences with conservation fencing and identify the costs and benefits of such conservation activities. While their conclusions are varied, it seems clear that the use of conservation fencing should only be employed when the benefits to biodiversity (i.e. protection from impending extinction) outweigh the costs (i.e. financial costs of construction; social costs of visual amenity; and ecological costs of isolation, fragmentation and mortality).

References

Bauer, H. & Van der Merwe, S. (2004) Inventory of free-ranging lions *Panthera leo* in Africa. *Oryx,* **38**, 26–31.

Berger, J. (2004) The last mile: how to sustain long-distance migration in mammals. *Conservation Biology,* **18**, 320–331.

Berger, J., Young, J.K. & Berger, K.M. (2008) Protecting migration corridors: challenges and optimism for Mongolian saiga. *PLoS Biology,* **6**, e165.

Caro, T.M. & Scholte, P. (2007) When protection falters. *African Journal of Ecology,* **45**, 233–235.

Davies-Mostert, H.T., Mills, M.G.L. & MacDonald, D.W. (2009) South Africa's wild dog *Lycaon pictus* meta-population management programme. *The Reintroduction of Top-order Predators* (eds M.W. Hayward & M.J. Somers), pp. 10–42. Wiley-Blackwell Publishing, Oxford, UK.

Donlan, J., Berger, J., Bock, C.E., Bock, J.H., Burney, D.A., Estes, J.A., Foreman, D., Martin, P.S., Roemer, G.W., Smith, F.A., Soule, M.E. & Greene, H.W. (2006) Pleistocene rewilding: an optimistic agenda for twenty-first century conservation. *American Naturalist,* **168**, 660–681.

Donlan, J., Greene, H.W., Berger, J., Bock, C.E., Bock, J.H., Burney, D.A., Estes, J.A., Foreman, D., Martin, P.S., Roemer, G.W., Smith, F.A. & Soule, M.E. (2005) Re-wilding North America. *Nature,* **436**, 913–914.

East M.L. et al. (2012) Does the vastness of the Serengeti limit human-wildlife conflicts? In: M.J. Somers and M.W. Hayward (eds) Fencing for Conservation. Springer, New York.

Gadd M.E. (2012) Barriers, the beef industry and unnatural selection: a review of the impacts of veterinary fencing on mammals in southern Africa. In: M.J. Somers and M.W. Hayward (eds) Fencing for Conservation. Springer, New York.

Gusset, M., Graf, J.A. & Somers, M.J. (2006) The re-introduction of endangered wild dogs into Hluhluwe-iMfolozi Park, South Africa: an update on the first 25years. *Re-introduction NEWS,* **25**, 31–33.

Harris, G., Thirgood, S.J., Hopcraft, J.G.C., Cromsigt, J.P.G.M. & Berger, J. (2009) Global decline in aggregated migrations of large terrestrial mammals. *Endangered Species Research,* **7**, 55–76.

Hayward, M.W. (2002) The ecology of the quokka (*Setonix brachyurus*) (Macropodidae: Marsupialia) in the northern jarrah forest of Australia. *Department of Biological, Earth and Environmental Science*. University of New South Wales, Sydney, Australia.

Hayward, M.W. (2009) Conservation management for the past, present and future. *Biodiversity and Conservation,* **18**, 765–775.

Hayward, M.W., Adendorff, J., O'Brien, J., Sholto-Douglas, A., Bissett, C., Moolman, L.C., Bean, P., Fogarty, A., Howarth, D., Slater, R. & Kerley, G.I.H. (2007a) Practical considerations for the reintroduction of large, terrestrial, mammalian predators based on reintroductions to South Africa's Eastern Cape Province. *The Open Conservation Biology Journal,* **1**, 1-11 doi 10.2174/1874-8392/07 http://www.bentham.org/open/toconsbj/openaccess2.html.

Hayward, M.W., Adendorff, J., O'Brien, J., Sholto-Douglas, A., Bissett, C., Moolman, L.C., Bean, P., Fogarty, A., Howarth, D., Slater, R. & Kerley, G.I.H. (2007b) The reintroduction of large carnivores to the Eastern Cape, South Africa: an assessment [Erratum]. *Oryx,* **41**, 413.

Hayward, M.W. & Kerley, G.I.H. (2009) Fencing for conservation: restriction of evolutionary potential or a riposte to threatening processes? *Biological Conservation,* **142**, 1–13.

Hetherington, D.A. (2006) The lynx in Britain's past, present and future. *Ecos,* **27**, 68–74.

Hofmeyr, M. (1997) *Operation Phoenix.* (ed P.A. Johnson), pp. 16. North West Parks Board, Rustenberg, South Africa.

Hunter, M.L. (2007) Climate change and moving species: furthering the debate on assisted colonization. *Conservation Biology,* **21**, 1356–1358.

Johnson, M.P. & Oring, L.W. (2002) Are nest exclosures an effective tool in plover conservation? *Waterbirds,* **25**, 184–190.

Kotchemidova, C. (2008) The culture of the fence: artifacts and meanings. *Counterblast - The Journal of Culture and Communication,* **2**, 1-4 http://www.nyu.edu/pubs/counterblast/fence.pdf.

Kowalczyk, R. et al. (2012) Do fences or humans inhibit the movements of large mammals in Białowieża Primeval Forest? In: M.J. Somers and M.W. Hayward (eds) Fencing for Conservation. Springer, New York.

KRWSSC (1994) Feasibility Report on the Native Wildlife Sanctuary Proposal for Karori Reservoir. Karori Reservoir Wildlife Sanctuary Trust Inc., Wellington, New Zealand.

Maddox, T. (2003) The ecology of cheetahs and other large carnivores in a pastoralist-dominated buffer zone. *Department of Anthropology*, pp. 365. University College & Institute of Zoology, London, U.K.

Marris, E. (2009) Reflecting the past. *Nature,* **462**, 31–32.

Mbaiwa, J.E. & Mbaiwa, O.I. (2006) The effects of veterinary fences on wildlife populations in Okavango Delta, Botswana. *International Journal of Wilderness,* **12**, 17–41.

Moseby, K.E. & O'Donnell, E. (2003) Reintroduction of the greater bilby, *Macrotis lagotis* (Reid) (Marsupialia: Thylacomyidae), to northern South Australia: survival, ecology and notes on reintroduction protocols. *Wildlife Research,* **30**, 15–27.

Owens, M.J. & Owens, D.D. (1985) *Cry of the Kalahari,* Collins, London.

de Tores, P.J. & Marlow, N. (2012) The relative merits of predator-exclusion fencing and repeated fox baiting for protection of native fauna: five case studies from Western Australia. In: M.J. Somers and M.W. Hayward (eds) Fencing for Conservation. Springer, New York.

Richards, J.D., Copley, P. & Morris, K.D. (2001) The wopilkara's return. *Nature Australia,* **26**, 52–61.

Richards, J.D. & Short, J. (2003) Reintroduction and establishment of the western barred bandicoot *Perameles bouganville* (Marsupialia: Peramelidae) at Shark Bay, Western Australia. *Biological Conservation,* **109**, 181–195.

Sandom, C. et al. (2012) Exploring the value of wolves (*Canis lupus*) in landscape-scale fenced reserves for ecological restoration in the Scottish Highlands. In: M.J. Somers and M.W. Hayward (eds) Fencing for Conservation. Springer, New York.

Short, J., Bradshaw, S.D., Giles, J., Prince, R.I.T. & Wilson, G.R. (1992) Reintroduction of macropods (Marsupialia: Macropodoidea) in Australia - a review. *Biological Conservation,* **62**, 189–204.

Short, J. & Turner, B. (2000) Reintroduction of the burrowing bettong *Bettongia lesueur* (Marsupialia: Potoroidae) to mainland Australia. *Biological Conservation,* **96**, 185–196.

Slotow, R. & Hunter, L.T.B. (2009) Reintroduction decisions taken at the incorrect social scale devalue their conservation contribution: African lion in South Africa. *The Reintroduction of Top-order Predators* (eds. M.W. Hayward & M.J. Somers), pp. 43–71. Wiley-Blackwell Publishing, Oxford, UK.

Stoner, C.J., Caro, T.M., Mduma, S., Mlingwa, C., Sabuni, G., Borner, M. & Schelten, C. (2007) Changes in large herbivore populations across large areas of Tanzania. *African Journal of Ecology,* **45**, 202–215.

Zimov, S.A. (2005) Pleistocene park: return of the mammoth's ecosystem. *Science,* **308**, 796–798.

Chapter 2
Perspectives on Fencing for Conservation Based on Four Case Studies: Marsupial Conservation in Australian Forests; Bushmeat Hunting in South Africa; Large Predator Reintroduction in South Africa; and Large Mammal Conservation in Poland

Matthew W. Hayward

Introduction

Effective conservation requires the separation of biodiversity from processes that threaten it (Hayward and Kerley 2009). Barrier fencing is one method that conservation managers can employ; however, more "metaphorical" barriers can also perform a similar function (Hayward and Kerley 2009). Fences also provide a defined management unit which minimises confusion about the destination of conservation actions aimed at stopping threats as diverse as intruders, poachers, land clearance, introduced predators, disease and weeds (Hayward and Kerley 2009).

The fundamental benefit of conservation fencing is that it separates biodiversity from threats to its existence and, hence, is critical to conservation actions, yet fencing for conservation also has some clear costs (Hayward and Kerley 2009). Fences have a high financial cost, as well as ecological costs such as inhibiting movement patterns, isolation, inbreeding, predation sinks, continuing management, visual costs and ethical issues (Hayward and Kerley 2009).

This chapter investigates some of the key benefits and problems of employing fencing for conservation purposes. These costs and benefits of fencing are discussed in light of four case studies from three countries in three continents. The first case study highlights how using poison, instead of a physical fence, as a barrier to

---

M.W. Hayward (✉)
Scotia Sanctuary, Australian Wildlife Conservancy, CARE P.O, Wentworth, NSW 2648, Australia

Centre for African Conservation Ecology, Nelson Mandela Metropolitan University, Port Elizabeth, South Africa

Mammal Research Institute, Polish Academy of Science, Białowieża, Poland

School of Biological, Earth and Environmental Science, University of New South Wales, Sydney, Australia
e-mail: hayers111@gmail.com

M.J. Somers and M.W. Hayward (eds.), *Fencing for Conservation: Restriction of Evolutionary Potential or a Riposte to Threatening Processes?*,
DOI 10.1007/978-1-4614-0902-1_2,

introduced predators may not be a successful long-term strategy for conserving the vulnerable macropodid marsupial, the quokka *Setonix brachyurus*. The second case study of bushmeat hunting in the Transkei region of South Africa illustrates how critical fence maintenance is to ensuring the effectiveness of fences for conservation. The third case study illustrates how effective fence maintenance has ensured the success of large predator reintroductions in South Africa's Addo Elephant National Park; however, this in turn has created management problems through population isolation. Finally, I look at how human influences around Poland's Białowieża Primeval Forest effectively act as a barrier fence to restrict animal movements.

Case Study 1 – Metaphorical Fences of Poison: Conserving the Quokka in Australia's Northern Jarrah Forest

The quokka is a 3–5-kg macropodid marsupial that is endemic to the south-western corner of Western Australia (Kitchener 1995). The species is one of the most abundant in the fossil-bearing sites of the region (Glauert 1948; Cook 1960), where its fossils peaked in abundance around 20,000 years ago (Balme et al. 1978). Aboriginal people regularly hunted them by burning their swampy habitats and spearing them as they fled the fire (Gardner 1957; Gould 1973; Nicholson 1981). Quokkas were also abundant when Europeans arrived in the region, such that they were classed as "vermin" by forestry officials (Stewart 1936). Reports of the quokkas' decline on the mainland in the 1930s are common in the literature (White 1952; Serventy et al. 1954; Barker et al. 1957). Conversely, the two island populations – Bald and Rottnest – remained stable through this period (Waring 1959; Storr 1965). Although the 1930s saw the major decline of the quokka, recent research suggests this decline has continued (Hayward et al. 2003) (Fig. 2.1).

Numerous factors were listed as causing the initial decline of the quokka, including disease, urbanisation, habitat alteration, competition with introduced herbivores and predation (White 1952; Main 1959; Short and Calaby 2001). Of these, there is only evidence for predation and habitat alteration playing distinguishable roles (Hayward et al. 2005a). There are several lines of evidence implicating the European red fox *Vulpes vulpes* in the quokkas' decline. Critically, the decline of the quokka on the mainland corresponded to the arrival of the fox in the south-west (King and Smith 1985); quokka populations on fox-free islands have remained stable; reintroduced quokka populations have failed due to predation by foxes (Short et al. 1992) and the fox remains the major mortality source today (Hayward et al. 2005a). Predation by feral cats *Felis catus* seems unlikely to be the cause of the quokkas decline as they coexist on Rottnest Island without any apparent population limitation (Main et al. 1959). It seems likely that the quokka is within the preferred weight range of prey of the red fox, but larger than the preferred prey weight range of cats (Hayward et al. 2006). There is no evidence of an increase in predation by native carnivores or of humans in the 1930s (Hayward 2002; Hayward et al. 2005a). Finally, reanalysis of the evidence for disease suggests that it was more likely surplus killing by newly invading red foxes (Short et al. 2002).

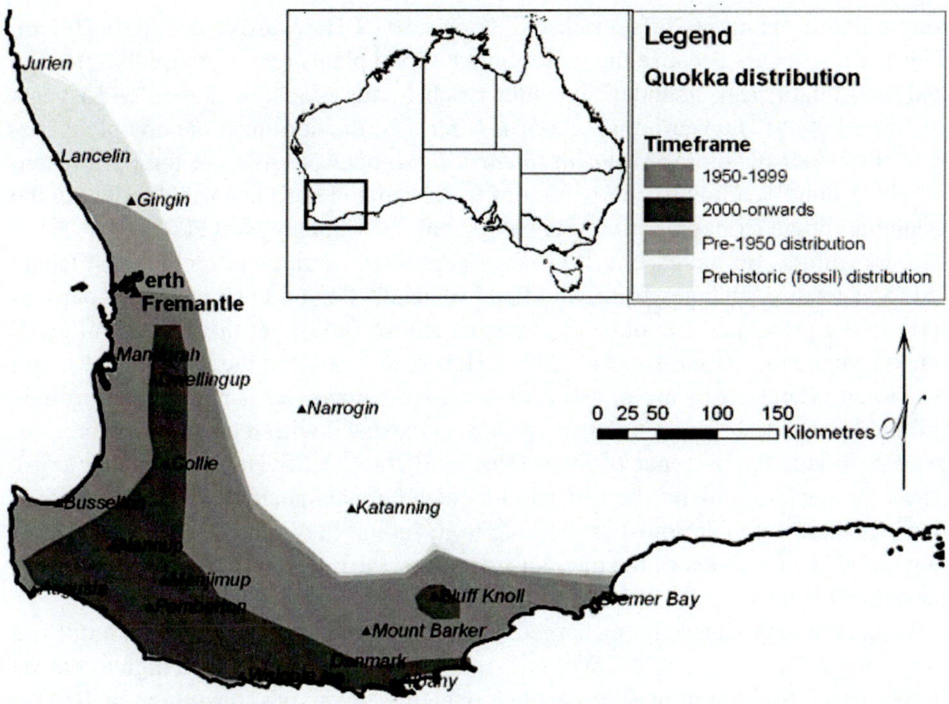

Fig. 2.1 Map showing the decline of the quokka from Hayward (2002). The area of quokka occupancy prehistorically is shown as the *dotted region*, prior to 1950 is hatched, from 1950 until 1990 is *light grey* and in 2000 is shown as *dark grey*

The quokkas' decline may also have been driven by habitat alteration. The Swan Coastal Plain has been largely cleared of natural vegetation for urban development and so few quokka populations remain there (de Tores et al. 2007). The fragments that do remain are small and susceptible to invasion by foxes (Hayward 2002). The jarrah forest is a multi-use area that includes forestry, bauxite mining, water storage (dams) and conservation. Forestry has not affected quokkas substantially as the 20 m buffer zone around quokka swamps appears sufficient to protect them (Calver and Dell 1998a, b; Hayward 2002). The large-scale clearance involved in mining has affected individual quokka swamps; however, this is a modern impact and mitigation measures are now put in place to protect native fauna, including quokkas (Hayward 2002). Quokka swamps have also been dammed to provide water, but again this has only affected a few individual populations since the 1930s (Hayward 2002). The biggest habitat alteration quokkas have faced is due to change in fire regimes. Quokkas evolved to cope with Aboriginal burning regimes of fires every 3–4 years in the jarrah forest that entered the swamps every 6–8 years (Wallace 1966; Burrows et al. 1995; Ward and Sneeuwjagt 1999). They then had to cope with complete fire exclusion during the early European occupation and more recently to low intensity, prescribed burns on a 7-year rotation (Hayward et al. 2005b). Fire can eradicate individual populations; however, quokkas recolonise swamps within months of a fire, reach peak densities at 8–15 years, before becoming absent in long-unburnt swamps (Christensen and Kimber 1975; Hayward et al. 2005b, 2007e). This pattern occurs because the major quokka food plants resprout rapidly after fire and they attain peak abundance within reach of foraging quokkas at 8–15 years (Hayward 2005; Hayward et al. 2005b). After this, the dominant canopy plant species of quokka swamps, *Taxandria linearifolia*, blocks out most of the light reaching the ground, leading to an opening of the shrub and herb layer and reducing the available refuge from predation (Hayward et al. 2005b; Hayward 2008).

Interestingly, an increase in fires since permanent settlement on Rottnest Island led to vegetation changes from woodland to heath (Pen and Green 1983), particularly in the presence of quokka overgrazing (Storr 1963), yet this has not affected quokka abundance (Johnson et al. 1989). Hence, it seems that the quokka is resilient to habitat alteration in the absence of red foxes; however, it requires the refuge offered by patches of long-unburnt swamp interspersed with recently burnt areas to provide food in the presence of foxes (Hayward 2002, 2005; Hayward et al. 2005b). Thus, the quokka may not be restricted to such habitat specificity on the mainland because these areas are most favourable to it; rather, they may be the habitat least "favoured" by the agent of the quokka's decline – the red fox. In essence, these may be refuge habitats.

Conservation managers are charged with a difficult task, particularly in multi-use forest areas (Hayward 2009a). Western Australia's native fauna has a high tolerance to the poison Sodium monofluoroacetate, which is the active constituent of 1080 ® (see summary by de Tores and Marlow 2012). Hence, managers tasked with conserving quokkas implemented a monthly poisoning campaign around known quokka swamps as part of the Western Shield programme (de Tores et al. 2004). After 6 years of fox poisoning, however, there was no evidence of a population response by quokkas, although two unbaited populations went extinct in that time (Hayward et al. 2003, 2005a). It was concluded that quokkas originally existed as a metapopulation that has collapsed since the arrival of the fox in the 1930s (Hayward et al. 2003), despite current movement patterns being sufficient to allow colonisation of adjacent, suitable patches (Hayward et al. 2004). Future management will include small-scale, precise burns of quokka swamps to create a mosaic of time-since-burn age classes (Hayward et al. 2005b, 2007e).

The plight of the quokka illustrates that "metaphorical" fences of poison are available and can potentially perform the same function as physical structures (see also de Tores and Marlow 2012). The continued vulnerability and decline of the quokka suggests that relying solely on poisoning may not be a long-term solution, particularly given the likelihood of foxes developing aversion or even tolerance to 1080 as rabbits *Oryctolagus cuniculus* have done in less than 100 years (Twigg et al. 2002).

This has severe ramifications for conservation management throughout Australia and New Zealand, where poison is frequently used to reduce the densities of introduced species. While physical fencing may be an extremely expensive option (Hayward and Kerley 2009), eventually the high construction costs will be allayed compared to the continued costs of poisoning, particularly if poisons become less effective (Fig. 2.2).

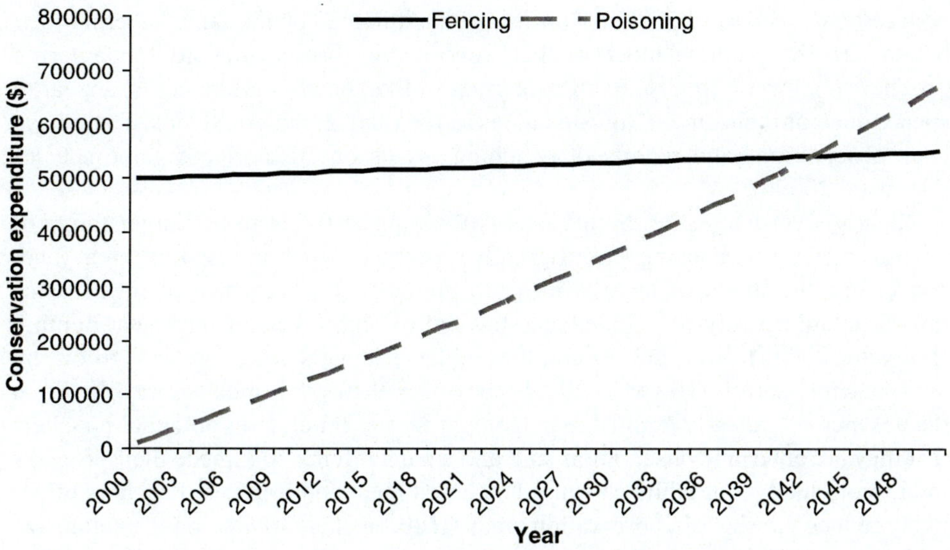

Fig. 2.2 Schematic representation of how continued annual poisoning will eventually cost more than the construction and maintenance of a physical fence. The hypothetical fence cost $500,000 to build, but only $1000 per annum for maintenance. The hypothetical poisoning campaign costs $10,000 per annum, but loses effectiveness by 1% per year as animals develop bait aversion and tolerance to the poison

Case Study 2 – Ineffective Fencing: Bushmeat Hunting in the Dwesa and Cwebe Nature Reserves, South Africa

Dwesa and Cwebe Nature Reserves are on the Wild Coast of South Africa's Transkei region between Port Elizabeth and Durban. This area has received relatively little attention from ecologists until recent years (Hayward et al. 2005c). The combined area of the reserves is 15,254 ha and they are covered by similar coastal indigenous rainforest communities, but are separated by the Mbashe River (Timmermans and Naicker 2002). Like all conservation areas in South Africa that house large and potentially dangerous wildlife species, Dwesa and Cwebe were fenced originally; however, this fencing has fallen into disrepair in recent years (Hayward 2009b).

It may be argued that there are two phases to the conservation of Dwesa and Cwebe. Following the creation of the reserves, several species were reintroduced including "plains game", such as white rhinoceros *Ceratotherium simum*, blue wildebeest *Connochaetes taurinus* and plains zebra *Equus quagga*, which did not originally occur in the region (Skead 1987; Feely 1999; Hayward et al. 2005c). These species did well on the grasslands that grew following clearing of some parts of the coastal forests, and they were frequently observed by tourists.

More recently, the fence has fallen into disrepair and has even been removed in places, which has led to a high level of human encroachment and poaching throughout both reserves (Hayward 2009b). This has resulted in the "plains game" altering their habitat use by spending much of their time inside forests to avoid encountering poachers (Hayward 2009b). Even bulk grazers, like the white rhinoceros, appear to spend much of their time in forested areas rather than on the grasslands. There have been observations and reports of poaching events on all the large game species (Hayward 2009b).

Clearly, over-harvesting by humans is the biggest threat to the larger fauna of Dwesa and Cwebe. Fencing was originally constructed to minimise this threat (and, conversely, the threat of large wildlife to humans). Lack of investment in fence infrastructure by way of maintenance has led to high levels of bushmeat hunting (Hayward 2009b). Both Dwesa and Cwebe have rangers who regularly go out on anti-poaching patrols (Hayward 2009b); however, their job is much more difficult in the absence of a clearly defined management unit to reiterate to potential poachers that they are entering conservation land and a barrier fence to impede their progress in or, more likely, out while carrying butchered meat and trophies. Poachers effectively reduce the size of conservation areas (Hilborn et al. 2006), and the small size of Dwesa and Cwebe means there is little scope for the fauna to utilise poaching refuges (areas too far from villages or in too dense habitat to make poaching an optimal strategy)(Wilkie and Carpenter 1999). Thus, for fencing to be an effective conservation action, it must be adequately maintained, and this requires adequate investment in the conservation body tasked with managing the area and, ideally, the local community to allow them to reap the financial rewards of living alongside a conservation area, which may reduce the amount of damage the fence receives.

Case Study 3 – Effective Fencing: The Reintroduction of Top-Order Predators to South Africa's Addo Elephant National Park

South Africa's Eastern Cape Province has experienced a rapid transition from marginally viable pastoral land use to both public and private conservation which yields 4 times the profit while employing 4 times as many people (Kerley and Boshoff 1997; Sims-Castley et al. 2004). As part of this process, large numbers of wildlife have been reintroduced to the province, including large predators (Hayward et al. 2007a–c).

The managers of Addo Elephant National Park reintroduced lions *Panthera leo*, leopards *P. pardus* and spotted hyaenas *Crocuta crocuta* in 2003 and 2004 (Hayward et al. 2007d). Although South African National Parks managers manipulated the sex ratio of the founder population to slow the potential population increase, the populations of lions and hyaenas increased rapidly. This is a feature of the reintroductions of lions and hyaenas to the 12 reserves in the Eastern Cape, such that overpopulation is now a major problem (Hayward et al. 2007b, d).

The success of large predator reintroductions in the Eastern Cape is largely attributable to fencing. Large predators were driven extinct in the region in the late nineteenth century by human persecution (Skead 1987). Large predators invariably come into conflict with humans and in most developed societies are increasingly restricted to isolated conservation areas (Bauer and van der Merwe 2004; Bauer et al. 2004). Encroachment and poaching act to further limit the area available to these species. Effective fencing greatly minimises human-animal conflict, which ensures large predators are free from retributive human persecution. Effective fencing also minimises encroachment and poaching. Hence, it is clear that the fencing around Addo Elephant National Park, and other reserves in the Eastern Cape, is effectively assisting in conserving large predators. The Eastern Cape reserves are not alone in linking their reintroduction successes with the presence of fencing. The reintroduction of African wild dogs *Lycaon pictus* has been attributed to the presence of fencing also (Gusset et al. 2008).

Fencing does raise some potential problems. Firstly, each population is isolated and there is no potential for natural mixing of populations to avert inbreeding, which has been identified as a problem about to arise (Frankham 2009). There is the potential, in the Eastern Cape, to bring down fences between adjacent reserves to create a mega-conservancy (Hayward et al. 2007b). If this does not occur, or where it is impossible, then continued management of animals will be required to avoid loss of genetic diversity.

A second potential problem is the effect of fencing on the spatial and social behaviour of the reintroduced species. Large predator home range sizes are generally negatively related to the biomass of preferred prey (Nilsen and Linnell 2006; Hayward et al. 2008). Thus, their territory size can be predicted using the biomass of preferred prey. Data from Addo Elephant National Park show that the range size of lions and spotted hyaenas (and a leopard) are not constrained by the fences and conform to predictions based on the availability of preferred prey (Hayward et al. 2008).

There were two male coalitions in Addo's lion population that initially battled for social dominance of the park's lionesses. After several months, one coalition became dominant and this forced the subordinate males to spend much of their time on the periphery of the dominant coalition's territory – that is, along the fence line (Hayward and Hayward 2007). Is the fence line limiting their movements? Their home range size is still as expected, based on the availability of food. In unfenced populations, subordinate lions (or nomads) are forced to live on the periphery of pride territories (Schaller 1972). Hence, if Addo's lions were surrounded by other prides rather than fences, subordinate or nomadic lions would be forced to live along the periphery of these territories and, thus, it seems unlikely that the fences are substantially affecting lion social behaviour.

The barrier fencing around Addo was financially costly, although this cost is likely to be rapidly recouped with the increase in tourism attributable to the reintroduction of large predators (Hayward et al. 2007b). While isolation is a factor that requires ongoing management and reduces the value of fenced reserves for conservation (Hunter et al. 2007), the return of species to the region for the first time in over 100 years seems to be a conservation advance for the species through distribution expansion, increased abundance and spreading risk to more populations.

Case Study 4 – Fence-Free Barriers: The Fauna of Białowieża Primeval Forest, Poland

Białowieża Primeval Forest is a remnant of the once, great temperate deciduous forests that covered much of Europe (Jędrzejewska and Jędrzejewski 1998). The forest spans the Polish and Belarussian border, but is largely isolated by a border fence (to the east) and cleared agricultural lands and urban areas to the north, south and west. The effect of the border fence is discussed elsewhere in this book (see Kowalczyk et al. 2012). Here, I discuss my impression of the "metaphorical" fence that limits animal movements and distribution to the north, south and west of the forest.

Figures illustrating the locations of radio tagged wolves *Canis lupus*, Eurasian lynx *Lynx lynx* and European bison *Bison bonasus* provide graphic evidence of the effectiveness of the "metaphorical" fence to the north, south and west of Białowieża (see Figs. 13.1–13.3 of Kowalczyk et al. 2012). There are almost no locations of any of these species outside the forest. Animals leaving the forest not only face the threat of accidental mortality from vehicle road kills, but also face human persecution (legal or illegal – see Kowalczyk et al. 2012) and harassment from dogs.

The fauna of Białowieża has faced several 100 years of selection at the hands of humans (Samojlik and Jędrzejewska 2005). Białowieża's forest survived largely because it served as a hunting ground to Polish kings and Lithuanian dukes. Managers were employed to protect the forest and ensure the king's wildlife was not poached. Areas outside the king's forest were cleared. Since the seventeenth century, humans have been killing animals that left the forest, while protecting those that remained inside (until the rare periods that the king chose to hunt) (Samojlik and Jędrzejewska 2005). This unnatural selection must have shaped the behavioural ecology of these species and probably explains the sedentary behaviour and forest preferences of the European bison (compared to its North American cousin).

Thus, even in the absence of physical barrier fencing, metaphorical barriers can have dramatic impacts on the ecology of wildlife. Management efforts are now focused on resurrecting movement corridors to increase connectivity to adjacent habitat patches.

Conclusion

The case studies discussed above illustrated a diverse range of benefits and costs of conservation fencing. A cost-benefit analysis may determine which of these were valuable tools in the conservation arsenal at these sites.

Would the Quokka Survive on Mainland Australia Without the Metaphorical Barrier of Poison?

Quokkas have persisted on the Australian mainland in sympatry with red foxes for over 70 years. The recent introduction of fox control (1996), has not led to a noticeable increase in quokka abundance or known populations, and all known populations are critically low in number and isolated from other elements of the collapsed metapopulation (Hayward et al. 2003). Foxes remain the biggest mortality source for quokkas (Hayward et al. 2005a). This suggests the metaphorical fence of poison has not been completely effective. Whether this is because foxes have developed a tolerance to 1080 (Hayward and Kerley 2009) or because mesopredator suppression of cats has been released with the lowering of fox density is unknown. In light of the recent precipitous decline of the formerly conservation-dependent woylie to becoming endangered (de Tores and Marlow 2012), there should be concern for the persistence of the quokka, such that consideration should be given to creating a physical fence around some known quokka populations and surrounding swamps.

Are the Fauna of Dwesa and Cwebe Nature Reserves Threatened Because of Ineffective Fencing?

Bushmeat hunting is increasingly becoming a conservation threat (IUCN 2007). Areas where people have relatively easy access to wildlife become halos of defaunation (Wilkie and Carpenter 1999) and the impact of poaching effectively decreases the size of conservation areas in proportion to the area accessible to poachers (Hilborn et al. 2006). These issues suggest Dwesa and Cwebe are providing little value for conservation. Yet many of the species present in Dwesa and Cwebe are grassland species that are unlikely to be there naturally (Hayward et al. 2005c). Thus, their local extinction may not be a critical assessment of the conservation value of the reserves. Other threatened species that do persist in Dwesa and Cwebe (e.g. blue duiker *Philantomba monticola* and samango monkey *Cercopithecus mitis labiatus*) are of more relevance. Blue duikers avoid the roads and tracks used by poachers suggesting they are affected by the ineffective fencing (Hayward 2009b). Conversely, samango monkeys appear unaffected by Dwesa and Cwebe's fencing issues (Hayward 2009b). Thus, there is a suggestion that fencing might be important to conserving blue duikers by minimising risk of poaching.

Furthermore, if the wildlife inside the reserves cause injury or substantial financial damage (bushpigs *Potamochoerus larvatus* already do) to local communities, there may be increasing pressure to reconsider the merits of having a conservation reserve in the area. In the absence of effective fencing, this is more likely to occur. This also suggests that having an effective conservation fence in place is important to the conservation of the fauna of Dwesa and Cwebe.

Would the Reintroduction of Large Predators to Addo Elephant National Park (and the Other Eastern Cape Reserves) Have Been Successful Without Fencing?

Human persecution drove large predators extinct in the Eastern Cape Province of South Africa in the late nineteenth century (Skead 1987). Human persecution has increasingly restricted large predators to isolated conservation areas today. In the absence of fencing, it is likely that the large predators would leave the conservation reserves, kill livestock or people and be killed in retribution. Thus, it is unlikely that the reintroduction of large predators to the Eastern Cape would have been successful in the absence of fencing.

Has Fencing Caused the Isolation of Wildlife in Białowieża's Primeval Forest?

The border fence between Poland and Belarus is a physical structure that unquestionably fragments the forest for large fauna (see Kowalczyk et al. 2012). The habitat clearance to the north, south and west of the forest effectively isolates the fauna of Białowieża from Western Europe. This illustrates how metaphorical fencing, without any conservation goals, can act the same way as a physical structure. Thus, both fencing and habitat clearance have effectively isolated the wildlife of Poland's Białowieża Primeval Forest.

Although fences may ultimately prove to be as much a threat to biodiversity as the threats they are meant to exclude, the biodiversity crisis we are facing means they are likely to be a critically important part of the landscape for many years to come (Hayward and Kerley 2009).

Acknowledgements The research discussed above was supported by Paul de Tores, Mike Augee, Peter Banks, Barry Fox, Graham Kerley, and Bogumiła and Włodzimierz Jędrzejewsky, and funded by the Australian (APA) and South African (NRF) governments, and the European Union Marie Curie Actions.

References

Balme, J., Merrilees, D. & Porter, J.K. (1978) Late Quaternary mammal remains, spanning 30,000 years, from excavations in Devil's Lair, Western Australia. *Journal of the Royal Society of Western Australia*, **61**, 33–65.

Barker, S., Main, A.R. & Sadlier, R.M. (1957) Recent capture of the quokka (*Setonix brachyurus*) on the mainland. *The Western Australian Naturalist*, **6**, 53–55.

Bauer, H. & van der Merwe, S. (2004) Inventory of free-ranging lions *Panthera leo* in Africa. *Oryx*, **38**, 26–31.

Bauer, H., Nowell, K. & IUCN (2004) *Panthera leo. 2004 Red List of Threatened Species*, pp. 1–11. IUCN, Gland, Switzerland.

Burrows, N.D., Ward, B. & Robinson, A.D. (1995) Jarrah forest fire history from stem analysis and anthropological evidence. *Australian Forestry*, **58**, 7–16.
Calver, M.C. & Dell, J. (1998a) Conservation status of mammals and birds in southwestern Australian forests. I. Is there evidence of direct links between forestry practices and species decline and extinction? *Pacific Conservation Biology*, **4**, 296–314.
Calver, M.C. & Dell, J. (1998b) Conservation status of mammals and birds in southwestern Australian forests. II. Are there unstudied, indirect or long-term links between forestry practices and species decline and extinction. *Pacific Conservation Biology*, **4**, 315–325.
Christensen, P.E.S. & Kimber, P.C. (1975) Effect of prescribed burning on the flora and fauna of south-western Australian forests. *Proceedings of the Ecological Society of Australia*, **9**, 85–106.
Cook, D.L. (1960) Some mammal remains found in caves near Margaret River. *The Western Australian Naturalist*, **7**, 107–108.
de Tores, P.J., Hayward, M.W. & Rosier, S.M. (2004) The western ringtail possum, *Pseudocheirus occidentalis*, and the quokka, *Setonix brachyurus*, case studies: Western Shield review - February 2003. *Conservation Science*, **5**, 235–257.
de Tores, P.J., Hayward, M.W., Dillon, M.J., Liddelow, G. & Brazell, R. (2007) Review of the distribution and conservation status of the quokka, *Setonix brachyurus*, (Macropodidae: Marsupialia). *Conservation Science Western Australia*, **6**, 13–73.
Feely, J. (1999) *Historical distribution of some larger mammals in Transkei*. Department of Economic Affairs, Environment and Tourism, Pretoria, South Africa.
Frankham, R. (2009) Genetic considerations in reintroduction programs for large terrestrial predators. *The Reintroduction of Top-order Predators* (eds. M.W. Hayward & M.J. Somers), pp. 371–387. Wiley-Blackwell Publishing, Oxford, UK.
Gardner, C.A. (1957) The fire factor in relation to the vegetation of Western Australia. *The Western Australian Naturalist*, **5**, 166–173.
Glauert, L. (1948) The cave fossils of the south-west. *The West Australian Naturalist*, **1**, 100–104.
Gould, J. (1973) *Kangaroos*. The Macmillan Company of Australia Pty Ltd, Melbourne.
Gusset, M., Ryan, S.J., Hofmeyr, M., Van Dyk, G., Davies-Mostert, H.T., Graf, J.A., Owen, C., Szykman, M., MacDonald, D.W., Monfort, S.L., Wildt, D.E., Maddock, A., Mills, M.G.L., Slotow, R. & Somers, M.J. (2008) Efforts going to the dogs? Evaluating attempts to re-introduce endangered wild dogs in South Africa. *Journal of Applied Ecology*, **45**, 100–108.
Hayward, M.W. (2002) The ecology of the quokka (*Setonix brachyurus*) (Macropodidae: Marsupialia) in the northern jarrah forest of Australia. *Department of Biological, Earth and Environmental Science*. University of New South Wales, Sydney, Australia.
Hayward, M.W. (2005) Diet of the quokka (*Setonix brachyurus*) (Macropodidae: Marsupialia) in the northern jarrah forest of Western Australia. *Wildlife Research*, **32**, 15–22.
Hayward, M.W. (2008) Home range overlap of the quokka *Setonix brachyurus* (Macropodidae: Marsupialia) suggests a polygynous mating system. *Conservation Science Western Australia*, **7**, 57–64.
Hayward, M.W. (2009a) Conservation management for the past, present and future. *Biodiversity and Conservation*, **18**, 765–775.
Hayward, M.W. (2009b) Bushmeat hunting in the Dwesa - Cwebe Nature Reserve, Eastern Cape, South Africa. *South African Journal of Wildlife Research*, **39**, 70–84.
Hayward, M.W. & Hayward, G.J. (2007) Activity patterns of reintroduced lion *Panthera leo* and spotted hyaena *Crocuta crocuta* in Addo Elephant National Park, South Africa. *African Journal of Ecology*, **45**, 135–141.
Hayward, M.W. & Kerley, G.I.H. (2009) Fencing for conservation: restriction of evolutionary potential or a riposte to threatening processes? *Biological Conservation*, **274**, 1–13.
Hayward, M.W., de Tores, P.J., Augee, M.L., Fox, B.J. & Banks, P. (2004) Home range and movements of the quokka *Setonix brachyurus* (Macropodidae: Marsupialia), and its impact on the viability of the metapopulation on the Australian mainland. *Journal of Zoology*, **263**, 219–228.

Hayward, M.W., de Tores, P.J., Augee, M.L. & Banks, P.B. (2005a) Mortality and survivorship of the quokka (*Setonix brachyurus*) (Macropodidae: Marsupialia) in the northern jarrah forest of Western Australia. *Wildlife Research*, **32**, 715–722.

Hayward, M.W., de Tores, P. J. & Banks, P. B. (2005b) Habitat use of the quokka *Setonix brachyurus* (Macropodidae: Marsupialia) in the northern jarrah forest of Australia. *Journal of Mammalogy*, **86**, 683–688.

Hayward, M.W., White, R.M., Mabandla, K.M. & Bukeye, P. (2005c) Mammalian fauna of indigenous forest in the Transkei region of South Africa, an overdue survey. *South African Journal of Wildlife Research*, **35**, 117–124.

Hayward, M.W., Henschel, P., O'Brien, J., Hofmeyr, M., Balme, G. & Kerley, G.I.H. (2006) Prey preferences of the leopard (*Panthera pardus*). *Journal of Zoology*, **270**, 298–313.

Hayward, M.W., Adendorff, J., Moolman, L.C., Hayward, G.J. & Kerley, G.I.H. (2007a) The successful reintroduction of leopard *Panthera pardus* to the Addo Elephant National Park. *African Journal of Ecology*, **45**, 103–104.

Hayward, M.W., Adendorff, J., O'Brien, J., Sholto-Douglas, A., Bissett, C., Moolman, L.C., Bean, P., Fogarty, A., Howarth, D., Slater, R. & Kerley, G.I.H. (2007b) Practical considerations for the reintroduction of large, terrestrial, mammalian predators based on reintroductions to South Africa's Eastern Cape Province. *The Open Conservation Biology Journal*, **1**, 1–11.

Hayward, M.W., Adendorff, J., O'Brien, J., Sholto-Douglas, A., Bissett, C., Moolman, L.C., Bean, P., Fogarty, A., Howarth, D., Slater, R. & Kerley, G.I.H. (2007c) The reintroduction of large carnivores to the Eastern Cape, South Africa: an assessment [Erratum]. *Oryx*, **41**, 413.

Hayward, M.W., Adendorff, J., O'Brien, J., Sholto-Douglas, A., Bissett, C., Moolman, L.C., Bean, P., Fogarty, A., Howarth, D., Slater, R. & Kerley, G.I.H. (2007d) The reintroduction of large predators to the Eastern Cape Province, South Africa: an assessment. *Oryx*, **42**, 205–214.

Hayward, M.W., de Tores, P.J., Dillon, M.J. & Banks, P.B. (2007e) Predicting the occurrence of the quokka, *Setonix brachyurus* (Macropodidae: Marsupialia), in Western Australia's northern jarrah forest. *WIldlife Research*, **34**, 194–199.

Hayward, M.W., de Tores, P.J., Dillon, M.J. & Fox, B.J. (2003) Local population structure of a naturally-occurring metapopulation of the quokka (*Setonix brachyurus* Macropodidae: Marsupialia). *Biological Conservation*, **110**, 343–355.

Hayward, M.W., Hayward, G.J., Druce, D. & Kerley, G.I.H. (2008) Do fences constrain predator movements on an evolutionary scale? Home range, food intake and movement patterns of large predators reintroduced to Addo Elephant National Park, South Africa. *Biodiversity and Conservation*, **18**, 887–904.

Hilborn, R., Arcese, P., Borner, M., Hando, J., Hopcraft, J.G.C., Loibooki, M., Mduma, S. & Sinclair, A.R.E. (2006) Effective enforcement in a conservation area. *Science*, **314**, 1266.

Hunter, L.T.B., Pretorius, K., Carlisle, L.C., Rickelton, M., Walker, C., Slotow, R. & Skinner, J.D. (2007) Restoring lions *Panthera leo* to northern KwaZulu-Natal, South Africa: short-term biological and technical success but equivocal long-term conservation. *Oryx*, **41**, 196–204.

IUCN. (2007) *2007 Red List of Threatened Species*. [*Accessed 28/09/2008*], International Union for the Conservation of Nature and Natural Resources (IUCN), Gland.

Jędrzejewska, B. & Jędrzejewski, W. (1998) *Predation in Vertebrate Communities*: *The Bialowieza Primeval Forest as a Case Study*, Springer, Berlin, Germany.

Johnson, K.A., Burbidge, A.A. & McKenzie, N.L. (1989) Australian Macropodoidea: status, causes of decline and future research and management. *Kangaroos, Wallabies and Rat-kangaroos*. (eds. G.C. Grigg, P.J. Jarman & I.D. Hume), pp. 641–657. Surrey Beatty & Sons Pty Ltd, Sydney, Australia.

Kerley, G.I.H. & Boshoff, A.F. (1997) A Proposal for a Greater Addo National Park: a regional and national conservation and development opportunity. pp. 1–63. Terrestrial Ecology Research Unit, Department of Zoology, University of Port Elizabeth, Port Elizabeth.

King, D.R. & Smith, L.A. (1985) The distribution of the European red fox (*Vulpes vulpes*) in Western Australia. *Records of the Western Australian Museum.*, **12**, 197–205.

Kitchener, D.J. (1995) Quokka (*Setonix brachyurus*). *Mammals of Australia* (ed R. Strahan), pp. 401–403. Reed Books, Sydney, Australia.

Main, A.R. (1959) Rottnest Island: the Rottnest Biological Station and recent scientific research. Rottnest Island as a location for biological studies. *Journal of the Royal Society of Western Australia*, **42**, 66–67.

Main, A.R., Shield, J.W. & Waring, H. (1959) Recent studies on marsupial ecology. *Biogeography and Ecology in Australia* (eds. A. Keast, R.L. Crocker & C.S. Christian), pp. 315–331. Dr. W. Junk, The Hague, Netherlands.

Nicholson, P.H. (1981) Fire and the Australian Aborigine - an enigma. *Fire and the Australian Biota* (eds. A.M. Gill, R.H. Groves & I.R. Noble), pp. 55–76. Australian Academy of Science, Canberra.

Nilsen, E.B. & Linnell, J.D.C. (2006) Intra-specific variation and taxa- sampling affects the home range - body mass relationship. *Acta Theriologica*, **51**, 225–232.

Pen, L.J. & Green, J. . (1983) Botanical exploration and vegetational changes on Rottnest Island. *Journal of the Royal Society of Western Australia*, **66**, 20–24.

Samojlik, T. & Jędrzejewska, B. (2005) The bison - rich treasure of the forests. *Conservation and Hunting: Bialowieza Forest in the Time of Kings*. (ed T. Samojlik), pp. 75–88. Mammal Research Institute, Polish Academy of Science, Bialowieza, Poland.

Schaller, G.B. (1972) *The Serengeti Lion*, University of Chicago Press, Chicago.

Serventy, D.L., Glauert, L., Carnaby, I.C., Loaring, W.H., Aitken, R. & Jones, A.D. (1954) The recent increase of the rarer native mammals. *The Western Australian Naturalist*, **4**, 128–141.

Short, J. & Calaby, J.H. (2001) The status of Australian mammals in 1922 - collections and field notes of museum collector Charles Hoy. *Australian Zoologist*, **31**, 533–562.

Short, J., Bradshaw, S.D., Giles, J., Prince, R.I.T. & Wilson, G.R. (1992) Reintroduction of macropods (Marsupialia: Macropodoidea) in Australia - a review. *Biological Conservation*, **62**, 189–204.

Short, J., Kinnear, J.E. & Robley, A. (2002) Surplus killing by introduced predators in Australia - evidence for ineffective anti-predator adaptations in native prey species? *Biological Conservation*, **103**, 283–301.

Sims-Castley, R., Kerley, G.I.H. & Geach, B. (2004) A questionnaire-based assessment of the socio-economic significance of ecotourism-based private game reserves in the Eastern Cape., pp. 31. Terrestrial Ecology Research Unit, University of Port Elizabeth, Port Elizabeth, South Africa.

Skead, C.J. (1987) *Historical Mammal Incidence in the Cape Province. Volume 2: The Eastern Half of the Cape Province, including the Ciskei, Transkei and East Griqualand*, Chief Directorate Nature and Environmental Conservation of the Provincial Administration of the Cape of Good Hope, South Africa., Cape Town.

Stewart, D.W.R. (1936) Notes on marsupial damage in pine plantations. *Australian Journal of Forestry*, **1**, 41–44.

Storr, G.M. (1963) Some factors inducing change in the vegetation of Rottnest Island. *The Western Australian Naturalist*, **9**, 15–22.

Storr, G.M. (1965) Notes of Bald Island and the adjacent mainland. *The Western Australian Naturalist*, **9**, 187–196.

Timmermans, H. & Naicker, K. (2002) The land. *From Conflict to Negotiation: Nature-based Development on South Africa's Wild Coast*. (eds. R. Palmer, H. Timmermans & D. Fay), pp. 2–14. Human Sciences Research Council, Pretoria, South Africa.

Twigg, L.E., Martin, G.R. & Lowe, T.J. (2002) Evidence of pesticide resistance in medium-sized mammalian pests: a case study with 1080 poison and Australian rabbits. *Journal of Applied Ecology*, **39**, 549–560.

Wallace, W.R. (1966) Fire in the jarrah forest environment. *Journal of the Royal Society of Western Australia*, **49**, 33–44.

Ward, D.J. & Sneeuwjagt, R. (1999) Aboriginal fire: its relevance to present day management of the jarrah forest of south-western Australia. *Proceedings of 'Bushfire 99' Conference* (eds. I. Lunt, D.G. Green & B. Lord), pp. 54–55. Charles Sturt University, Albury, New South Wales, Australia.

Waring, H. (1959) Rottnest Island: the Rottnest Biological Station and recent scientific research. Introduction. *Journal of the Royal Society of Western Australia*, **42**, 65–66.

White, S.R. (1952) The occurrence of the quokka in the south-west. *The Western Australian Naturalist*, **3**, 101–103.
Wilkie, D.S. & Carpenter, J.F. (1999) Bushmeat hunting in the Congo Basin: an assessment of impacts and options for mitigation. *Biodiversity and Conservation*, **8**, 927–955.
de Tores, P.J. & Marlow, N. (2012) The relative merits of predator-exclusion fencing and repeated fox baiting for protection of native fauna: five case studies from Western Australia. In: M.J. Somers and M.W. Hayward (eds) Fencing for Conservation. Springer, New York.
Kowalczyk, R. et al. (2012) Do fences or humans inhibit the movements of large mammals in Białowieża Primeval Forest? In: M.J. Somers and M.W. Hayward (eds) Fencing for Conservation. Springer, New York.

Chapter 3
The Relative Merits of Predator-Exclusion Fencing and Repeated Fox Baiting for Protection of Native Fauna: Five Case Studies from Western Australia

Paul J. de Tores and Nicky Marlow

Introduction

Many mammal species in Western Australia (WA) are susceptible to predation from the introduced European red fox *Vulpes vulpes* and the feral cat *Felis catus*. Predation by these species is now formally recognised as a "key threatening process" (DEWHA 2008a, b). There is also wide recognition by conservation land managers in Australia that repeated baiting for fox control through use of the toxin sodium monofluoroacetate (known as 1080) can result in enhanced survival of many species of native fauna (Braysher 1993; Saunders et al. 1995). Similarly, predator-exclusion fencing has been shown to give protection from fox and cat predation (Long and Robley 2004).

Here we provide an overview of the achievements from repeated fox baiting in WA and identify the requirements for effective fencing. We then review five case studies from WA; two studies where "predator-exclusion fencing" and three where repeated 1080 baiting was used for protection of native fauna. By necessity, each case study was *post hoc* and we were unable to compare the chosen approach with the alternative under the same conditions.

Fencing may appear to be an easy short-term solution to what appears to be an intractable problem of predation by introduced species. However, it is not the solution for protection of Australian native fauna from predation by foxes and cats, nor is repeated 1080 baiting for foxes alone. We outline why this is the case and identify the different premises on which each strategy is based.

P.J. de Tores (✉) • N. Marlow
Department of Environment and Conservation and Invasive Animals Cooperative Research Centre, Western Australian Wildlife Research Centre, PO Box 51, Wanneroo, Western Australia 6946, Australia
e-mail: paul.detores@dec.wa.gov.au

M.J. Somers and M.W. Hayward (eds.), *Fencing for Conservation: Restriction of Evolutionary Potential or a Riposte to Threatening Processes?*,
DOI 10.1007/978-1-4614-0902-1_3,

Background to 1080 Baiting in Western Australia: A Demonstrated Need for Control of the Fox

A number of landmark studies in WA in the 1980s and 1990s demonstrated the effectiveness of repeated 1080 baiting. Successes included:

- Increased trap success of woylies, or brush-tailed bettongs, *Bettongia penicillata* at Dryandra Woodland and Tutanning Nature Reserve (Kinnear et al. 2002) and the Perup Forest (Christensen 1980a, b).
- Increased sightings of numbats *Myrmecobius fasciatus* (Friend 1990; Friend and Thomas 2003) at Dryandra Woodland.
- Increased density of the black-footed rock-wallaby *Petrogale lateralis* at several small granite outcrop reserves within the WA wheatbelt (Kinnear et al. 1988, 1998).
- Increased trap success of the chuditch, or western quoll, *Dasyurus geoffroii* at Batalling Forest Block (Morris et al. 2003) (Fig. 3.1).

Examples of initial translocation success where repeated fox baiting was carried out include woylie translocations to Batalling Forest and Julimar Forest (Orell 2004), western ringtail possum *Pseudocheirus occidentalis* translocations to Leschenault Peninsula Conservation Park (de Tores et al. 1998a) and chuditch translocations to Julimar Forest (Orell 2004) (Fig. 3.1).

Following these initial successes, the WA Department of Conservation and Land Management (CALM, now the Department of Environment and Conservation, DEC) commenced two large-scale aerial 1080 baiting programs in south-west WA. The first of these, "Operation Foxglove", commenced in 1994. This was a collaborative research program between CALM and the then Vertebrate Biocontrol Cooperative Research Centre (VB CRC, which became the Pest Animal CRC and subsequently the Invasive Animals CRC). The core area of the Operation Foxglove study site was 550,000 ha within the northern jarrah forest (Fig. 3.1). The study focussed on assessing the response of native fauna to different frequencies of fox baiting. The second large-scale aerial baiting program, the "Western Shield Introduced Predator Control and Fauna Recovery Program", became the umbrella program for all of CALM's 1080 baiting within south-west WA. Western Shield commenced in 1996 and now covers approximately 3.6 million ha of state forest and conservation reserves in south-west WA.

The large-scale use of 1080 baits to control foxes has been possible in WA due to the divergent levels of tolerance to this poison by introduced and native fauna (King et al. 1978; McIlroy 1981, 1986; Twigg et al. 2003). Fluoroacetate, the active component of 1080, is present naturally in plants from the genus *Gastrolobium* which occur extensively in south-west WA. Many species of native fauna in WA have very high levels of 1080 tolerance due to their evolutionary exposure to the naturally occurring sources of this poison (King et al. 1978). Foxes, feral cats, dogs *Canis familiaris* and dingoes *C. lupus* have a very low tolerance to 1080 (McIlroy 1981). Researchers from the WA Agriculture Protection Board exploited this vulnerability through development of a dried meat bait for fox control. Baits consisted of a 120-g piece of kangaroo meat, injected with 4.5 mg of 1080 in solution, then

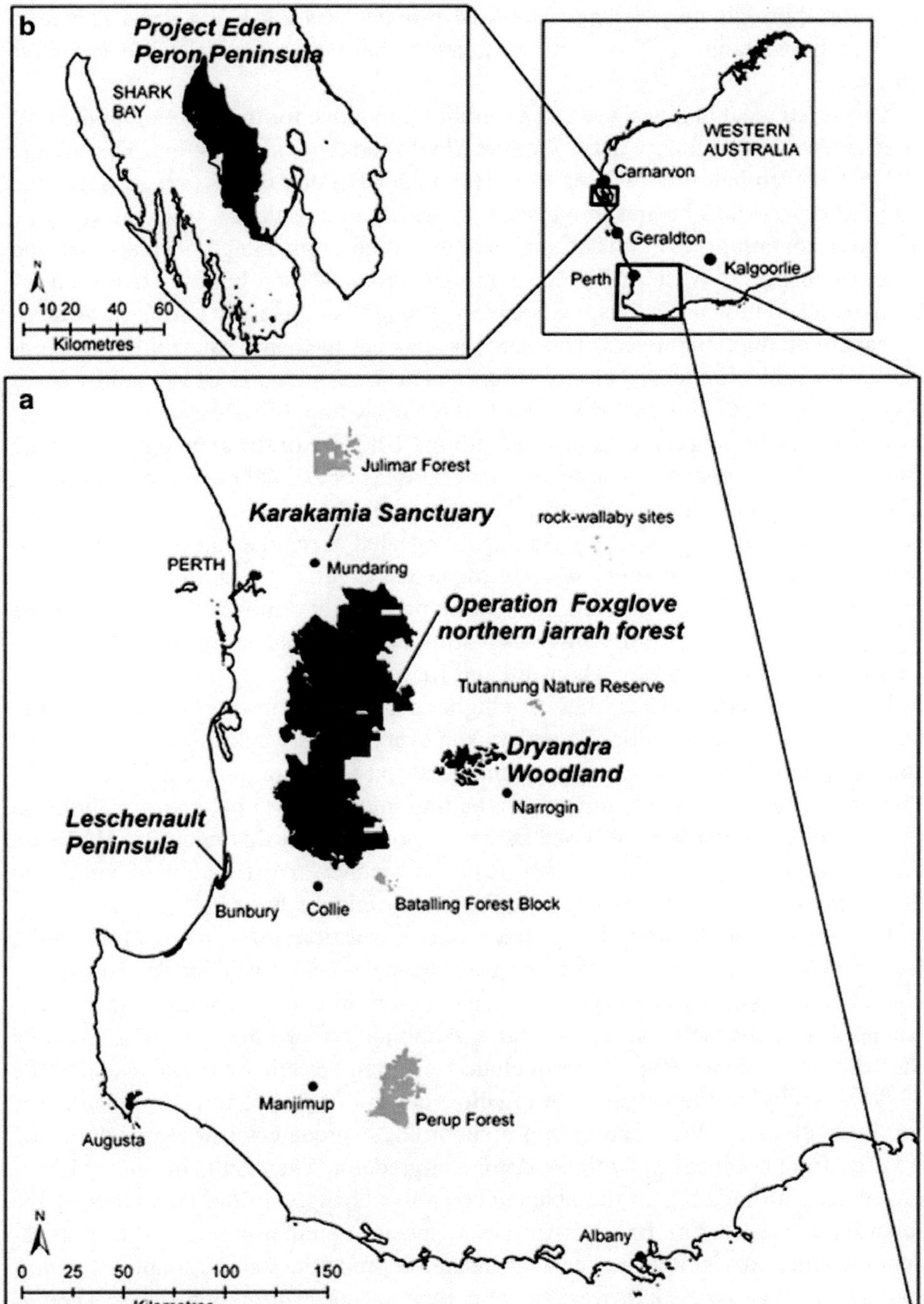

Fig. 3.1 (**a**) The location of landmark studies in Western Australia where control of the introduced European red fox led to demonstrable increases in native fauna capture success. Also shown is the location of four of the five case studies where exclusion fencing or repeated 1080 baiting was used to protect native fauna. The case studies are labelled in bold and italicised. (**b**) The location of the fifth case study, Project Eden at Peron Peninsula, where a barrier fence was constructed

dried for approximately 4 days. The final product was a 40–50g dried kangaroo meat bait, palatable to foxes and sufficiently hard to be unpalatable to native species.

This dried meat bait was used by Operation Foxglove for the duration of the program and was used initially in the Western Shield aerial baiting program. The amount of 1080 in each bait was reduced to 3.0 mg in 2001 (Peter Orell[1] pers. comm.) and from 2005 there has been a progressive transition to a "salami" style sausage bait (Probait) containing 3 mg 1080 and manufactured from kangaroo meat, fat and flavour enhancers. Western Shield aerial and ground-based baiting (bait delivery from vehicles) now use Probait exclusively. Probait is physically hard as a result of the salami production process. This hardness reduces the bait's palatability to potentially vulnerable non-target species (Martin et al. 2002). Uptake of Probait by foxes is comparable to that of the dried meat bait (Marlow unpublished data).

In WA, fox baiting regimes vary according to the size of the area to be protected, with aerial baiting generally adopted only for areas of 20,000 ha or more (de Tores 1994). Large areas with a low perimeter to area ratio have historically been baited every 3 months (de Tores 1994). Baiting is repeated at regular intervals in recognition that foxes can disperse considerable distances in relatively short periods of time (Saunders et al. 1995). This repeated baiting potentially removes foxes which have immigrated to a site since the previous baiting event and is analogous to the "metaphorical fence" described by Hayward and Kerley (2009).

In smaller reserves, where there is a higher perimeter to area ratio and potentially more rapid reinvasion, baiting is undertaken every 4 or 5 weeks and baits are hand delivered (thrown from a vehicle or tethered at fixed bait stations) (de Tores 1994). Bait uptake by foxes is maximised at a baiting intensity of 5 baits km^{-2} (Thomson and Algar 2000) and vehicle based baiting is nominally at this intensity. However, the road network within most reserves results in an uneven distribution of baits, with the baiting intensity invariably greater than the nominal 5 baits km^{-2}.

Despite ongoing baiting of large tracts of conservation estate, many of the initial fauna recoveries and translocation successes have not been sustained (de Tores et al. 2004; Orell 2004). Various hypotheses have been proposed to explain these declines but none is universally accepted and a combination of causal factors is likely (de Tores et al. 2004). Hypotheses include mesopredator release (Crooks and Soulé 1999) whereby, in the presence of effective control of the dominant predator (the fox in south-west WA), one or more subordinate predators (mesopredators) are released from competition with the dominant predator. The results of this "release" can include an increase in the abundance and/or changes in the behaviour of the subordinate predator(s). In south-west WA, species potentially released from competition with foxes include the feral cat and native predators such as goannas *Varanus* spp., chuditch and the south-west carpet python *Morelia spilota imbricata*. There is now increasing evidence from WA supporting the hypothesis of mesopredator release of cats in the presence of repeated fox baiting (de Tores et al. 2005; de Tores and Berry 2007).

[1] Peter Orell: Zoologist (Western Shield), Species and Communities Branch, Nature Conservation Division, Department of Environment and Conservation, Perth, WA.

Characteristics of Fences Used to Protect Species of High Conservation Value

In Australia, predator-exclusion fencing is used primarily to protect listed "threatened" species from threatening processes such as fox and cat predation. Fencing is also widely used in Australia to protect species translocated to areas from which they have become locally extinct. For a predator-exclusion fence to be effective, it must be designed to exclude the predator(s) of concern. The fence design must also consider native species potentially enclosed (Long and Robley 2004), irrespective of whether these are species for which the protection was intended. Environmental features such as water courses and rocky outcrops may reduce the effectiveness or durability of the fence and need to be considered (Long and Robley 2004). There may also be implications from exclusion of the target predator and non-target native species. Exclusion of the desired predator may result in increased predation outside the fence. Fence construction may obstruct or interfere with normal patterns of behaviour or movement for species such as large macropods, it may prevent their use of existing routes (roo pads) and may result in fragmentation of populations of small, less mobile species.

A fundamental requirement of a predator-exclusion fence is that during, or immediately after construction, all individuals of the target introduced predator(s) species confined within the fenced enclosure will be eradicated. For eradication inside the fenced area to be effective the target species must be vulnerable to the control method(s) used e.g. baiting, trapping or shooting. Therefore, the decision to adopt fencing is not independent of the need for other control measures. Subject to the size of the fenced area and the terrain and vegetation enclosed, the removal of introduced predators can be a complex and long process. There must be no pockets where the predator has remained (e.g. in areas where baits cannot be laid) and it must not breed more quickly than it can be controlled (survival and fecundity rates may increase as the density of the target species decreases). The probability of the fence being breached must be sufficiently low to enable control of incursions (Parkes 2002) and there must be ongoing monitoring to detect incursions and maintenance to control such incursions (Clapperton and Day 2001).

Electric fences will deter foxes (Minsky 1980) but foxes are capable of jumping electrified fences 105 cm high (Poole and McKillop 2002). The integrity of the electrical connectivity also needs to be checked regularly (Helen Crisp[2] pers. comm.). It may need to be daily, as is the case for a 14 km fence at "Arid Recovery" in Roxby Downs, South Australia (Helen Crisp pers. comm.). Foxes will also readily walk around the "open" ends of electric fences constructed to create a barrier instead of an enclosure (Poole and McKillop 2002). Despite fence ends extending up to 300 m into coastal waters at Peron Peninsula, WA, occasional very low tidal

[2] Helen Crisp: Ecologist, Arid Recovery, Roxby Downs, South Australia.

events have, at times, exposed the ends above low water level and made incursion around them possible (Colleen Sims[3] pers. comm.).

Long and Robley (2004) highlighted the difficulties of making comparisons of the costs of building fences under different biophysical conditions. There are differences in the costs of fences depending upon their designs, the cost of freight, labour costs and the availability of raw materials. Similarly, Clapperton and Day (2001) cited many factors which influence the cost of fencing, including the size of the area to be fenced, its shape, the number of water courses present, access requirements, the soil type and the subsidence risk of the terrain. Exclusion fences were considered to be more cost effective if the area to be protected is large relative to the length of fencing required (Clapperton and Day 2001). However, with increased size, there are increased costs for initial eradication and ongoing maintenance. Coman and McCutchan (1994) claimed it is only feasible to fence relatively small areas in Australia due to the costs involved. Nonetheless, predator-exclusion over large areas can be achieved, as demonstrated at the Australian Wildlife Conservancy's (AWC) Scotia Sanctuary, western New South Wales. Scotia Sanctuary contains two large fenced areas each 4,000 ha and 80 km of fence-line is checked every second day. In the 5 year period 2005–2010 there were two rabbit incursions only and the fenced areas have remained free of foxes, cats and goats *Capra hircus* (Matthew Hayward[4] pers. comm.).

Case Studies of the Effectiveness of Predator-Exclusion Fencing and Repeated Fox Baiting for Protection of Native Species

We provide five case studies from WA where we compare the protection afforded to vulnerable species from fencing and repeated 1080 baiting. Each case study was examined for:

- Its effectiveness in excluding, or reducing the density of foxes and feral cats.
- The outcomes with respect to the species identified as requiring protection.
- Other realised and potential positive and negative biodiversity conservation outcomes.

Case Study 1: Karakamia Wildlife Sanctuary

Karakamia Wildlife Sanctuary, approximately 50 km north-east of Perth (Fig. 3.1), is a 275-ha, privately owned fenced sanctuary managed by AWC. Karakamia supports a suite of naturally occurring native mammal and reptile species and is also a translocation release site for several threatened mammal species.

[3] Colleen Sims: Project Eden Manager, Department of Environment and Conservation, Shark Bay District, Denham.

[4] Matthew Hayward: South East Regional Ecologist, Australian Wildlife Conservancy.

Table 3.1 Mammal species known to be present at Karakamia Wildlife Sanctuary prior to construction of predator-exclusion fencing

Common name	Scientific name
Yellow-footed antechinus	*Antechinus flavipes*
At least one species of dunnart	*Sminthopsis* sp.
Western pigmy-possum	*Cercartetus concinnus*
Common brushtail possum	*Trichosurus vulpecula hypoleucus*
Western brush wallaby	*Macropus irma*
Western grey kangaroo	*Macropus fuliginosus*
Short-beaked echidna	*Tachyglossus aculeatus*
House mouse[a]	*Mus musculus*
Black rat[a]	*Rattus rattus*

Species are inclusive of those considered extant and known from distribution records only
Source: AWC (2006)
[a] Indicates non-native species

The primary objective of the sanctuary was *to re-estabish the medium-sized mammals that had either declined significantly in the region, or had become regionally extinct* (AWC 2006). This was to be achieved through *the protection and effective management of critical habitat, and the exclusion of all feral animals* (AWC 2006).

A fox and cat proof fence was established in 1992 (Schmitz and Copley 1997). The fence is a "mesh/electric wire composite fence" with closely spaced wires which form an electrical deterrent. It has wire netting to prevent animals pushing through the base of the fence and a buried mesh apron which deters animals from pushing or digging under the fence (Long and Robley 2004). Baiting and trapping were used to remove feral predators from the fenced area after construction, followed by intensive monitoring (using sandplots and spotlighting) for several months to ensure all foxes and cats were eradicated (Manda Page[5] pers. comm.).

Mammal species known to be present at Karakamia Wildlife Sanctuary at the time of fence construction and the species subsequently translocated to the Sanctuary are listed in Tables 3.1 and 3.2, respectively. At least one of the species re-introduced (the woylie) was considered a candidate species to provide founders for future translocations to other areas.

Monitoring of the effectiveness of the fencing was initially intense, with regular inspections to detect incursions. Only four incursions were detected in the 16 year period 1992–2008, two incursions by cats and two by foxes. On each occasion cats were trapped within days and foxes baited within weeks of detection. Incursions were again recorded in 2008 and 2009. In 2008, two dogs were found inside the fenced area and were shot within hours. In 2009, based on evidence from tracks, spotlighting sightings and camera traps, two foxes and one cat were confirmed present. These proved more difficult to remove than previous incursions (Manda Page pers. comm.). Baiting was restricted because of the risk to non-target native species

[5] Manda Page: South West Regional Ecologist, Australian Wildlife Conservancy.

Table 3.2 Mammal species translocated to Karakamia Wildlife Sanctuary subsequent to completion of predator-exclusion fencing

Common name	Scientific name	Date of initial translocation	Category of translocation (based on historic distributions)
Southern brown bandicoot, or quenda	*Isoodon obesulus*	September 1994	Re-introduction
Numbat	*Myrmecobius fasciatus*	December 1994	Re-introduction
Woylie, or brush-tailed bettong	*Bettongia penicillata*	January 1995	Re-introduction
Western ringtail possum	*Pseudocheirus occidentalis*	August 1995	Re-introduction
Quokka	*Setonix brachyurus*	October 1996	Re-introduction
Tammar wallaby	*Macropus eugenii*	November 1998	Re-introduction
Common brushtail possum (eight individuals only)	*Trichosurus vulpecula hypoleucus*	1994	Augmentation
Western brush wallaby (two individuals only)	*Macropus irma*	1994	Augmentation

Source: AWC (2006)

and its effectiveness was limited because of the abundance of alternative food (native prey species). Both species avoided leg-hold traps which also posed a lethal threat to non-target fauna. Shooting proved to be difficult as spotlighting seemed to disturb the animals and sent them into "hiding" in thickly vegetation areas. Sniffer dogs were brought in on two occasions and were not effective. One fox was eventually shot. The other fox and the cat were assumed to have taken baits or escaped as no carcasses were recovered. The incursion point(s) were never identified. "Boundary runs" are now carried out every 2 days and access gates and the fence are being improved and updated. Additional surveillance measures (using spotlighting and dragging tracks) to detect incursions are now undertaken monthly (Manda Page pers. comm.).

Native fauna were monitored using a range of techniques including radio telemetry, trapping, spotlighting and opportunistic sightings. There are unequivocal data from capture success rates for the woylie, tammar wallaby *Macropus eugenii* and common brushtail possum *Trichosurus vulpecula* to conclude populations of these species have increased and have become self-sustaining.

Karakamia Wildlife Sanctuary has now provided founding animals for translocations of woylies, southern brown bandicoots *Isoodon obesulus* and common brushtail possums. Over 795 animals have been translocated from the Sanctuary to other locations in Australia. In excess of 600 woylies were translocated from Karakamia by December 2006 (AWC 2006; Richards et al. 2009). Common brushtail possums have been translocated to Lorna Glen (a DEC managed pastoral lease 1,100 km north-east of Perth) (Manda Page pers. comm.). The quokka *Setonix brachyurus* and western ringtail possum are regularly observed but data are sparse and there are no population estimates. The brush-tailed phascogale *Phascogale tapoatafa* was recorded for the first time at Karakamia in 2010 (Zimmermann[6] unpublished). The numbat is possibly no longer present.

We have interpreted the original stated objective for Karakamia Wildlife Sanctuary as meaning the goals were to (1) ensure populations of extant mammal fauna are self-sustaining; and (2) establish self-sustaining populations of locally extinct mammal fauna. Assessed against the latter, Karakamia Wildlife Sanctuary has clearly established self-sustaining populations of the woylie, tammar wallaby and common brushtail possum (Manda Page pers. comm. and unpublished annual trapping data at 2009). The founding population of woylies was 15 individuals, sourced from Dryandra Woodland in January 1995 (AWC 2006; de Tores unpublished) and supplemented over the following decade with approximately 25 additional individuals, primarily sourced from wildlife carers (AWC 2006). Recent examination of the genetic diversity of the Karakamia woylie population confirmed it has lower variability than two of the three extant, naturally occurring populations (Dryandra Woodland and Perup Forest) and greater variability than the third (Tutanning Nature Reserve) (Carlo Pacioni[7] pers. comm., unpublished).

The genetic structure of other species has not been assessed. The woylie and tammar wallaby populations may now be at carrying capacity (Manda Page pers. comm.). Trapping data from resident mammal fauna (i.e. not translocated) show the southern brown bandicoot is consistently present but numbers fluctuate and interpretation of trapping data is confounded by trap saturation from woylies. The common brushtail possum population may also be at carrying capacity and the western grey kangaroo *Macropus fuliginosus* population has been regularly culled. The physical condition of western brush wallabies *Macropus irma* has been reported to be poor (AWC 2006).

Case Study 2: Project Eden (Peron Peninsula)

Project Eden is a large-scale conservation project aimed at *translocation and reconstruction of the pre-European fauna* of the Peron Peninsula (Morris et al. 2004). Peron Peninsula is the eastern most peninsula of the Shark Bay region of WA and is 900 km north of Perth (Fig. 3.1). It is a 105,000-ha peninsula comprised of Francois Peron National Park and Unallocated Crown Land within the Shark Bay World Heritage Area. Prior to European settlement, the Shark Bay area supported 37 native terrestrial species (Baynes 1990 as cited by Morris et al. 2004). Peron Peninsula is reported to have a rich herpetofauna assemblage (McKenzie et al. 2000). It is a translocation release site for four threatened mammal species and one threatened bird species. Peron also supports a captive breeding facility for three of the four threatened mammal species released on the peninsula (Morris et al. 2004). The specific objectives of Project Eden were to (1) control introduced predators; (2) re-establish

[6] Laura Zimmermann: Honours student, University of Western Australia.

[7] Carlo Pacioni: PhD student, Murdoch University.

Fig. 3.2 Predator-exclusion fence at Project Eden, Peron Peninsula. The road provides public access and is a potential limitation to effective exclusion, despite the presence of a stock grid and an ultrasonic noise emitter. Photograph: Western Australian Department of Environment and Conservation, Shark Bay District

a range of threatened animals, particularly mammals; and (3) foster and promote sustainable nature-based tourism. The project was seen as an "operational experiment" with three phases: phase one – introduced animal control; phase two – fauna translocations; and phase three – nature based tourism. Commencement of phase two was subject to successful completion of phase one and commencement of phase three was similarly subject to successful completion of phase two (Morris et al. 2004). However, as the project developed it became clear complete feral eradication was difficult and it was recognised there would be some degree of overlap of the three stages (Colleen Sims pers. comm.).

Eight native species of non-volant mammal and five species of introduced mammal, including the fox and feral cat, were known to be extant at Peron Peninsula at the commencement of Project Eden (Morris et al. 2004; Colleen Sims pers. comm.). Cyanide baiting, using the methods outlined by Algar and Kinnear (1992), estimated the density foxes as 2.5 km^{-2} before aerial baiting with fox baits. A 3.4-km electrified barrier fence was established in 1996, after the initial aerial baiting for foxes. The fence was constructed across the isthmus at the southern end of the peninsula with the objective of preventing reinvasion by foxes and feral cats (Morris et al. 2004). The fence is comprised of wire netting (~2 m high) to provide a physical barrier and has a horizontal 30 cm wide rabbit mesh overhang, with two electric wires underneath the overhang to deter climbing animals. It has a buried mesh apron on the "outside" to deter animals from pushing or digging under the fence (Long and Robley 2004; Colleen Sims pers. comm.). A recording of a dog barking, an ultrasonic noise emitter and stock grid were installed at the point where the 5–6 m wide main public access road breaches the barrier fence (Fig. 3.2). The recording and ultrasonic noise is activated when an infrared beam spanning the road is broken (Morris et al. 2004).

The initial aerial fox baiting in 1995 was estimated to have reduced the fox population by 95%, as determined by track counts. Follow-up baiting and trapping produced "effective eradication" or at least absence of detection of foxes (Colleen Sims pers. comm.). However, fox numbers were considered to be at "very low levels" (Morris et al. 2004) which indicates eradication had not been achieved and/or there had been incursions. Infrequent incursions were reported over the 15 year period from 1995 to 2009 and fox track counts were observed on 17 of 125 standardised track count monitoring sessions. However, these counts did not exceed six foxes 100 km^{-1} (Colleen Sims pers. comm.). Formal track counting sessions are carried out every 1–3 months. When fox tracks are observed they are immediately targeted with vehicle based baiting and checked for continued presence after baiting. The presence of tracks usually ceases within 2–3 days of baiting (Colleen Sims pers. comm.). Fortnightly monitoring and baiting of the barrier fence is also carried out.

Cat numbers were considered to have increased after fox baiting. However, the methodology used to determine this was unclear, as prior to implementing fox baiting, cat numbers were described as "difficult to determine" because of the high number of rabbit *Oryctolagus cuniculus*, fox and goat *Capra hircus* tracks (Morris et al. 2004). Cat control was attempted through various baiting techniques but results were highly variable. Aerial baiting targeting cats was carried out in 1996 and, based on track counts, was estimated to have resulted in an 80% reduction in cat numbers. Similarly, aerial cat baiting in 2002 resulted in a 65–85% reduction in cat numbers based on track counts, use of rhodamine (biomarker) baiting and known mortality of radio-collared cats. The 2002 baiting and follow-up trapping resulted in cat track counts of less than 10 cats 100 km^{-1}. This was kept at less than 20 cats 100 km^{-1} for 9 months, before gradually increasing (Colleen Sims pers. comm.). Various baiting protocols in intervening and subsequent years failed to produce a detectable impact on feral cat populations (Colleen Sims pers. comm. and unpublished data). The ability of cats to persist in the presence of baiting may have been due to the timing of the baiting and the concurrent seasonal availability of alternative food sources, in particular rabbits, spinifex hopping-mouse *Notomys alexis* and house mouse *Mus musculus* (Morris et al. 2004; Colleen Sims pers. comm.). Trapping became "the main control" option for cat control and numbers were estimated at 20–30 cats 100 km^{-1} of track count (Morris et al. 2004).

Monitoring of resident native mammal fauna was reported in terms of capture success only (Morris et al. 2004) and indicated no response to predator control or the barrier fence. A goal of recording less than 10 cats and foxes 100 km^{-1} of track count was set as the pre-requisite for commencing translocations. Despite the target track count not being met, woylies were translocated to Peron Peninsula between 1997 and 2000. Captive bred bilbies *Macrotis lagotis*, banded hare-wallabies *Lagostrophus fasciatus* and rufous hare-wallabies *Lagorchestes hirsutus* were released between 2000 and 2001.

Overall survivorship of woylies was poor (quantitative estimates of survivorship were not reported), monitoring was considered inadequate and the fate of un-collared animals was unknown (Morris et al. 2004). Subsequent monitoring found 33% of woylie deaths were a result of predation by cats (Morris et al. 2004).

Fig. 3.3 The predator-exclusion fence at Project Eden extends into Shark Bay to minimise the possibility of introduced predator incursion at low tide. Photograph: Linda Reinhold, Western Australian Department of Environment and Conservation, Shark Bay District

Monitoring of radio-tagged bilbies (approximately 50% of 55 captive bred released individuals) showed 81% to be alive 6 months after release, however only 7% were alive after 12 months (Morris et al. 2004). Monitoring further releases of radio-tagged bilbies (44 radio-tagged from a total of 171 releases) over a period ranging from 2 days to 16 months recorded only 2 mortalities, neither attributed to predation. However, there was a high loss of radio tags. Although there are no quantitative data analyses available, 10 years after the initial release the bilby population is considered self-sustaining (Colleen Sims pers. comm.) and this is thought unlikely to have been possible if the initial high mortality rate had continued (Colleen Sims pers. comm.).

Monitoring of banded hare wallaby and rufous hare-wallaby was more intense than the woylie and bilby monitoring. Results from radio-telemetry monitoring revealed a high mortality rate for both species of hare wallaby. Cat predation was considered the primary cause of death (Morris et al. 2004).

The results clearly indicated cat predation to be the major impediment to translocation success for most species at Peron. The use of existing cat baits and intensive trapping was unable to achieve eradication of feral cats after installation of the barrier fence. The cat population is likely derived from individuals present before the fence was constructed and/or from domestic and stray cats from within the township, augmented by recruitment of cats from one or more incursions through or around the fence (Colleen Sims, pers. comm.) (Fig. 3.3).

Peron Peninsula is a complex area to manage. It presents a possibly unique circumstance of a barrier fence through which the public gain vehicle access to a township within the fenced area. Cat ownership is permitted within the town. The fenced area is extensive and the track system does not provide complete access to all areas where cats can be present. It has seasonally high prey populations of rabbits and native mammals which support cat populations and reduce the effectiveness of cat baiting, but seemingly not fox baiting.

Case Study 3: Leschenault Peninsula Conservation Park

Leschenault Peninsula Conservation Park is approximately 150 km south of Perth (Fig. 3.1), is approximately 11 km long and has an area of 1,070 ha. The primary purpose of management of Leschenault, and all conservation parks in WA, is maintenance and restoration of the natural environment (Conservation and Land Management Act 1984).

Leschenault Peninsula has extant populations of several native mammal and reptile species and is a translocation release site for the western ringtail possum. Western ringtail possum translocation commenced in 1991 (de Tores et al. 1998a). A cyanide baiting program (see Algar and Kinnear 1992) was conducted to estimate the number of foxes present before the initial translocation. Eleven foxes and one feral cat were removed after three consecutive days of cyanide baiting. A monthly 1080 baiting program commenced in September 1991 and has been maintained, albeit with changes to the baiting regime and missed baiting sessions (de Tores et al. 2004). A follow-up cyanide baiting program was conducted in the mid 1990s and resulted in removal of only one fox, indicating the on-going 1080 baiting program was effective at reducing fox density. Foxes are now rarely observed at Leschenault Peninsula but, as anticipated from repeated baiting, they have not been eradicated.

By 1998, radio telemetry monitoring of survivorship, home range, dispersal patterns and habitat use, combined with spotlight surveys indicated translocated western ringtail possums had established a self-sustaining population (de Tores et al. 1998a). A follow-up spotlighting program conducted in 2002 revealed the population had declined to a level where presence only was able to be detected (de Tores et al. 2004). The resident population of common brushtail possums increased after commencement of 1080 baiting and did not show the subsequent decline shown by western ringtail possums. Various hypotheses were proposed to explain the western ringtail possum decline, including changes to the baiting regime, drought, mesopredator release of cats and pythons, prey switching (rabbit numbers had been reduced as a result of a concurrent 1080 oat baiting program), unsuitable habitat at the release sites, competition with the sympatric common brushtail possum and anthropogenic factors such as fire and disease (de Tores et al. 2004).

Further release of western ringtail possums and intensive radio telemetry monitoring of survivorship commenced in 2004 at Leschenault and at an unbaited control site within Yalgorup National Park, 30 km north of Leschenault (Fig. 3.1). Predation by feral cats and the native south-west carpet python were shown to be limiting translocation success at Leschenault, with negligible predation events attributable to foxes.

Of 25 radio collared ringtails known to have died as a result of predation, nine were attributed to cat predation, 11 to predation by the south-west carpet python and only one to fox predation. The remaining four predation events were unable to be attributed to a predator species. Conversely, at the Yalgorup National Park unbaited control site, only one of 15 predation events was attributed to cat predation, none to pythons, three or four to raptors and three or four attributed to fox predation. Again, the remaining predation events were unable to be attributed to a predator species (de Tores unpublished).

Although inconclusive, the results are consistent with the mesopredator release hypothesis, where as a result of the effective fox baiting program, feral cats and the south-west carpet python have been released from competition with foxes. The effective fox baiting may also be responsible for the recolonisation (immigration) by two native predator species, the chuditch and brush-tailed phascogale, both previously thought to be locally extinct. The lack of a detectable increase in these two predators may be a result of competition with cats.

Case Study 4: Dryandra Woodland

Dryandra Woodland is situated 26 km north-west of Narrogin, in the wheatbelt region of south-west WA. Dryandra is considered an iconic reserve and is managed for conservation and recreation purposes. It is one of only three locations supporting naturally occurring populations of the woylie and one of only two locations supporting naturally occurring populations of the numbat. Dryandra Woodland is surrounded by farmland, is heavily dissected and has a large perimeter to area ratio. Fox baiting commenced at Dryandra in 1989 and it is ground baited every 4 weeks.

Trap success rates for woylies and common brushtail possums increased dramatically after baiting commenced (Kinnear et al. 2002; Orell 2004). Two species which were once locally extinct, the southern brown bandicoot and chuditch, have naturally recolonised the site. Populations of the tammar wallaby, western brush wallaby, red-tailed phascogale *Phascogale calura* and malleefowl *Leipoa ocellata* have been sustained.

Despite ongoing four-weekly baiting, trap success of the woylie has declined and returned to its pre-1989 level. The success of translocated populations of the bilby and burrowing bettong *Bettongia lesueur* at Dryandra has also been variable. The first translocation of bilbies, carried out between May 2000 and October 2001, failed. A subsequent release in spring 2003 appears to have been successful, although bilby density is low. All releases of the burrowing bettong have been unsuccessful with most individuals suffering predation from foxes and cats, with some predation by south-west carpet pythons (Neil Thomas[8] pers. comm.).

At Dryandra Woodland the effectiveness of fox baiting is still equivocal and, as expected with repeated baiting, foxes have not been eradicated. However, the current

---

[8] Neil Thomas: Principal Technical Officer, Department of Environment and Conservation, Western Australian Wildlife Research Centre.

fox baiting regime is sufficient to protect common brushtail possums and reintroduced bilbies. The trap success of species which are susceptible to cat predation, such as woylies, is generally low and feral cats have been identified as being a major predator of radio-collared woylies (Marlow unpublished). This may be a consequence of mesopredator release of cats. There has been no evidence of increased predation on woylies by the south-west carpet python or raptors (Marlow unpublished).

Case Study 5: Operation Foxglove (The Northern Jarrah Forest)

The northern jarrah forest spans the Darling Plateau and Darling Scarp from Mundaring to Collie (Fig. 3.1). A large-scale research project, Operation Foxglove, was undertaken in the northern jarrah forest from 1994 to 2000. The project assessed native fauna response to different frequencies of 1080 baiting. The abundance of resident native fauna was monitored, as was survivorship of translocated populations of the woylie. The track count technique for deriving indices of fox and cat activity (Allen et al. 1996; Engeman 2005) was considered inappropriate to infer values for density and a site specific technique was developed and validated to derive estimates of fox and cat density (de Tores 1999).

1080 baiting was effective in reducing fox density, with annual estimates of density lowest in the most frequently baited treatment and highest in the unbaited control. Conversely, the annual estimate of cat density was higher in the baited sites (de Tores 1999). Modelling of woylie survivorship gave most support for models where survivorship was a function of a treatment effect (baiting frequency) and a covariate for a seasonal effect. The most frequently baited treatment showed the highest level of survivorship (de Tores et al. 1998b), with survivorship lower across all treatments in winter. The survivorship study also indicated the number of woylie mortalities attributed to predation by cats was highest where fox density had been reduced (de Tores 1999). This suggests a reduction in the number of predation events attributed to foxes may be compensated for by an increase in the number of predation events attributed to cats. This pattern is consistent with the principle of mesopredator release. Other supporting evidence for mesopredator release, albeit equivocal, was the increased density and survivorship of chuditch in the most frequently baited treatment (de Tores 1999). Similarly, the lack of recovery of the quokka at sites in the northern jarrah forest where fox baiting was implemented was hypothesised to be the result of predation by cats (Hayward et al. 2003).

Discussion

Fencing and baiting clearly have their respective merits and limitations. The case studies from WA have highlighted some known issues and benefits of each approach. Some of these are relevant to fencing only, some to repeated baiting only, and some are common to both. The decision as to which option to adopt should be determined by how well each option is likely to meet the objectives. Implicit in this is the assumption that the objectives are well defined. Parkes (1990) noted the goals for goat control in New Zealand were often confused with the means of control and this resulted in ambiguity as to whether the goals were achieved. In Australia, the goals and methodologies for introduced predator control are often similarly confused.

Predator-Exclusion Fencing: General

In most cases in Australia where predator-exclusion fencing is used, there is an awareness of the need to eradicate foxes and cats from within the fenced area. Small fenced enclosures provide opportunities for what is effectively captive breeding of species translocated to, and released in, the fenced area. These individuals may then be used, through translocation programs, to establish new populations. The Karakamia Wildlife Sanctuary case study provides a good example of what can be achieved with predator-exclusion fencing.

Fenced enclosures also provide the wherewithal to protect remnant populations of threatened species which may otherwise become locally extinct. An example is the critically endangered Gilbert's potoroo *Potorous gilbertii*. Fencing, combined with the existing captive breeding program, may be necessary to prevent its extinction and may be an essential part of the recovery process.

However, the term "predator-exclusion fencing" is a misnomer. In most cases where "predator-exclusion fencing" has been adopted in WA, there have been repeated incursions by introduced predators, as evident in the case studies. This is not a groundbreaking finding and has been identified as an issue by numerous authors (Coman and McCutchan 1994 and studies cited therein; Short and Turner 2000). However, the consequences are not trivial.

Exceptions to the pattern of repeated incursions include the 14 km long, predator-exclusion fence at Roxby Downs, South Australia. Moseby and Read (2006) found foxes, cats and rabbits were unable to breach this fence. The fence is checked daily to ensure there is no "shorting out" of the electric wires (Helen Crisp, pers. comm.). When assessing the potential for incursion of a 47-km exclusion fence on Maungatautari Ecological Island in New Zealand, Connolly et al. (2009) found there was a high probability a breach will be detected and exploited by mammalian predators (in this case rodents) within 24 h of the breach occurring.

Where fencing does not result in exclusion of introduced predators and complete eradication of introduced species is not achieved, it will be of limited conservation value. This appears to be the case in the Project Eden case study, where the consequences of not achieving eradication of foxes and cats were compounded by lack of eradication of rabbits.

Repeated Baiting: General

Issues relevant to repeated 1080 baiting include the potential threat to domestic and working dogs, the development of bait shyness and/or bait aversion and an evolved tolerance to 1080. Risk assessment protocols to reduce the risk to working dogs are usually incorporated when conservation estate managers are planning baiting programs. These risk assessments less often incorporate measures to reduce the effects from bait shyness and bait aversion.

We differentiate bait shyness from bait aversion by defining the former as a response to a sub-lethal dose of 1080, where the target species may experience a non-lethal ill effect from ingestion of a bait and, in doing so, become "shy" of consuming another bait. We define the latter as an individual animal's preference (or more accurately, lack of preference) to take a bait. The development of bait shyness is a potential risk from a one-off baiting event and long-term repeated exposure to 1080 baits. Repeated 1080 baiting programs and experimental trials in New Zealand have demonstrated bait shyness by brushtail possums (Morgan et al. 1996; O'Connor and Matthews 1999; Ogilvie et al. 2000).

The issue of evolved tolerance to 1080 is a clear and present concern. In WA, rabbits have shown an increased tolerance to 1080 when repeatedly exposed to it (Twigg et al. 2002). Some fox baiting regimes, such as those in small fragmented Wheatbelt reserves in WA, deliver baits up to 13 times year^{-1} at baiting intensities in excess of the nominal 5 baits km^{-2}. If maintained over the long-term, such baiting regimes will inevitably lead to evolution of 1080 tolerance by foxes. To some extent this is being addressed by research aimed at developing alternative toxins. Management measures could include alternating use of 1080 baiting with trapping and shooting and use of baiting regimes which alternate the toxin (once options are available) to prevent continued, long term exposure to a single toxin.

Issues Common to Exclusion Fencing and Repeated Baiting: Monitoring

Identified weaknesses common to predator-exclusion fencing and repeated 1080 baiting include a lack of a commitment to meaningful, quantifiable monitoring of introduced predators in fenced and baited areas. In most of the WA case studies, the level and type of monitoring of the effectiveness of baiting and fencing, in terms of the abundance of introduced predators, was inadequate to enable meaningful inference. Similarly, the level and type of monitoring of native fauna was inadequate to detect changes in abundance unless these changes were of an order of magnitude or more.

Issues Common to Exclusion Fencing and Repeated Baiting: Gene Flow

There are recognised effects from lack of immigration and emigration created by the construction of a fence and the subsequent lack of gene flow (Frankham et al. 2004). Where immigration and emigration is prevented, for example, when a threatened species is translocated and released in a predator-proof fenced enclosure, this can, and invariably will, result in a loss of gene flow between the translocated population and the source population. Other species inadvertently "fenced in" are subject to the same constraints on immigration and emigration and may be unable to maintain gene flow with congeners outside the fenced area. However, the issue of restricted gene flow is also applicable to small, isolated, unfenced conservation reserves. Maintenance of the genetic stock should be of equal importance as fence and baiting maintenance. It can be achieved through periodic population supplementation but, as with other management functions, requires monitoring.

Issues Common to Exclusion Fencing and Repeated Baiting: Overabundance

Overabundance of some native species was also evident from the WA case studies. Again, this is not a new finding and applies to fenced enclosures and baited areas. The historic response by managers has been to cull, or to capture and remove excess animals. Overabundance of woylies at Karakamia led to difficulties with effective trapping-based monitoring of other species – the large number of woylie captures resulted in trap saturation (AWC 2006). The common brushtail possum also appeared to be overabundant at Karakamia and in excess of 60 individuals were trapped and translocated to AWC's Paruna Sanctuary (AWC 2006). At Karakamia, common brushtail possums were frequently found using unusual diurnal rest sites and the behaviour suggested the sanctuary had reached carrying capacity. The overabundance of western grey kangaroos at Karakamia has necessitated a regular culling program (AWC 2006). Overabundance is not limited to fenced populations. Tutanning Nature Reserve, in the WA wheatbelt, has been baited since 1984 (Kinnear et al. 2002). The baiting program was designed and implemented to protect one of only three extant populations of the woylie. The tammar wallaby was also present at low density at Tutanning when baiting commenced. The tammar wallaby population increased to the extent where it was necessary to remove excess animals (Peter Orell pers. comm.). These removed tammars have been used as founders for translocations to other sites.

Issues Common to Exclusion Fencing and Repeated Baiting: Predation by Cats

The inability to control feral cats was the major impediment to fauna recovery in three of the five case studies and identified as an issue in the remaining two. Cat predation has long been recognised as a factor involved in the decline of native fauna in south-west WA and Australia generally (see e.g. Dickman 1996). The impact of cats on native fauna is compounded as fox baiting does not protect susceptible fauna from predation by feral cats. Confounding this further is that cats prefer to feed on live prey (DEWHA 2008b) and, even in the event of development of a cat bait, will remain more difficult to bait than foxes. Although there has been considerable investment of funds by the WA and Commonwealth governments in the development of a cat bait, the Project Eden case study should highlight concerns as to the efficacy of existing cat baits. There are also many technical impediments still to be overcome before cat baiting can be implemented operationally.

Conclusion

We believe enclosures can be made fox- and cat-free but only if the type of fencing required is adequately researched (*i.e.* selection of fencing appropriate for the biophysical conditions and the predators to be excluded) and installed. It is essential to eradicate foxes and cats after fence installation and there must be a serious commitment to effective maintenance and incursion control. Labour intensive trapping and/or use of non target-specific baits will be required to eradicate the cat population from within each fenced enclosure. Equally essential is the requirement to eradicate rabbits which not only present a risk to the environment but have the potential to act as the primary prey source for cats. In doing so, they increase the difficulty of cat eradication.

Repeated 1080 baiting, or "metaphorical fencing" (Hayward and Kerley 2009), is based on a different premise from exclusion fencing. Unlike fenced areas, there should be no expectation for baited areas to be free of introduced predators. However, repeated 1080 baiting provides the opportunity to reduce the effects of fox predation at a local and landscape scale. It provides the opportunity for in situ and translocated populations of native species to increase in number and establish or maintain self-sustaining populations. By definition, these self-sustaining populations must be able to withstand some degree of predation, whether this is from native and/or introduced predators. In the absence of predator-exclusion fencing there is the potential for immigration and emigration of all species, predators and prey alike. The long-term biodiversity conservation implications of unrestricted immigration and emigration of native prey and native predator species, combined with effective introduced predator control (foxes and cats) through repeated baiting are profound in two ways. Firstly, introduced predators can have significantly greater negative effects than native predators (Salo et al. 2007). This effect can be mitigated if foxes and cats are controlled. Secondly, as native predator diversity increases, food web stability is enhanced (Gross et al. 2009) and intra-guild predator interactions increase (Finke and Denno 2004). This in turn can dampen the negative effects from trophic cascades (Finke and Denno 2004).

The hybrid option of establishing predator-proof enclosures at the core of larger baited areas has the potential to provide the significant positive outcomes for biodiversity conservation. Overabundant species and successful breeding within the introduced-predator free enclosure would provide founders for the larger baited area. This scenario would provide the opportunity, over an evolutionary timeframe, for predation sensitive native species to evolve predator avoidance strategies.

We conclude by hypothesising in the absence of a long-term, well considered approach to introduced predator control, recognising the evolutionary potential of introduced predators and native fauna, further extinctions of Australian native fauna are inevitable.

Acknowledgements We thank Manda Page, Colleen Sims and Helen Crisp for their willingness to provide information from Karakamia Wildlife Sanctuary, Peron Peninsula and Roxby Downs, respectively. Thanks also to Peter Orell for providing information on the tammar wallaby population at Tutanning Nature Reserve, Carlo Pacioni for providing unpublished genetic data on woylie populations and Neil Thomas for providing information on the fate of translocated burrowing bettongs at Dryandra Woodland. We also thank Matt Hayward and the Society for Conservation Biology for the invitation to present the case studies from WA. Thanks to Suzanne Rosier for editing and advice on drafts and Matt Hayward and two anonymous referees for constructive comments on a draft.

References

Algar, D. & Kinnear, J.E. (1992) Cyanide baiting to sample fox populations and measure changes in relative abundance. *Bureau of Rural Resources Proceedings*, **11**, 135–138.

Allen, L., Engeman, R. & Krupa, H. (1996) Evaluation of three relative abundance indices for assessing dingo populations. *Wildlife Research*, **23**, 197–206.

AWC (2006) Karakamia Wildlife Sanctuary. Jan–Dec 2006. Unpublished report prepared by the Australian Wildlife Conservancy.

Baynes, A. (1990) Report on a contract to search for Bilby remains on Peron Peninsula, Shark Bay, Western Australia. Unpublished report prepared for the Department of Conservation and Land Management, Western Australia.

Braysher, M. (1993) *Managing Vertebrate Pests: Principles and Strategies.* Bureau of Resource Sciences, Department of Primary Industries and Energy, Canberra.

Christensen, P.E.S. (1980a) A sad day for native fauna. *Forest Focus*, **23**, 2–12.

Christensen, P.E.S. (1980b) The biology of *Bettongia penicillata* Gray, 1837, and *Macropus eugenii* (Desmarest, 1817) in relation to fire. Bulletin 91. Forests Department of Western Australia. Perth.

Clapperton, B.K. & Day, T.D. (2001) *Cost effectiveness of exclusion fencing for stoat and other pest control compared with conventional control.* Department of Conservation. New Zealand, Wellington.

Coman, B. & McCutchan, J. (1994) *Predator exclusion fencing for wildlife management in Australia.* Australian Nature Conservation Agency, Canberra.

Connolly, T.A., Day, T.D. & King, C.M. (2009) Estimating the potential for reinvasion by mammalian pests through pest exclusion fencing. *Wildlife Research*, **36**, 410–421.

Crooks, K.R. & Soulé, M.E. (1999) Mesopredator release and avifaunal extinctions in a fragmented system. *Nature*, **400**, 563–566.

de Tores, P.J. (1994) *Fox Control Manual. Operational guidelines for control of the Red Fox,* Vulpes vulpes, *through the use of Sodium monofluoroacetate or '1080' on CALM managed estate and in other CALM programs.* Unpublished report prepared for the Western Australian Department of Conservation and Land Management. Perth.

de Tores, P.J. (1999) *1997–1998 Report to Environment Australia National Feral Animal Control Program ISF06638. Control and Ecology of the Red Fox in Western Australia - Prey Response to 1080 Baiting Over Large Areas.* Unpublished report prepared for Environment Australia, National Feral Animal Control Program. ISF06638. Department of Conservation and Land Management. Perth.

de Tores, P.J. & Berry, O. (2007) Will curiosity kill the cat? *Landscope*, **22**, 49–55.

de Tores, P.J., Rosier, S.M. & Paine, G. (1998a) Conserving the western ringtail possum. *Landscope*, **13**, 28–35.

de Tores, P.J., Himbeck, K., Dillon, M., Cocking, J., MacArthur, B. & Rosier, S. (1998b) Large scale fox control in the northern jarrah forest of southwest Western Australia. In: *11th Australian Vertebrate Pest Conference*. (pp. 275–281), Western Australia. Bunbury.

de Tores, P.J., Hayward, M.W. & Rosier, S.M. (2004) The western ringtail possum, *Pseudocheirus occidentalis*, and the quokka, *Setonix brachyurus*, case studies: *Western Shield* review - February 2003. *Conservation Science Western Australia*, **5**, 235–257.

de Tores, P.J., Rosier, S., Guthrie, N., Jackson, J. & Bertram, I. (2005) The western ringtail possum - a resilient species or another taxon on the decline? (Part 2) *Western Wildlife*, **9**, 1 & 4–5.

DEWHA (2008a) *Threat abatement plan for predation by the European red fox*. Department of the Environment, Water, Heritage and the Arts, Canberra.

DEWHA (2008b) *Threat abatement plan for predation by feral cats*. Department of Environment, Water, Heritage and the Arts, Canberra.

Dickman, C.R. (1996) *Overview of the impacts of feral cats on Australian native fauna*. Australian Nature Conservation Agency, Canberra.

Engeman, R.M. (2005) Indexing principles and a widely applicable paradigm for indexing animal populations. *Wildlife Research*, **32**, 203–210.

Finke, D.L. & Denno, R.F. (2004) Predator diversity dampens trophic cascades. *Nature*, **429**, 407–410.

Frankham, R., Ballou, J. D. & Briscoe, D. A. (2004) *A Primer of Conservation Genetics*. Cambridge University Press. Cambridge.

Friend, J.A. (1990) The Numbat *Myrmecobius fasciatus* (Myrmecobiidae): history of decline and potential for recovery. In: Australian Ecosystems: 200 years of Utilization, Degradation and Reconstruction. Geraldton, Western Australia. Ecological Society of Australia. *Proceedings of the Ecological Society of Australia*, **16**, 369–377.

Friend, J.A. & Thomas, N.D. (2003) Conservation of the numbat (*Myrmecobius fasciatus*). *Predators with pouches: The Biology of Carnivorous Marsupials*. (eds. M. Jones, C. Dickman & M. Archer). pp. 446–457. CSIRO Publishing. Melbourne.

Gross, T., Rudolf, L., Levin, S. & Dieckman, U. (2009) Generalised Models Reveal Stabilizing Factors in Food Webs. *Science*, **325**, 747–750.

Hayward, M.W. & Kerley, G.I.H. (2009) Fencing for conservation: Restriction of evolutionary potential or a riposte to threatening processes? *Biological Conservation,* **142**, 1–13.

Hayward, M.W., de Tores, P.J., Dillon, M.J. & Fox, B.J. (2003) Local population structure of a naturally occurring metapopulation of the quokka (*Setonix brachyurus* Macropodidae: Marsupialia). *Biological Conservation*, **110**, 343–355.

King, D.R., Oliver, A.J. & Mead, R.J. (1978) The adaptation of some Western Australian mammals to food plants containing fluoroacetate. Australian *Journal of Zoology*, **26**, 699–712.

Kinnear, J.E., Onus, M.L. & Bromilow, R.N. (1988) Fox control and rock-wallaby population dynamics. *Australian Wildlife Research*, **15**, 435–450.

Kinnear, J.E., Onus, M.L. & Sumner, N.R. (1998) Fox Control and rock-wallaby population dynamics - II. An update. *Wildlife Research*, **25**, 81–88.

Kinnear, J.E., Sumner, N.R. & Onus, M.L. (2002) The red fox in Australia - an exotic predator turned biocontrol agent. *Biological Conservation*, **108**, 335–359.

Long, K. & Robley, A. (2004) *Cost Effective Feral Animal Exclusion Fencing for Areas of High Conservation Value in Australia*. Australian Government Department of the Environment and Heritage, Canberra.

Martin, G.R., Twigg, L.E., Marlow, N.J., Kirkpatrick, W.E., King, R.D. & Gaikhorst, G. (2002) The acceptability of three types of predator baits to captive non-target animals. *Wildlife Research*, **29**, 489–502.

McIlroy, J.C. (1981) The sensitivity of Australian animals to 1080 poison. II. Marsupial and eutherian carnivores. *Australian Wildlife Research*, **8**, 385–399.

McIlroy, J.C. (1986) The sensitivity of Australian animals to 1080 poison. IX. Comparisons between the major groups of animals, and the potential danger non-target species face from 1080-poisoning campaigns. *Australian Wildlife Research*, **13**, 39–48.

McKenzie, N.L., Rolfe, J.K., Aplin, K.P., Cowan, M.A. & Smith, L.A. (2000) Herpetofauna of the southern Carnarvon Basin, Western Australia. *Biodiversity of the southern Carnarvon Basin.* (eds. A.H. Burbidge, M.S. Harvey & N.L. McKenzie), pp. 335–360. Records of the Western Australian Museum Supplement No. 61.

Minsky, D. (1980) Preventing fox predation at a least tern colony with an electric fence. *Journal of Field Ornithology*, **50**, 80–81.

Morgan, D.R., Morriss, G. & Hickling, G.J. (1996) Induced 1080 bait-'shyness' in captive brushtail possums and the implications for management. *Wildlife Research*, **23**, 207–212.

Morris, K., Johnson, B., Orell, P., Gaikhorst, G., Wayne, A. & Moro, D. (2003) Recovery of the threatened chuditch (*Dasyurus geoffroii*): A case s study. *Predators With Pouches. The biology of carnivorous marsupials.* (eds. M. Jones, C. Dickman & M. Archer). CSIRO Publishing. Collingwood, Victoria.

Morris, K., Sims, C., Himbeck, K., Christensen, P., Sercombe, N., Ward, B. & Noakes, N. (2004) *Project Eden* - fauna recovery on Peron Peninsula, Shark Bay: *Western Shield* review - February 2003. *Conservation Science Western Australia*, **5**, 202–234.

Moseby, K.E. & Read, J.L. (2006) The efficacy of feral cat, fox and rabbit exclusion fence designs for threatened species protection. *Biological Conservation*, **127**, 429–437.

O'Connor, C.E. & Matthews, L.R. (1999) 1080-induced bait aversions in wild possums: influence of bait characteristics and prevalence. *Wildlife Research,* **26**, 375–382.

Ogilvie, S.C., Thomas, M.D., Morriss, G.A., Morgan, D.R. & Eason, C.T. (2000) Investigation of sodium monofluoroacetate (1080) bait shyness in wild brushtail possum (*Trichosurus vulpecula*) populations. *International Journal of Pest Management,* **46**, 77–80.

Orell, P. (2004) Fauna monitoring and staff training: *Western Shield* review - February 2003. *Conservation Science Western Australia*, **5**, 51–95.

Parkes, J. (1990) Feral goat control in New Zealand. *Biological Conservation*, **54**, 335–348.

Parkes, J.P. (2002) Potential use of rabbit haemorrhagic disease virus (RHDV) as a biological agent for European rabbits (*Oryctolagus cuniculus*) on Clarion Island, Mexico. Landcare Research Contract report LC0203/015.

Poole, D.W. & McKillop, I.G. (2002) Effectiveness of two types of electric fence for excluding the Red Fox (*Vulpes vulpes*). *Mammal Review*, **32**, 51–57.

Richards, J., Gardner, T. & Copley, M. (2009) Bringing back the animals. *Landscope*, **24**, 55–61.

Salo, P., Korpima, E., Banks, P.B., Nordstro, M. & Dickman, C.R. (2007) Alien predators are more dangerous than native predators to prey populations. *Proceedings of the Royal Society B*, **274**, 1237–1243.

Saunders, G., Coman, B., Kinnear, J. & Braysher, M. (1995) *Managing Vertebrate Pests: Foxes.* Australian Government Publishing Services. Canberra.

Schmitz, A. & Copley, M. (1997) Karakamia Sanctuary. *Landscope*, **13**, 17–22.

Short, J. & Turner, B. (2000) Reintroduction of the burrowing bettong *Bettongia lesueur* (Marsupialia: Potoroidae) to mainland Australia. *Biological Conservation,* **96**, 185–196.

Thomson, P. C. & Algar, D. (2000) The uptake of dried meat baits by foxes and investigations of baiting rates in Western Australia. *Wildlife Research*, **27**, 451–456.

Twigg, L.E., Martin, G. ., Eastman, A.F., King, D.R. & Kirkpatrick, W.E. (2003) Sensitivity of some Australian animals to sodium fluoroacetate (1080): additional species and populations, and some ecological considerations. *Australian Journal of Zoology*, **51**, 515–531.

Twigg, L.E., Martin, G.R. & Lowe, T.J. (2002) Evidence of pesticide resistance in medium-sized mammalian pests: a case study with 1080 poison and Australian rabbits. *Journal of Applied Ecology*, **39**, 549–560.

Chapter 4
Fences or Ferals? Benefits and Costs of Conservation Fencing in Australia

Chris R. Dickman

Introduction

The arrival of white settlers on Australia's east coast in 1788 set in train a series of events that had very large effects on the nation's first people and on the native flora and fauna. The grand sweep of these events and their repercussions have been documented at some length (e.g. Clark 1981) but, for some elements of the Australian vertebrate fauna, arguably the most important change wrought by the new settlers was the introduction of new species. Domestic animals were imported to provide meat, dairy produce, leather and other materials, companionship, and a sense of the familiar in a new environment (Rolls 1969). Some were introduced for sporting and hunting purposes or, like house cats *Felis catus*, deliberately to replace the native fauna; others such as black rats *Rattus rattus* and house mice *Mus musculus* were carried as stowaways on the early ships (Low 1999; Long 2003). From the mid-nineteenth century acclimatisation societies systematically imported many new species of birds and mammals to Australia. At the same time state governments began developing policies such as the Marsupial Destruction Acts to extirpate native species that were considered to interfere with the interests of the new settlers (Hrdina 1997; Dickman 2007). Although these attitudes no longer prevail, the activities of the early European settlers facilitated the spread of many new species in the Australian environment. Two introduced carnivores – the European red fox *Vulpes vulpes* and the domestic house cat – have spread widely, and are now of particular concern owing to their damaging effects on a wide range of native Australian vertebrates (Salo et al. 2007).

C.R. Dickman (✉)
Institute of Wildlife Research, School of Biological Sciences, University of Sydney, Sydney, NSW 2006, Australia
e-mail: chris.dickman@sydney.edu.au

M.J. Somers and M.W. Hayward (eds.), *Fencing for Conservation: Restriction of Evolutionary Potential or a Riposte to Threatening Processes?*,
DOI 10.1007/978-1-4614-0902-1_4, © Springer Science+Business Media, LLC 2012

The first cats were brought to Australia in the late eighteenth century, but appear to have remained close to villages and townships in the early years of settlement. They spread into less disturbed areas from points of introduction along the coast between 1824 and 1886 (Gaynor 2000; Abbott 2002), and now occur in all terrestrial habitats in all parts of the continent as well as on over 40 islands offshore (Dickman 1992). Cat populations have been categorised as domestic if their resource requirements are wholly and intentionally met by humans, and feral if the population is self-sustaining and not dependent on access to human-derived resources (Moodie 1995). A third category, stray or semi-feral, includes cats that use remote human resources such as those at rubbish tips (Denny et al. 2002). Although cats may move between categories within their lifetimes or between generations, most attention has focused on the impacts of feral cats on native fauna (Denny and Dickman 2010). The major impact of feral cats is almost certainly that of direct predation, with historical evidence suggesting that this predator may have driven seven species of small native mammals to extinction and contemporary experimental evidence suggesting that cats can severely deplete local populations of extant small mammals and birds (Dickman 1996a, b; Risbey et al. 2000; de Tores and Marlow 2012). Cats may also affect native vertebrates by acting as vectors for diseases such as toxoplasmosis and sparganosis (Moodie 1995) and are suspected to have further deleterious impacts on ecologically equivalent native species such as quolls *Dasyurus* spp. via competition (Glen and Dickman 2005, 2008). Predation by feral cats has been listed by the Australian government as a key threatening process under the Environment Protection and Biodiversity Conservation (EPBC) Act 1999.

The red fox was introduced to Australia on several occasions in the mid-nineteenth century, but became established in the early 1870s near Melbourne following the earlier spread of the European rabbit *Oryctolagus cuniculus* (Rolls 1969; Abbott 2011). Foxes spread rapidly – up to 160 km a year – in southern coastal and sub-coastal areas, probably with human assistance; the rate of spread was slower in arid regions and in the tropical north where human settlement was sparse (Saunders et al. 1995). Foxes now occupy the southern three quarters of Australia. They appear transiently in some of the more arid regions after heavy rainfalls have stimulated eruptions of small mammals (Letnic and Dickman 2010), and are moving slowly into tropical areas in both Queensland and Western Australia (Long 1988; Saunders et al. 2010). A wealth of historical, circumstantial and experimental evidence confirms that foxes have strongly deleterious effects on a wide range of native Australian species, with depredation affecting mammals, birds and reptiles weighing 35–5,500 g most severely (Burbidge and McKenzie 1989; Kinnear et al. 2010; Saunders et al. 2010). Foxes may exert further effects via transmission of diseases such as sarcoptic mange and sparganosis, competition and an array of indirect effects (Saunders et al. 2010), but the relative importance of these non-predatory interactions remains unclear. Foxes are also a recognised pest for sheep farmers, with rates of predation on lambs ranging from 1 to 30% (Lugton 1993; Saunders et al. 2010). As for feral cats, predation by the European red fox has been listed as a key threatening process under the EPBC Act 1999.

Several methods can be employed to reduce the detrimental effects of feral cats and foxes. Shooting and trapping are often used but are costly, labour-intensive and potentially effective only in small areas. In agricultural landscapes the impacts of foxes are sometimes controlled by drives, den fumigation, or the deployment of guard animals in the paddocks (Saunders and McLeod 2007). However, the most commonly used control method in both conservation reserves and agricultural areas is the setting of baits laced with the poison sodium fluoroacetate, or "1080" (Saunders et al. 2010). In New South Wales about a million baits are laid each year over several hundred thousand hectares (Saunders and McLeod 2007), while equally large or larger areas are baited annually in Victoria and Western Australia (Armstrong 2004; Department of Sustainability and Environment 2005). The consequent reductions in numbers of predators, mostly foxes, allow the persistence of a wide range of native vertebrates (Kinnear et al. 2002; Dexter and Murray 2009; Mahon 2009). Despite these benefits, there are general concerns that 1080 may place some non-target species at risk of poisoning, that its mode of action is not humane (Marks et al. 2009), and that in the long term selection for 1080-resistance will occur (Twigg et al. 2002). In addition to these concerns, feral cats seldom take 1080-baits and their control over large areas remains an almost intractable problem (Risbey et al. 1997; Short et al. 1997; but cf. Algar et al. 2007). Current research on new toxins such as para-aminopropiophenone (PAPP) and new methods of bait delivery give some hope for the future, but will need to meet concerns about possible effects on non-target species before new baits can be deployed at a broad scale (Denny and Dickman 2010). Several authors have proposed that populations of foxes and cats could be reduced over large areas by dingoes *Canis lupus dingo*, but to be effective this would require cessation of the active culling and suppression programs of the dingo by the pastoral community (Johnson et al. 2007; Dickman et al. 2009; Letnic et al. 2009). Biological control of both predator species has also been suggested, but research on this possibility has been unsuccessful so far (Strive et al. 2007; Denny and Dickman 2010).

Against this background, an enterprising but contentious approach has been developed in recent years to provide complete and potentially long term protection for native fauna against the depredations of both feral cats and foxes. This is the conservation fence. Fences have long been used to retain livestock or provide protection against pests or wildlife intruders (Hayward and Kerley 2009), but fences designed to secure large and viable populations of high-value native species are a more recent concept. In Australia the first such fence appears to have been completed by John Wamsley at Warrawong in South Australia in 1975 (De Alessi 2003/04). Such fences have considerable advantages over other methods of fauna protection, but also are very expensive and have been criticised as being little more than captive breeding areas (Pickard 2007).

In this chapter I describe different kinds of conservation fences that are used in Australia, their rationale, effectiveness and costs, and conclude by discussing ways in which fence designs could be extended and improved. Because of the disproportionately negative effects of feral cats and foxes on native biota, I focus primarily on fences that are intended to exclude these predators. I also focus on built structures.

The importance of natural barriers such as water bodies and peninsulas in limiting or excluding feral cats and foxes has been explored elsewhere (Dickman 1992; Short et al. 1994; Burbidge and Manly 2002) and, with the exception of buffers created by the deployment of 1080 baits (Armstrong 2004), there is little evidence that metaphorical barriers (sensu Hayward and Kerley 2009) are effective for these predators.

Kinds of Conservation Fence

Australia has the distinction of having built some of the earliest and longest fences that may have had some positive outcomes for conservation, even if the beneficial consequences were inadvertent. In the 1880s, a fence was constructed to limit incursions of dingoes into the sheep-grazing lands of south-eastern Australia; at 3,374 km, it is second in length only to the Great Wall of China (Breckwoldt 1988). Initially this fence would have helped to reduce predation on kangaroos and emus (Caughley et al. 1980), although it is also an effective barrier to their movement (Letnic 2007). Soon after work began on the dingo fence, thousands of kilometres of rabbit-proof fences were constructed in eastern, central and western regions of Australia to limit the spread of this pest species (Broomhall 1991). This almost certainly delayed the destructive effects of rabbit grazing and excavation in some districts, and perhaps also slowed the local rate of increase of fox numbers (Newsome et al. 1997). Despite the great effort involved in the construction and maintenance of fences for rabbits and dingoes, however, neither was designed to exclude feral cats or foxes and there is little evidence that either of these predators was deterred by them.

Virtually all fences constructed in the last 25–30 years to exclude feral cats and foxes have used wire mesh or netting with a buried apron running along the base of the fence on both sides, with or without an overhanging top, and sometimes with wires carrying charge sufficient to shock and repel animals attempting to cross (Fig. 4.1). Once construction is complete, foxes, feral cats and sometimes other invasive species are removed from the area to be protected; introduced predators may even leave of their own accord once prey such as rabbits have been removed. In most cases the fence surrounds an irregularly shaped area and constitutes the entire barrier against predators that might otherwise move in from outside. However, innovative designs at Heirisson Prong and Peron Peninsula in Western Australia and at Venus Bay in South Australia have used geography to advantage, with physical fences erected across the narrow necks of peninsulas preventing land-based incursion of predators and natural seawater barriers preventing ingress elsewhere (Short et al. 1994; Morris et al. 2004; Department for Environment and Heritage 2006).

Current conservation fence designs in Australia use ideas developed in Europe and the United States (e.g. Forster 1975; Lokemoen et al. 1982), but with many refinements that improve their reliability for local conditions. In the first instance, a barrier fence that is intended to keep out particular species must be designed with the physical characteristics and abilities of those species in mind. These include the sizes of the animals and their ability to jump, dig, climb and bite through given fence materials.

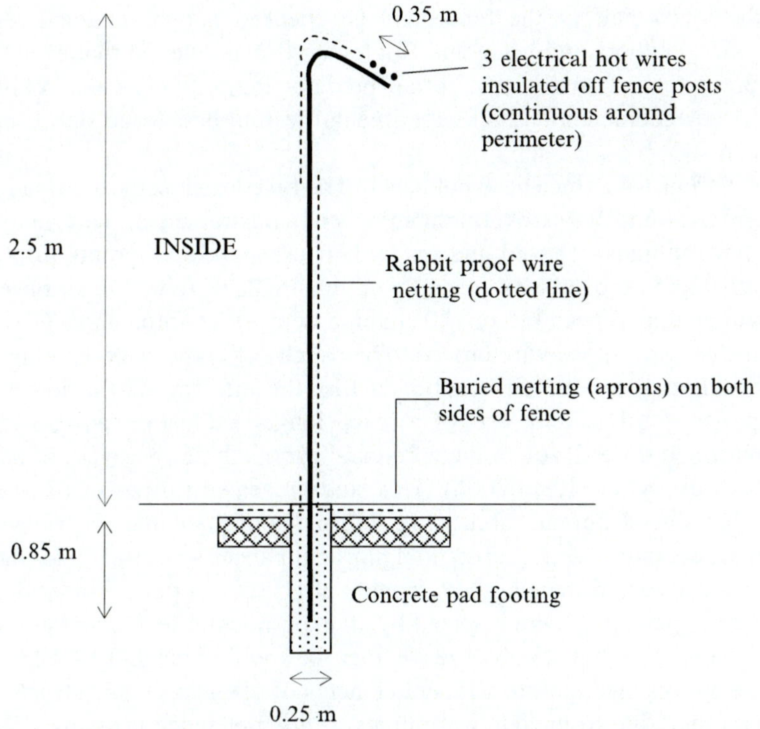

Fig. 4.1 Side view of a predator-proof fence. This example was developed by the Western Australian Department of Environment and Conservation to protect a small mammal breeding enclosure near Narrogin, Western Australia. At 2.5 m this design is higher than usual and has a rigid rather than floppy top to the fence. See text for a discussion of the many different fence designs that have been trialled and of the designs that are in most current use (redrawn from Saunders and McLeod 2007)

Feral cats weigh 3–7 kg and have field-measured head widths that vary from 5.5 to 9.0 cm (Denny 2005); strength of jaw closure is moderate but not sufficient to cut standard gauge fence wire. Standing jumps of 1.4 m have been observed and, provided that they can obtain good purchase on a substrate, cats can use their footpads and sharp claws to climb high above ground (Coman and McCutchan 1994). Cats are reluctant diggers, but can excavate short burrows in soft substrates. Foxes weigh 4–8 kg, zygomatic width (≈head width) ranges from 7.0 to 9.1 cm (Lloyd 1980) and the strength of jaw closure is sufficient to tear through some types of woven wire fences (Coman and McCutchan 1994). Foxes are generally considered to be poor climbers, but have been observed scaling netting and chain mesh fences over 2 m high and to be capable of leaping fences in excess of 1.3 m (Coman and McCutchan 1994; Moseby and Read 2006). They are proficient diggers.

With these attributes, Coman and McCutchan (1994) recommended that barrier fences for these predators should be at least 1.4 m high, have wire netting or chain mesh aprons to prevent digging, have mechanical or electrical barriers such as netting overhangs and/or pulsed offset wires to dissuade climbing, mesh or netting in at least the lower half of the fence and be checked and maintained regularly. However, these authors and Long and Robley (2004) also noted that even the best predator-proof fences usually just curtail predator ingress rather than completely prevent it, and recommended that experimental testing be carried out to optimise fence design.

Two such trials have since been published. The first, by Moseby and Read (2006), placed feral cats and foxes overnight into pens constructed of wire netting and recorded how animals behaved and made their escape. Weak points in the fence design were then improved incrementally until animals were no longer observed to escape. Wire netting fences 115 or 180 cm high with a foot apron and a 60 cm-wide internal overhang of floppy wire proved to be equally effective at containing captive predators, although the authors cautioned that the efficacy of the lower fences needed further testing. Field tests of a much longer exclusion fence, 180 cm in height, confirmed the efficacy of the apron and overhang design over periods of up to 6 years (Moseby and Read 2006). This study revealed further that metal fence posts greatly reduced animals' ability to climb and confirmed that electric wires had a deterrent effect only when a substantial physical barrier was already in place.

In the second experimental study, Robley et al. (2007) placed individual feral cats or foxes in pens that were bisected by an experimental fence, and left animals there for 3 days and nights to observe whether they could breach the fence. Six different fence designs were used, with enticements of open space and structured refuge on the other side to encourage animals to attempt fence crossings. With the exception that both predators were found to breach fences that were 1.2 m high, the authors' conclusions were remarkably similar to those of Moseby and Read (2006). Thus, they concluded that effective predator-proof fences should be 1.8 m high with an apron of hard wire mesh to prevent foxes chewing through and a recurved overhang at least 60 cm wide at the top to deter animals from climbing over the fence from underneath (Robley et al. 2007). In contrast to other studies (e.g. Poole and McKillop 2002), but like Moseby and Read (2006), Robley et al. (2007) found that electric wires could add slightly to the effectiveness of an appropriate physical fence but could not be justified in terms of the extra capital and maintenance costs they entailed.

Rationale

The major objective of predator-proof fences is, of course, conservation. Despite this, it is possible to distinguish several different motivations for fence construction. These are listed in Table 4.1, together with examples of fenced areas that were developed with different conservation imperatives in mind.

In the first instance fences may be constructed to conserve threatened species in situ. Although this would seem to be an obvious motivation for the construction of fences, it has been invoked relatively infrequently and even then only when populations that are critical to a species' survival have been identified. In the case of the

Table 4.1 Examples of predator-proof fences in Australia, showing the main conservation rationale for construction, location, biome, area enclosed and key species protected

Location	Biome	Area (ha)	Key species protected
In situ protection of threatened species			
Epping Forest National Park, Qld	Grassy woodland	2,500	*Lasiorhinus krefftii*
Twin Swamps Nature Reserve, WA	Swan coastal plain	150	*Pseudemydura umbrina*
Ellen Brook Nature Reserve, WA	Swan coastal plain	34	*Pseudemydura umbrina*
Reintroduction of threatened species			
Mulligans Flat Woodland Sanctuary, ACT	Box gum grassy woodland	484	*Climacteris picumnus* (further species are to be reintroduced)
Scotia Sanctuary, stage 1, NSW	Mallee woodland, spinifex, scrub	4,000	*Manorina melanocephalus*, *Leporillus conditor*, *Myrmecobius fasciatus*, *Macrotis lagotis*, *Bettongia lesueur*, *B. penicillata*, *Onychogalea fraenata*, *Lagorchestes hirsutus* (extra-limital)
Scotia Sanctuary, stage 2, NSW	Mallee woodland, spinifex, scrub	4,000	*Myrmecobius fasciatus*, *Bettongia penicillata*, *Onychogalea fraenata* (further species are to be reintroduced)
Eraring Power Station, Newcastle, NSW	Regenerating forest	160	*Phascolarctos cinereus*
Uluru-Kata Tjuta National Park, NT	Hummock grassland	170	*Lagorchestes hirsutus*
Watarrka National Park, NT	Hummock grassland	100	*Lagorchestes hirsutus*
Currawinya National Park, Qld	Semi-arid shrubland and woodland	2,500	*Macrotis lagotis*
Richard Underwood Nature Refuge, Qld	Grassy woodland	105	*Lasiorhinus krefftii*
Arid Recovery, SA	Chenopod shrubland, woodland, sand dunes	6,000	*Aspidites ramsayi*, *Myrmecobius fasciatus*, *Macrotis lagotis*, *Perameles bougainville*, *Bettongia lesueur*, *Leporillus conditor*
Venus Bay, SA	Coastal heath, grassland, woodland	1,460	*Burhinus grallarius*, *Macrotis lagotis*, *Bettongia penicillata*, *Leporillus conditor*
Banrock Station, SA	Riparian woodland	1,600	*Macrotis lagotis* (at least two more species are to be reintroduced)
Woodlands Historic Park, Vic	Woodland, open grassland	300	*Perameles gunnii*
Hamilton Community Parklands, Vic	Woodland, open grassland	100	*Perameles gunnii*
Heirisson Prong, WA	Coastal heath, open low shrubland	1,200	*Perameles bougainville*, *Bettongia lesueur*, *Leporillus conditor*, *Pseudomys fieldi*
Peron Peninsula, WA	Open low shrubland, dense closed shrubland, some hummock grassland	105,000	*Leipoa ocellata*, *Isoodon obesulus*, *Macrotis lagotis*, *Bettongia penicillata*, *Lagostrophus fasciatus*, *Lagorchestes hirsutus*
Karakamia Sanctuary, WA	Swan coastal plain	275	*Myrmecobius fasciatus*, *Isoodon obesulus*, *Pseudocheirus occidentalis*, *Bettongia penicillata*, *Macropus eugenii*, *Setonix brachyurus*

(continued)

Table 4.1 (continued)

Location	Biome	Area (ha)	Key species protected
Lorna Glen, WA	Semi-arid rangeland	1,100	*Macrotis lagotis*, *Trichosurus vulpecula* (nine more species are to be reintroduced)
Wadderin Sanctuary, WA	Semi-arid woodland	430	*Phascogale calura*, *Isoodon obesulus*, *Trichosurus vulpecula*, *Bettongia penicillata*
Education/ecotourism			
Australia Walkabout Wildlife Park, NSW	Mixed forest	32	Wide range of local native fauna
Living Desert Flora and Fauna Sanctuary, NSW	Arid shrubland	180	Unspecified large kangaroos and wallabies
Warrawong Sanctuary, SA	Mixed forest and scrub	34	*Ornithorhynchus anatinus*, *Isoodon obesulus*, *Bettongia penicillata*, *Macropus eugenii*
Yookamurra Sanctuary, SA	Mallee woodland	1,100	*Myrmecobius fasciatus*, *Macrotis lagotis*, *Bettongia lesueur*, *B. penicillata*
Little Desert Nature Lodge, Nhill, Vic	Mallee woodland	140	*Burhinus grallarius*, *Leipoa ocellata*, *Isoodon obesulus*, *Bettongia penicillata*
Moonlit Sanctuary, Vic	Mixed forest and scrub	10	*Dasyurus viverrinus*, *Petaurus norfolcensis*, *Bettongia gaimardi*
Woodland Reserve, Whiteman Park, WA	Banksia woodland	?	*Bettongia penicillata*
Yelverton Brook Conservation Sanctuary, WA	Mixed forest and scrub	40	*Isoodon obesulus*, *Bettongia penicillata* (at least two more species are to be reintroduced)
Education/research			
Tidbinbilla Nature Reserve, ACT	Forest and swamp	17	Waterfowl
La Trobe Melbourne Wildlife Sanctuary, Vic	Mixed forest and wetland	30	Waterfowl, local fauna native to the area
Mount Rothwell Biodiversity and Interpretive Centre, Vic	Woodland, grassy woodland	400	*Dasyurus viverrinus*, *Perameles gunnii*, *Petrogale penicillata*
Harry Waring Marsupial Reserve, WA	Mixed forest, woodland and swamp	254	*Isoodon obesulus*, *Trichosurus vulpecula*, *Macropus irma*, *Setonix brachyurus*

Sites protecting <10 ha are not shown

Sources: Osborne and Williams (1991), Copley (1994), Short et al. (1992, 1994), Kuchling (2000), Richards (2003), Mawson (2004), Morris et al. (2004), Department for Environment and Heritage (2006), Moseby and Read (2006), Finlayson et al. (2008), Pohl (2008), Winnard and Coulson (2008), Broken Hill City Council (2009), Dunlop and Morris (2009), Moseby et al. (2009), Short and Stone (2009), Anon (2010), Finlayson et al. (2010), Mills (2010), M.W. Hayward and J. Short (personal communications) western swamp tortoise *Pseudemydura umbrina*, early research by Burbidge (1967, 1981) confirmed that just two small populations remained and that recovery was likely to be limited by predation from foxes. Together with a captive breeding program, tortoise numbers have increased following the construction of fox-proof fences around both the remnant populations (Kuchling 2000). The northern hairy-nosed wombat *Lasiorhinus krefftii* likewise persists in a confined area; the single remaining natural colony of this species lives behind a fence that was constructed to prevent predation, especially by dingoes (Triggs 2009). In other situations, existing fence structures have been augmented or extended to protect in situ populations against marauding foxes and feral cats. For example, double fences were used to increase shelter for eastern barred bandicoots *Perameles gunnii* at Hamilton in Victoria (Arnold et al. 1990) and wire mesh fences to protect long-nosed bandicoots *P. nasuta* at North Head in New South Wales (Scott et al. 1999; Banks 2004). In both cases, however, the fences were not designed to exclude predators and attacks on bandicoots continued.

The second and most prevalent rationale for constructing predator-proof fences has been to use the newly-protected areas as sites to reintroduce species that once occurred there (Table 4.1). Such reintroductions usually involve one or more species of mammals or birds that now occur in small parts of their former ranges, and hence represent attempts to establish further populations to improve species' status. However, reintroductions sometimes include charismatic reptiles such as the woma *Aspidites ramsayi* or widespread but declining species such as the common brush-tail possum *Trichosurus vulpecula*. At least two of the sites listed in Table 4.1 have had variable success in removing introduced predators. Banrock Station in South Australia uses the Murray River as a "fence" boundary and has been unable to prevent foxes swimming in. Conversely, Peron Peninsula is so large that it has not been possible to eradicate all introduced predators; while fox numbers have been reduced to virtually zero, feral cats remain (Morris et al. 2004). Although de Tores and Marlow (2012) argue that introduced predators must be eradicated for a conservation fence to be effective, malleefowl *Leipoa ocellata*, bilbies *Macrotis lagotis* and perhaps brush-tailed bettongs *Bettongia penicillata* appear likely to persist on Peron following their release (Morris et al. 2004).

Conservation fences for species reintroductions are the most extensive such fences that have been constructed, and they have facilitated extensive research on reintroduction protocols and the biology of the species involved. At Arid Recovery in South Australia, for example, numerous studies have explored the diets, habitats, population dynamics and effects of reintroduced populations on the environment (e.g. Finlayson and Moseby 2004; Bice and Moseby 2008; Read et al. 2008; Bester and Rusten 2009). Several similar studies have been carried out at Scotia Sanctuary in western New South Wales (e.g. Pizzuto et al. 2007; Vieira et al. 2007; Finlayson et al. 2008), at Heirisson Prong in Western Australia (e.g. Sander et al. 1997; Risbey et al. 2000; Robley et al. 2001, 2002; Richards 2003) and elsewhere. Although not all reintroductions have been successful, predator-proof areas provide unique opportunities to reconstruct the species composition and ecological processes that prevailed in earlier times. Most of the sites selected for species reintroductions are in open biomes such as grassland, open woodland, heathland and shrubland (Table 4.1) where species losses have historically been greatest (Burbidge and McKenzie 1989). Fences are also easier to build and maintain in open environments than in heavily forested or topographically diverse ones (Mills 2010).

A third rationale for constructing predator-proof fences is to display native species within protected surroundings for educational and ecotourism purposes (Table 4.1). These fenced areas are usually relatively small to allow reliable viewing of wildlife, and often have residential accommodation, cafes and other facilities to enhance the experience for visitors (Green et al. 1999). Some offer facilities for the rehabilitation of sick or injured wildlife. Except for the Living Desert Flora and Fauna Sanctuary near Broken Hill in western New South Wales, many ecotourist facilities in higher rainfall areas contain patches of forest and advertise the presence of iconic arboreal species such as gliders or koalas *Phascolarctos cinereus*. Although these facilities often support small breeding populations of threatened species that have been reintroduced, some also contain populations of species that never occurred naturally in the local area but have been introduced to enhance the visitor experience. Warrawong Sanctuary, for example, supports red-necked pademelons *Thylogale thetis* and rufous bettongs *Aepyprymnus rufescens*, attractive and visitor-friendly species whose geographical ranges lie far to the east of the sanctuary (R. and A. Williams, personal communication).

Finally, several conservation fences have been built to serve dual educational and research functions (Table 4.1). The Harry Waring Marsupial Reserve, near Perth, provides a good example. Owned and run by the University of Western Australia, this facility has long been a venue for instructing undergraduate students in marsupial biology as well as for a wide range of investigations into the ecology, physiology and behaviour of a wide range of native fauna and flora (e.g. Broughton and Dickman 1991; Abensperg-Traun et al. 1993). The reserve is not open to the public. The other areas listed provide greater opportunities for general visitor education via signage, guided walks, school programs and active community engagement. They also support research (Lambie 1988); for example, the Mount Rothwell site has been used to assess methods for the reintroduction and population persistence of *P. gunnii* on the mainland of Australia (Winnard and Coulson 2008).

Although the above categories delineate the main motivations for constructing predator-proof fences, they are not necessarily as exclusive as Table 4.1 might indicate. For example, the site at Heirisson Prong was developed collaboratively between government agencies, a mining company and local people to make it a resource for environmental education and scientific research as well as a sanctuary for threatened species (Short et al. 1994). The Mulligans Flat Woodland Sanctuary near Canberra likewise was developed with extensive community consultation, and rangers now run guided tours as part of a broader education program (Mills 2010). A further benefit of conservation fences, although not necessarily a prime justification for their construction, is that many non-target native species will respond positively when invasive predators have been removed. This in turn permits scarce but non-threatened species to be studied more readily (e.g. Moseby et al. 2009).

A further rationale for constructing predator-proof fences, not shown in Table 4.1, is that of captive breeding. For this purpose fenced areas are usually small (<10 ha) and often incorporated within the larger protected areas that support self-sustaining populations (e.g. Short et al. 1994; Morris et al. 2004; Finlayson et al. 2008). However, several small, fenced facilities exist separately, such as Secret Creek in New South Wales and Barnia Mia in Western Australia (Hughes et al. 2005). These ventures often provide captive wildlife experiences for visitors and use funds to help support the breeding programs. The line between in situ and ex situ conservation can become blurred as conditions within protected areas become more and more confined (Dickie et al. 2008), and I do not consider fenced enclosures for breeding any further here.

Effectiveness and Costs

An effective conservation fence should ensure the persistence of populations of different target species within its boundaries. Ideally, it should do this for long enough to ensure that it achieves the original objectives for which it was set up; although, in the case of providing protection from feral cats and foxes, this could be a very long time. Fenced populations should also be self-sustaining over long periods with little need for demographic intervention, addition of extra food, water or shelter resources or manipulation of genetic stock (Frankham 1994). However, as the first conservation fence was established in Australia only in 1975 and most fences within the last 20 years, their long-term effectiveness is difficult to judge. Nonetheless, several studies have reviewed the early results of fence erection on populations of resident, introduced or reintroduced species, and suggest that fences can work very effectively.

In an early review, Short et al. (1992) investigated factors contributing to the successful reintroduction of macropodid marsupials and showed that success depended on the absence of introduced predators: 8% of reintroductions were successful if predators were present, compared with 82% if they were not. Most sites without predators in this review were offshore islands rather than fenced reserves, and one site with a conservation fence – the Harry Waring Marsupial Reserve – was considered to be an area with predators as the fence did not constitute an effective barrier (Short et al. 1992). Finlayson et al. (2010) reported similar results for bettongs. Thus, of 24 translocation attempts that could be judged a success or failure for the brush-tailed bettong *B. penicillata*, only one success occurred in the presence of introduced predators compared with six where predators had been excluded by fences or water barriers. The same authors reported four successful translocations of burrowing bettongs *Bettongia lesueur* into fenced reserves or islands. For both species failures occurred largely where animals had been moved to mainland areas, even if intensive efforts had been made to reduce the impacts of predators prior to the bettongs being reintroduced (Finlayson et al. 2010). Winnard and Coulson (2008) reported 3/3 successful releases of eastern barred bandicoots into fenced sites in Victoria compared with just 1/5 into sites without fences. Broadly similar results were documented by Mawson (2004) for mammals and other vertebrates in Western Australia, and by Copley (1994) for vertebrates in South Australia, although both authors commented that success or failure was difficult to judge in many cases owing to either lack of records or the recency of translocation attempts. Averaging across all these reviews suggests that the success rate of vertebrates released into predator-free areas – either fenced reserves or islands – is about 80%, whereas the success rate for releases into unprotected sites can vary from 0 to 60% depending on the intensity of predator control that is maintained, the release protocol and the identity, number, sex ratio and quality of animals released.

These results indicate that conservation fences can effectively support wildlife populations; given the limitation that most fences are relatively new, long-term population persistence should be possible provided that fence integrity is maintained and introduced predators are kept at bay. But what are the costs of fence construction and maintenance?

The material cost of conservation fencing has been estimated to range from Aus$6,000 to Aus$30,000/km (Coman and McCutchan 1994), with estimates of around Aus$7000 to Aus$12,500/km cited for designs that exclude feral cats and foxes (Moseby and Read 2006; Bode and Wintle 2010). These costs do not include transport of materials, clearing straight lines for fence placement, labour, or differences in cost associated with substrate or topography. Beyond the initial capital investment, ongoing monitoring also is required to ensure that fences remain intact and gates are closed properly and secured after use. Small areas can be patrolled on foot, but larger areas may be checked efficiently only by the use of quad bikes or other vehicles to cover the greater distances involved. Some use also may be made of alarms that sound if any section of the fence is damaged. Electrical monitors with radio or satellite links are used in some remote areas and are recommended if regular direct checks cannot be carried out (Coman and McCutchan 1994). Fences can be undermined by determined burrowing animals such as echidnas *Tachyglossus aculeatus* or common wombats *Vombatus ursinus*, damaged during severe storms, by falling trees, or deliberate vandalism; in areas with public access, gates also can be left open inadvertently (Winnard and Coulson 2008). The actual costs of maintaining a predator-proof fence have not been reported, but Pickard (2007) estimated conservatively that they could be about $1,000/km/year.

Regular monitoring and swift repair of damage to fences is clearly important, but even the best fence designs will be unlikely to keep out feral cats and foxes in perpetuity. Indeed, protected sites with dense populations of native species may act as "magnets" to predators outside the fence and provide motivation for determined individuals to gain access. Recent research by Connolly et al. (2009) in New Zealand confirms that mammalian pest species constantly "test" the perimeters of predator-proof conservation fences, with incursions very likely to occur within just 24 h of a breach opening up. Animals also may be deliberately dumped into protected areas by people. In the mid-1980s, for example, cats were introduced on at least three occasions to the Harry Waring Marsupial Reserve near Perth (B.T. Clay, personal communication). Foxes also gained access, most likely by detecting and entering

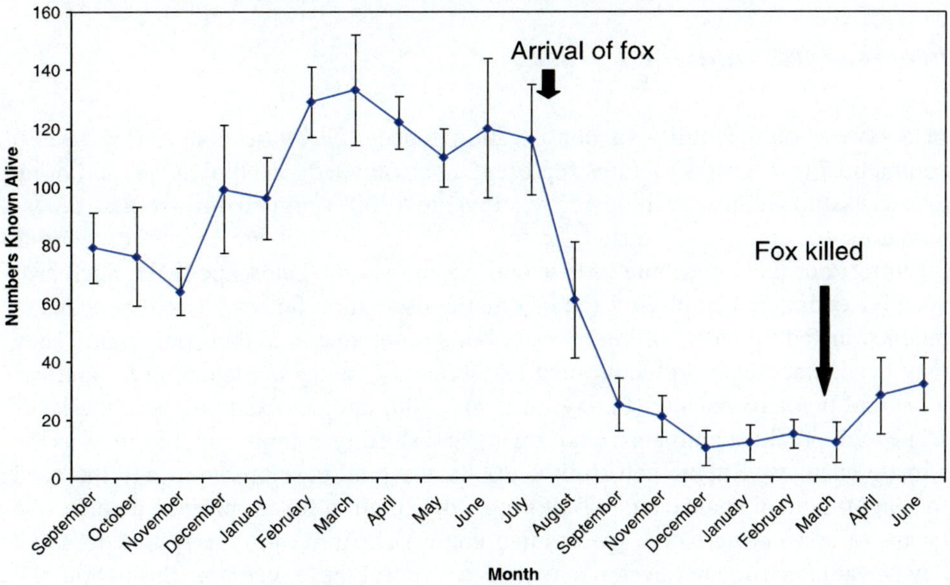

Fig. 4.2 Estimated population numbers of southern brown bandicoots *Isoodon obesulus* (Jolly-Seber estimates ± SE) during a monthly live-trapping program in the Harry Waring Marsupial Reserve, Western Australia. A single fox was known to have gained entry to the reserve from tracks and sightings; its carcase was found about 6 months after the incursion began (C.R. Dickman, unpublished)

via weak points in the fence. During work in this reserve at that time, I was able to track the rapid depletion of a population of the southern brown bandicoot *Isoodon obesulus* that followed the arrival of a single fox (Fig. 4.2). Bandicoots were hunted and eaten, hunted and cached and also bitten to death and left on the ground in what appeared to be an example of surplus killing (Kruuk 1972). If so, this would add to many other examples of such surplus killing of native mammals by foxes (Short et al. 2002). Thus, monitoring using camera traps, tracks or other signs is essential to detect incursions by predators so that control actions can be initiated quickly. Several authors have also suggested that regular predator control be undertaken in areas outside the conservation fence to further reduce the chance of incursions taking place (Coman and McCutchan 1994; Short et al. 1994; Pickard 2007; Moseby et al. 2009). Bode and Wintle (2010) outlined a return-on-investment framework that minimises the long-term management costs of fences by explicitly balancing the capital expenses of secure designs against the costs of coping with more frequent failures. If adopted, this approach can be expected to improve the efficiency of conservation fencing more broadly.

Fences, Ferals and the Future

In a review of the utility of conservation fences, Hayward and Kerley (2009) commented that these structures represent an acknowledgement that we are failing to successfully conserve biodiversity. Pickard (2007) argued further that unless predators are controlled outside the fences, the enclosures are essentially captive breeding zoos that contribute little to conservation at the landscape scale. Such provocative comments emphasise the fact that conservation fences provide a means to retain some components of biodiversity but are not an end in their own right. They may even exacerbate problems such as inbreeding, genetic adaptation to confinement and predator naivety (Finlayson et al. 2008) and, as fixed points in the landscape, will preclude animals from tracking and shifting to more suitable areas as the climate changes (Ritchie and Bolitho 2008). Despite these problems and the need for ongoing maintenance and monitoring, conservation fences provide an effective means of conserving native species that are at risk from introduced predators and buy us valuable time to develop new tools to control these predators throughout the landscape. Could they be made more cost-effective?

Virtually all unconfined populations likely occupy geographical areas where the risk of predation varies, with individuals selecting particular sites at different times according to their resource needs and perception of hazards (Brown 1988). In landscapes where resources and risk are spatially predictable, source-sink or metapopulation dynamics may develop and confer some measure of resilience for species over large areas (Fryxell 2001). Even where predators are pervasive, both theory and empirical results suggest that predator-impacts can be counteracted by improving prey survival through habitat manipulation (Sinclair et al. 1998), and that viable populations can be sustained if they have access to pockets of absolute refuge (Arthur et al. 2005). If these lines of thinking are applied to fenced reserves, it may be possible to devise ways to extend the local benefits provided by the fences to the broader region (Smith and Quin 1996). Two conditions would have to be satisfied. Firstly, prey species would need to be able to move into and out of the protected areas, and secondly their survival would need to be improved while outside the fence. If these conditions were met, prey species could potentially use conservation reserves as absolute refuges but also occur in a halo region around them.

To achieve the first condition, fences would have to be selectively leaky and allow ingress and egress of prey species but not of their predators. This may occur already with small native species such as phascogales or proficient climbers such as possums. However, for medium-sized and ground-dwelling prey species, gates may be required that allow them passage while denying access to predators. Camera technology is now being used to identify livestock and allow them to move through gates to access watering points while preventing other species from doing so (Finch et al. 2006); similar technology could potentially be used to allow specific ingress of prey species into fenced reserves. However, even without this, broad scale population persistence could still be achieved using one-way gates to allow egress from reserves provided that the second condition is met. Trials to make effective one-way gates for medium-sized mammals such as bilbies *M. lagotis* and bettongs *Bettongia* spp. are currently under way (Crisp and Moseby 2010).

To improve prey survival outside the fenced reserves, predator activity could be reduced by poison baiting, shooting or other methods, or – better still – their access to prey reduced by increasing the complexity of the outside habitat. For already established reserves, complexity could be enhanced beyond the fence by seeding with shrubs and other ground cover species, addition of logs, burrows or other refugia. For conservation fences that have yet to be constructed, position within the landscape could be an important factor in the planning and site acquisition stages. For example, rather than position a fence around dense stands of vegetation, perhaps it would be better to construct it in more open habitat adjacent to those stands so that prey species dispersing out would have increased chances of survival. If conservation fences were integrated into the broader landscape more efficiently, perhaps their shape also could be modified. They would not be necessarily square or rectangular, but could be configured so that they provide the best protection between patches of complex habitat outside. They could, for example, provide linear barriers along travelling stock routes or other roadside reserves so that prey could move securely from one patch of woodland to another; they could also connect key breeding habitats that sustain metapopulations. Regular shaped designs have some advantages in that they are cheaper to construct and provide a relatively shorter fence perimeter for the area that is protected compared to irregular designs. This may be important in reducing barriers to the movement of larger species outside the exclosures, such as kangaroos and emus, and also reduce use of the fence by raptors and other predators outside that may use the fences as "drives" for smaller prey (van Dyk and Slotow 2003; M.W. Hayward, personal communication). Thus, the costs and benefits of novel fence configurations would need to be considered before construction.

Alternatively, if many small fenced areas were interspersed between patches of complex vegetation throughout the broader landscape, they could provide the basis for establishing prey metapopulations that do not currently exist. Many small fences would have the additional advantages that, if one or two were breached, the remaining fences and the habitat beyond them would continue to sustain the target prey species, they would minimise opportunities for external predators to use the fences as drives, and would minimise barriers to movement for large species outside.

In the last 25 years we have become proficient at building very effective fences to exclude feral cats, foxes and other undesirable species and have shown that populations of many native Australian prey species can flourish within their confines. It is now time to marry this hard-won proficiency with systematic planning considerations about fence position, number and configuration in the landscape, as well as with knowledge of the biology of the species that we wish to conserve. A good start has been made, but much remains to be done!

Acknowledgements I thank Graeme Finlayson for helping to stimulate my thinking about the biology of fenced populations, Matt Hayward and other contributors at the fencing symposium in Chattanooga for thoughtful discussions on the topic, Aaron Greenville for drawing Fig. 4.1, Carol

McKechnie for continued support and for critical reading of the draft manuscript, and Matt Hayward, Katherine Moseby and Jeff Short for helpful review comments on the final version.

References

Abbott, I. (2002) Origin and spread of the cat, *Felis catus*, on mainland Australia, with a discussion of the magnitude of its early impact on native fauna. *Wildlife Research,* **29**, 51–74.

Abbott, I. (2011) The importation, release, establishment, spread, and early impact on prey animals of the red fox *Vulpes vulpes* in Victoria and adjoining parts of south-eastern Australia. *Australian Zoologist*, **35**, 463–533.

Abensperg-Traun, M., Black, R. & Bunn, S. (1993) Selection of woody food by termites (Isoptera) of the Harry Waring Marsupial Reserve, near Jandakot, Western Australia. *Western Australian Naturalist,* **19**, 247–254.

Algar, D., Angus, G.J., Williams, M.R. & Mellican, A.E. (2007) Influence of bait type, weather and prey abundance on bait uptake by feral cats (*Felis catus*) on Peron Peninsula, Western Australia. *Conservation Science Western Australia*, **6**, 109–149.

Anon. (2010) Delivering effective conservation for Australia's wildlife. *Wildlife Matters* **Summer**, 2009/10.

Armstrong, R. (2004) Baiting operations: *Western Shield* review – February 2003. *Conservation Science Western Australia,* **5**, 31–50.

Arnold, A.H., Goldstraw, P.W., Hayes, A.G. & Wright, B.F. (1990) Recovery management of the eastern barred bandicoot in Victoria: the Hamilton conservation strategy. In *Management and Conservation of Small Populations* (eds. T.W. Clark and J.H. Seebeck), pp. 179–192. Chicago Zoological Society: Brookfield, Illinois.

Arthur, A.D., Pech, R.P. & Dickman, C.R. (2005) Effects of predation and habitat structure on the population dynamics of house mice in large outdoor enclosures. *Oikos,* **108**, 562–572.

Banks, P.B. (2004) Population viability analysis in urban wildlife management: modelling management options for Sydney's quarantined bandicoots. In *Urban Wildlife: More than Meets the Eye* (eds. D. Lunney and S. Burgin), pp. 70–77. Royal Zoological Society of New South Wales: Sydney, Australia.

Bester, A.J. & Rusten, K. (2009) Trial translocation of the numbat (*Myrmecobius fasciatus*) into arid Australia. *Australian Mammalogy,* **31**, 9–16.

Bice, J. & Moseby, K.E. (2008) Diets of the reintroduced greater bilby (*Macrotis lagotis*) and burrowing bettong (*Bettongia lesueur*) in the Arid Recovery reserve, northern South Australia. *Australian Mammalogy,* **30**, 1–12.

Bode, M. & Wintle, B. (2010) How to build an efficient conservation fence. *Conservation Biology,* **24**, 182–188.

Breckwoldt, R. (1988) *The Dingo: a very Elegant Animal.* Angus and Robertson: Sydney, Australia.

Broken Hill City Council. (2009) Annual Report 2008/2009. Broken Hill City Council: Broken Hill, Australia.

Broomhall, F.H. (1991) *The Longest Fence in the World: a History of the No. 1 Rabbit Proof Fence from its Beginning until Recent Times*. Hesperian Press: Perth, Australia.

Broughton, S.K. & Dickman, C.R. (1991) The effect of supplementary food on home range of the southern brown bandicoot, *Isoodon obesulus* (Marsupialia: Peramelidae). *Australian Journal of Ecology,* **16**, 71–78.

Brown, J.S. (1988) Patch use as an indicator of habitat preference, predation risk, and competition. *Behavioral Ecology and Sociobiology,* **22**, 37–47.

Burbidge, A.A. (1967) The biology of south-western Australian tortoises. PhD thesis, University of Western Australia: Perth, Australia.

Burbidge, A.A. (1981) The ecology of the western swamp tortoise *Pseudemydura umbrina* (Testudines: Chelidae) *Australian Wildlife Research,* **8**, 203–223.
Burbidge, A.A. & Manly, B.F.J. (2002) Mammal extinctions on Australian islands: causes and conservation issues. *Journal of Biogeography,* **29**, 465–473.
Burbidge, A.A. & McKenzie, N.L. (1989) Patterns in the modern decline of Western Australia's vertebrate fauna: causes and conservation implications. *Biological Conservation,* **50**, 143–198.
Caughley, G., Grigg, G.C., Caughley, J. & Hill, G.J.E. (1980) Does dingo predation control the densities of kangaroos and emus? *Australian Wildlife Research,* **7**, 1–12.
Clark, C.M. (1981) *A Short History of Australia*, 2nd edition. Macmillan: Melbourne, Australia.
Coman, B.J. &. McCutchan, J. (1994) *Predator Exclusion Fencing for Wildlife Management in Australia*. Australian Nature Conservation Agency: Canberra, Australia.
Connolly, T.A., Day, T.D. & King, C.M. (2009) Estimating the potential for reinvasion by mammalian pests through pest-exclusion fencing. *Wildlife Research,* **36**, 410–421.
Copley, P.B. (1994) Translocations of native vertebrates in South Australia: a review. In *Reintroduction Biology of Australian and New Zealand Fauna* (Ed. Serena M.), pp. 35–42. Surrey Beatty & Sons: Sydney, Australia.
Crisp, H. & Moseby, K. (2010) One-way gates: initial trial of a potential tool for preventing overpopulation within fenced reserves. *Ecological Management and Restoration*, **11**, 139–141.
De Alessi, M. (2003/04) Conservation Inc. The business of saving wildlife. *Policy,* **19**, 19–24.
de Tores, P.J. & Marlow, N. (2012) The relative merits of predator-exclusion fencing and repeated fox baiting for protection of native fauna: five case studies from Western Australia. In *Fencing for Conservation* (eds. M.J. Somers and M.W. Hayward). Springer: New York.
Denny, E.A. (2005) Ecology of free-living cats exploiting waste disposal sites: diet, morphometrics, population dynamics and population genetics. PhD thesis, University of Sydney: Sydney, Australia.
Denny, E.A. & Dickman, C.R. (2010) *Review of Cat Ecology and Management Strategies in Australia*. Invasive Animals Cooperative Research Centre: Canberra, Australia.
Denny, E., Yakovlevich, P., Eldridge, M.D.B. & Dickman, C.R. (2002) Social and genetic analysis of a population of free-living cats (*Felis catus* L.) exploiting a resource-rich habitat. *Wildlife Research,* **29**, 405–413.
Department for Environment and Heritage (2006) *Venus Bay Conservation Park Management Plan*. Department for Environment and Heritage: Adelaide, Australia.
Department of Sustainability and Environment (2005) *Southern Ark Operations Plan*. Department of Sustainability and Environment: Melbourne, Australia.
Dexter, N. & Murray, A. (2009) The impact of fox control on the relative abundance of forest mammals in East Gippsland. *Wildlife Research,* **36**, 252–261.
Dickie, L., Bonner, J.P. & West, C.D. (2008) *In situ* and *ex situ* conservation: blurring the boundaries between zoos and the wild. In *Zoos in the 21st Century: Catalysts for Conservation?* (eds. A. Zimmermann, M. Hatchwell, L.A. Dickie and C.D. West), pp. 220–235. Cambridge University Press: Cambridge, U.K.
Dickman, C.R. (1992) Conservation of mammals in the Australasian region: the importance of islands. In *Australia and the Global Environmental Crisis* (eds. J.N. Coles and J.M. Drew), pp. 175–214. Academy Press: Canberra, Australia.
Dickman, C.R. (1996a) Impact of exotic generalist predators on the native fauna of Australia. *Wildlife Biology,* **2**, 185–195.
Dickman, C.R. (1996b) *Overview of the Impacts of Feral Cats on Australian Native Fauna.* Australian Nature Conservation Agency: Canberra, Australia.
Dickman, C.R. (2007) *A Fragile Balance: the Extraordinary Story of Australian Marsupials.* Mallon Publishing: Melbourne, Australia.
Dickman, C.R., Glen, A.S. & Letnic, M. (2009) Reintroducing the dingo: can Australia's conservation wastelands be restored? In *Reintroduction of Top-order Predators* (eds. M.W. Hayward and M.J. Somers), pp. 238–269. Wiley-Blackwell Publishing: Oxford, U.K.
Dunlop, J. & Morris, K. (2009) Into the wild: restoring rangelands fauna. *Landscope,* **24**, 52–58.
Finch, N.A., Murray, P.J., Dunn, M.T. & Billingsley, J. (2006) Using machine vision classification to control access of animals to water. *Australian Journal of Experimental Agriculture,* **46**, 837–839.

Finlayson, G.R. & Moseby, K.E. (2004) Managing confined populations: the influence of density on the home range and habitat use of reintroduced burrowing bettongs (*Bettongia lesueur*). *Wildlife Research,* **31**, 457–463.

Finlayson, G.R., Vieira, E.M., Priddel, D., Wheeler, R. & Dickman, C.R. (2008) Multi-scale patterns of habitat use by re-introduced mammals: a case study using medium-sized marsupials. *Biological Conservation,* **141**, 320–331.

Finlayson, G.R., Finlayson, S.T. & Dickman, C.R. (2010) Returning the rat-kangaroos: translocation attempts in the Family Potoroidae (superfamily Macropodoidea) and recommendations for conservation. In *Macropods: the Biology of Kangaroos, Wallabies and Rat-kangaroos* (eds. G. Coulson and M. Eldridge), pp. 245–262. CSIRO Publishing: Melbourne, Australia.

Forster, J.A. (1975) Electric fencing to protect sandwich terns against foxes. *Biological Conservation,* **7**, 85.

Frankham, R. (1994) Genetic management of captive populations for reintroduction. In *Reintroduction Biology of Australian and New Zealand Fauna* (ed. M. Serena), pp. 31–34. Surrey Beatty & Sons: Sydney, Australia.

Fryxell, J.M. (2001) Habitat suitability and source-sink dynamics of beavers. *Journal of Animal Ecology,* **70**, 310–316.

Gaynor, A. (2000) Report on the history of the arrival of the feral cat population in Western Australia. *CALMScience,* **3**, 140–179.

Glen, A.S. & Dickman, C.R. (2005) Complex interactions among mammalian carnivores in Australia, and their implications for wildlife management. *Biological Reviews,* **80**, 387–401.

Glen, A.S. & Dickman, C.R. (2008) Niche overlap between marsupial and eutherian carnivores: does competition threaten the endangered spotted-tailed quoll? *Journal of Applied Ecology,* **45**, 700–707.

Green, R., Higginbottom, K. & Jones, D. (1999) A tourism classification of Australian wildlife and their habitats. In *Wildlife Tourism Discussion Document* (eds. K. Higginbottom and M. Hardy), pp. 52–68. CRC for Sustainable Tourism: Gold Coast, Australia.

Hayward, M.W. & Kerley, G.I.H. (2009) Fencing for conservation: restriction of evolutionary potential or a riposte to threatening processes? *Biological Conservation,* **142**, 1–13.

Hrdina, F.C. (1997) Marsupial destruction in Queensland 1877–1930. *Australian Zoologist,* **30**, 272–286.

Hughes, M., Newsome, D. & Macbeth, J. (2005) Case study: visitor perceptions of captive wildlife tourism in a Western Australian natural setting. *Journal of Ecotourism,* **4**, 73–91.

Johnson, C.N., Isaac, J.L. & Fisher, D.O. (2007) Rarity of a top predator triggers continent-wide collapse of mammal prey: dingoes and marsupials in Australia. *Proceedings of the Royal Society, B* **274**, 341–346.

Kinnear, J.E., Sumner, N.R. & Onus, M.L. (2002) The red fox in Australia – an exotic predator turned biocontrol agent. *Biological Conservation,* **108**, 335–359.

Kinnear, J.E., Krebs, C.J., Pentland, C., Orell, P., Holme, C. & Karvinen, R. (2010) Predator-baiting experiments for the conservation of rock-wallabies in Western Australia: a 25-year review with recent advances. *Wildlife Research,* **37**, 57–67.

Kruuk, H. (1972) Surplus killing by carnivores. *Journal of Zoology,* **166**, 233–244.

Kuchling, G. (2000) Conservation strategies for remnant turtle populations: the Western Australian swamp turtle *Pseudemydura umbrina* recovery programme. *Zugleich Kataloge des OÖ. Landesmuseums, Neue Folge,* **149**, 179–188.

Lambie, B. (1988) Breeding captive fauna at Tidbinbilla Nature Reserve. *Australian Ranger Bulletin,* **4**, 20–21.

Letnic, M. (2007) The impacts of pastoralism on the fauna of arid Australia. In *Animals of Arid Australia: out on their Own?* (eds. C. Dickman, D. Lunney and S. Burgin), pp. 65–75. Royal Zoological Society of New South Wales: Sydney, Australia.

Letnic, M. & Dickman, C.R. (2010) Resource pulses and mammalian dynamics: conceptual models for hummock grasslands and other Australian desert habitats. *Biological Reviews,* **85**, 501–521.

Letnic, M., Koch, F., Gordon, C., Crowther, M.S. & Dickman, C.R. (2009) Keystone effects of an alien top-predator stem extinctions of native mammals. *Proceedings of the Royal Society, B* **276**, 3249–3256.

Lloyd, H.G. (1980) *The Red Fox*. B.T. Batsford: London, U.K.

Lokemoen, J.T., Doty, H.A., Sharp, D.E. & Neaville, J.E. (1982) Electric fences to reduce mammalian predation on waterfowl nests. *Wildlife Society Bulletin*, **10**, 318–323.

Long, J.L. (1988) *Introduced Birds and Mammals in Western Australia*. Agriculture Protection Board of Western Australia: Perth, Australia.

Long, J.L. (2003) *Introduced Mammals of the World*. CSIRO Publishing: Melbourne, Australia.

Long, K. & Robley, A. (2004) *Cost Effective Feral Animal Exclusion Fencing for Areas of High Conservation Value in Australia*. Department of the Environment and Heritage: Canberra, Australia.

Low, T. (1999) *Feral Future*. Viking Books: Melbourne, Australia.

Lugton, I.W. (1993) Fox predation on lambs. In *Proceedings of the Australian Sheep Veterinary Society* (ed. D.A. Hucker), pp. 17–26. Australian Veterinary Association Conference: Gold Coast, Australia.

Mahon, P.S. (2009) Targeted control of widespread exotic species for biodiversity conservation: the red fox (*Vulpes vulpes*) in New South Wales, Australia. *Ecological Management and Restoration*, **10**, S59-S69.

Marks, C.A., Gigliotti, F. & Busana, F. (2009) Assuring that 1080 toxicosis in the red fox (*Vulpes vulpes*) is humane. II. Analgesic drugs produce better welfare outcomes. *Wildlife Research*, **36**, 98–105.

Mawson, P.R. (2004) Translocations and fauna reconstruction sites: *Western Shield* review – February 2003. *Conservation Science Western Australia*, **5**, 108–122.

Mills, P. (2010) Strategies to reduce conflict; protecting wildlife, engaging the community – the Mulligans Flat Woodland Sanctuary experience. In *Proceedings of the RSPCA Australia Scientific Seminar - Convergence or Conflict: Animal Welfare in Wildlife Management and Conservation* (eds. M. Tensen & B. Jones), pp. 35–40. RSPCA: Canberra, Australia.

Moodie, E. (1995) *The Potential for Biological Control of Feral Cats in Australia*. Australian Nature Conservation Agency: Canberra, Australia.

Morris, K., Sims, C., Himbeck, K., Christensen, P., Sercombe, N., Ward, B. & Noakes, N. (2004) *Project Eden* – fauna recovery on Peron Peninsula, Shark Bay: *Western Shield* review – February 2003. *Conservation Science Western Australia*, **5**, 202–234.

Moseby, K.E. & Read, J.L. (2006) The efficacy of feral cat, fox and rabbit exclusion fence designs for threatened species protection. *Biological Conservation*, **127**, 429–437.

Moseby, K.E., Hill, B.M. & Read, J.L. (2009) Arid Recovery – a comparison of reptile and small mammal populations inside and outside a large rabbit, cat and fox-proof exclosure in arid South Australia. *Austral Ecology*, **34**, 156–169.

Newsome, A.E., Pech, R., Smyth, R., Banks, P. & Dickman, C.R. (1997) *Potential Impacts on Australian Native Fauna of Rabbit Calicivirus Disease*. Environment Australia: Canberra, Australia.

Osborne, W. & Williams, K. (1991) Conservation problems with cats in the Australian Capital Territory. In *The Impact of Cats on Native Wildlife* (Ed. Potter, C.), pp. 14–17. Australian National Parks and Wildlife Service: Canberra, Australia.

Pickard, J. (2007) Predator-proof fences for biodiversity conservation: some lessons from dingo fences. In *Animals of Arid Australia: out on their Own?* (eds. C. Dickman, D. Lunney and S. Burgin), pp. 197–207. Royal Zoological Society of New South Wales: Sydney, Australia.

Pizzuto, T.A., Finlayson, G.R., Crowther, M.S. & Dickman, C.R. (2007) Microhabitat use of the brush-tailed bettong (*Bettongia penicillata*) and burrowing bettong (*B. lesueur*) in semi-arid New South Wales: implications for reintroduction programs. *Wildlife Research*, **34**, 271–279.

Pohl, M. (2008) *Yookamurra Sanctuary Education Camp Information: a Guide for Teachers and Science Educators*. Australian Wildlife Conservancy: Sedan, Australia.

Poole, D.W. & McKillop, I.G. (2002) Effectiveness of two types of electric fence for excluding the red fox (*Vulpes vulpes*). *Mammal Review*, **32**, 51–57.

Read, J.L., Carter, J., Moseby, K.M. & Greenville, A. (2008) Ecological roles of rabbit, bettong and bilby warrens in arid Australia. *Journal of Arid Environments,* **72**, 2124–2130.

Richards, J.D. (2003) Report on threatened Shark Bay marsupials, western barred bandicoot *Perameles bougainville bougainville*, burrowing bettong *Bettongia lesueur lesueur*, banded hare-wallaby *Lagostrophus fasciatus fasciatus*, and rufous hare-wallabies *Lagorchestes hirsutus bernieri* and *Lagostrophus fasciatus dorreae*. Department of the Environment and Heritage: Canberra, Australia.

Risbey, D.A., Calver M. & Short, J. (1997) Control of feral cats for nature conservation. I. Field tests of four baiting methods. *Wildlife Research,* **24**, 319–326.

Risbey, D.A., Calver, M.C., Short, J., Bradley, J.S. & Wright, I.W. (2000) The impact of cats and foxes on the small vertebrate fauna of Heirisson Prong, Western Australia. II. A field experiment. *Wildlife Research,* **27**, 223–235.

Ritchie, E.G. & Bolitho, E.E. (2008) Australia's savanna herbivores: bioclimatic distributions and an assessment of the potential impact of regional climate change. *Physiological and Biochemical Zoology,* **81**, 880–890.

Robley, A., Purdey, D., Johnston, M., Lindeman, M., Busana, F. & Long, K. (2007) Experimental trials to determine effective fence designs for feral cat and fox exclusion. *Ecological Management and Restoration,* **8**, 193–198.

Robley, A.J., Short J. & Bradley, A.J. (2001) Dietary overlap between the burrowing bettong (*Bettongia lesueur*) and the European rabbit (*Oryctolagus cuniculus*) in semi-arid coastal Western Australia. *Wildlife Research*, **28**, 341–349.

Robley, A.J., Short J. & Bradley, A.J. (2002) Do European rabbits (*Oryctolagus cuniculus*) influence the population ecology of the burrowing bettong (*Bettongia lesueur*)? *Wildlife Research,* **29**, 423–429.

Rolls, E.C. (1969) *They All Ran Wild: the Story of Pests on the Land in Australia.* Angus and Robertson: Sydney, Australia.

Salo, P., Korpimäki, E., Banks, P.B., Nordström, M. & Dickman, C.R. (2007) Alien predators are more dangerous than native predators to prey populations. *Proceedings of the Royal Society, B* **274**, 1237–1243.

Sander, U., Short, J. & Turner, B. (1997) Social organisation and warren use of the burrowing bettong, *Bettongia lesueur* (Macropodoidea: Potoroidae). *Wildlife Research,* **24**, 143–157.

Saunders, G. & McLeod, L. (2007) *Improving Fox Management Strategies in Australia.* Bureau of Rural Sciences: Canberra, Australia.

Saunders, G., Coman, B., Kinnear, J. & Braysher, M. (1995) *Managing Vertebrate Pests: Foxes.* Australian Government Publishing Service: Canberra, Australia.

Saunders, G.R., Gentle, M.N. & Dickman, C.R. (2010) The impacts and management of foxes *Vulpes vulpes* in Australia. *Mammal Review*, **40**, 181–211.

Scott, L.K., Hume, I.D. & Dickman, C.R. (1999) Ecology and population biology of long-nosed bandicoots (*Perameles nasuta*) at North Head, Sydney Harbour National Park. *Wildlife Research,* **26**, 805–821.

Short, J., Bradshaw, S.D., Giles, J., Prince, R.I.T. & Wilson, G.R. (1992) Reintroduction of macropods (Marsupialia: Macropodoidea) in Australia – a review. *Biological Conservation,* **62**, 189–204.

Short, J., Kinnear, J.E. & Robley, A. (2002) Surplus killing by introduced predators in Australia – evidence for ineffective anti-predator adaptations in native prey species? *Biological Conservation,* **103**, 283–301.

Short, J. & Stone, M. (2009) Farmers befriend phascogale. *Landscope*, **25**, 22–28.

Short, J., Turner, B., Parker, S. & Twiss, J. (1994) Reintroduction of endangered mammals to mainland Shark Bay: a progress report. In *Reintroduction Biology of Australian and New Zealand Fauna* (Ed. Serena M.), pp. 183–188. Surrey Beatty & Sons: Sydney, Australia.

Short, J., Turner, B., Risbey, D.A. & Carnamah, R. (1997) Control of feral cats for nature conservation. II. Population reduction by poisoning. *Wildlife Research,* **24**, 703–714.

Sinclair, A.R.E., Pech, R.P., Dickman, C.R., Hik, D., Mahon, P. & Newsome, A.E. (1998) Predicting effects of predation on conservation of endangered prey. *Conservation Biology*, **12**, 564–575.

Smith, A.P. & Quin, D.G. (1996) Patterns and causes of extinction and decline in Australian conilurine rodents. *Biological Conservation,* **77**, 243–267.

Strive, T., Hardy, C.M. & Reubel, G.H. (2007) Prospects for immunocontraception in the European red fox (*Vulpes vulpes*). *Wildlife Research,* **34**, 523–529.

Triggs, B. (2009) *Wombats*. CSIRO Publishing: Melbourne, Australia.

Twigg, L.E., Martin, G.R. & Lowe, T.J. (2002) Evidence of pesticide resistance in medium-sized mammalian pests: a case study with 1080 poison and Australian rabbits. *Journal of Applied Ecology,* **39**, 549–560.

van Dyk, G. & Slotow R. (2003) The effects of fences and lions on the ecology of African wild dogs introduced to Pilanesberg National Park, South Africa. *African Zoology,* **38**, 79–84.

Vieira, E.M., Finlayson, G.R. & Dickman, C.R. (2007) Habitat use and density of numbats (*Myrmecobius fasciatus*) reintroduced in an area of mallee vegetation, New South Wales. *Australian Mammalogy,* **29**, 17–24.

Winnard, A.L. & Coulson, G. (2008) Sixteen years of eastern barred bandicoot *Perameles gunnii* reintroductions in Victoria: a review. *Pacific Conservation Biology*, **14**, 34–53.

Chapter 5
The Use and Potential of Pest-Proof Fencing for Ecosystem Restoration and Fauna Conservation in New Zealand

Bruce Burns, John Innes, and Tim Day

Introduction

In New Zealand, fences are commonplace. In a country in which pastoral agriculture is one of the dominant land uses, and in which rotational grazing is a strong management paradigm, barriers to control the distribution of farm animals are present in most landscapes. It is not surprising, therefore, that New Zealand conservation managers have extended and adapted this basic and familiar technology to try and control the distribution of exotic mammalian pests (The term "pest" here is used to denote wild animals that adversely affect indigenous biota and ecosystems; Parkes and Murphy 2003). By this means, they have succeeded in isolating native ecosystems from the impacts of these pests and have made significant conservation gains. Also, the application of pest-proof fencing in New Zealand to local projects has focused and galvanized public involvement in conservation in a profound manner.

Since human colonization of the New Zealand archipelago, 31 species of mammals have established naturalized populations following deliberate or accidental introduction by humans (Parkes and Murphy 2003). Many of these introduced mammals have fundamentally changed the nature of New Zealand natural ecosystems as predators and competitors, causing national and local extinctions of many native species, and continuing to cause population declines in others (Forsyth et al. 2010; Innes et al. 2010; Lee et al. 2010). Many of the most damaging introduced mammals

B. Burns (✉)
School of Biological Sciences, University of Auckland, Private Bag,
92019, Auckland 1142, New Zealand
e-mail: b.burns@auckland.ac.nz

J. Innes
Landcare Research, Private Bag 3127, Hamilton 3240, New Zealand

T. Day
Xcluder Pest Proof Fencing, P.O. Box 7135, Rotorua 3042, New Zealand

M.J. Somers and M.W. Hayward (eds.), *Fencing for Conservation: Restriction of Evolutionary Potential or a Riposte to Threatening Processes?*,
DOI 10.1007/978-1-4614-0902-1_5,

are almost ubiquitous in New Zealand and continue to degrade even large, legally protected areas such as National Parks, e.g., ship rat *Rattus rattus*, brushtail possum *Trichosurus vulpecula* and stoat *Mustela erminea* (King 2005). Introduced pests are by far the largest conservation problem in New Zealand (Craig et al. 2000).

The use of fences designed to exclude all introduced mammals from natural habitats is a recent conservation strategy being applied in New Zealand for ecosystem restoration. It is an extension of the strategy of using pest-free islands for conservation by attempting to create such "islands" on the mainland, with fences rather than the sea as a barrier to pest reinvasion. The lack of native, terrestrial, flightless mammals in New Zealand (Gibbs 2006; Tennyson 2010) means that fences can selectively exclude this introduced guild with limited apparent disruption to native organisms. The areas established are therefore expected to reverse degrading ecological processes, such as flowering, pollination and regeneration, and regain species diversity (often through conservation translocations) and abundance. Projects centred on these areas hope for the re-emergence of an ecosystem structure close to that of a pre-human New Zealand, although they concede that such a goal is strictly unattainable given the lack of knowledge of what such an ecosystem looked like, extinction of some species, and changes in environment (e.g., climate) since that time (e.g., Karori Wildlife Sanctuary (recently renamed Zealandia), Campbell-Hunt 2002) (Simberloff 1990; Atkinson 2001).

This chapter reviews the development of conservation fencing, the extent and distribution of conservation areas based on fences, the experience of conservation managers with fences and the current contribution of fences to conservation in New Zealand. Areas with pest-proof fences are now considered an important resource for ongoing conservation efforts, although their long-term effectiveness (both practical and financial) is still being evaluated. Discussion is largely limited to specialist conservation ("pest-proof") fences rather than situations in which standard stock proof fences are used in a conservation setting.

History of Conservation Fencing in New Zealand

Fences as barriers to animal movement are a characteristic feature of New Zealand landscapes and reflect its agricultural heritage. As well, many innovations in fencing technology have been developed in New Zealand (e.g., types of electric fencing, Brook 1939). Adapting fences to conservation problems is therefore congruous with existing cultural traditions for land management. Initially, standard ungulate fences were used in New Zealand to protect reserved natural areas from domesticated browsers such as cattle *Bos taurus* and sheep *Ovis aries* (e.g., Brown 1994). Browsing by these farm animals is known to cause significant damage to native forest composition, structure and regeneration (Timmins 2002; Smale et al. 2005; Dodd and Power 2007), and stock-proof fencing (6–9-wire post and batten, or netting) is widely implemented around conservation areas to improve vegetation condition (Smale et al. 2005). Fencing native forest fragments from stock is a necessary condition of conservation easements on private lands of such organizations as the Queen Elizabeth II National Trust in New Zealand (Porteous 1993).

In the 1980s and 1990s, fences started to be investigated and used as a means of protecting populations of highly endangered species from specific predators in cases where the species of conservation interest could not be transferred to predator-free habitat. For example, fences were developed around vegetation containing the endangered giant flax snail *Placostylus ambagiosus* at Te Paki to keep out pigs *Sus scrofa*, an important predator, using netting reinforced with metal stakes (Parrish et al. 1995). In another initiative, the Cromwell Chafer Reserve was fenced to exclude rabbits *Oryctolagus cuniculus*, with the fence extending 1.1 m above ground and 0.6 m belowground, to prevent vegetation degradation that would threaten the critically endangered Cromwell chafer beetle *Prodontria lewisi* (Aviss and Roberts 1994). Also, a population of Maud Island frog *Leiopelma pakeka* on Maud Island was surrounded by a fence designed to exclude the native tuatara *Sphenodon punctatus*, its main predator on that island (Brown 1994).

At about this same time, conservation in New Zealand underwent a revolution fuelled by the success of pest eradication projects on small offshore islands (Taylor and Thomas 1993; Jansen 1993). Islands have long been a key component of New Zealand's conservation strategy with those that are pest-free acting as refuges for species vulnerable to pest mammals. In the 1980s, eradication technologies were developed to clear islands previously invaded by pest mammals to increase the area available for conservation purposes (Pryde 1997). The key advances achieved by these means was the ability to remove a suite of mammal species from islands, and the realization that even seemingly entrenched pests such as Norway rats *Rattus norvegicus* and mice *Mus musculus* could be eliminated. By 2008, 71 islands had all non-native mammals eradicated (Bellingham et al. 2010) and 65 had been cleared of at least rats including at least 2 islands >2,000 ha (Towns and Broome 2003). The residual biodiversity on these islands has often flourished following these pest removals (Bellingham et al. 2010).

These island experiences led some conservation managers to consider whether programmes targeting multiple pests could be applied in mainland situations. One of the first of these achieved reductions in a range of pest mammals, particularly brushtail possums and ship rats, to near zero densities over several years in central North Island forests, in order to restore populations of North Island kokako *Callaeas cinerea wilsoni* (Innes et al. 1999). Kokako populations increased dramatically as a result (Innes et al. 1999), but as well, other ecological responses were noted including improvements in populations of invertebrates, plants and other forest birds (Saunders and Norton 2001). The success of this project caused a shift in focus of some conservation management in New Zealand from species-centred to ecosystem-centred, and led to the creation of a series of "mainland islands" based on trying to achieve sustainable management regimes by which all critical pests were controlled to extremely low levels and attempts made to limit re-invasion (Saunders and Norton 2001). The Department of Conservation initiated six Mainland Island projects in 1995 and 1996 throughout New Zealand ranging in size from 117 to 6,000 ha using this approach (Saunders and Norton 2001).

Unfenced Mainland Island projects have resulted in numerous conservation success stories, although few have so far been published; an exception is from Trounson Kauri Park (Gillies et al. 2003). Behind these highlights, however, predator control in some years failed to prevent damage by surviving or invading pests; some pests such as mice and hedgehogs *Erinaceus europaeus* were mostly not targeted at all; concern about toxin residues in non-target wildlife remained (Hoare and Hare 2006) and interactions between pest species have become apparent. These interactions include: rabbits increasing after stoat and cat control (Gillies et al. 2003); mice increasing after ship rat removal (Innes et al. 1995); ship rats increasing after possum removal (Sweetapple and Nugent 2007), and stoats eating more birds when ship rats are absent (Murphy and Bradfield 1992). Fences provided the alluring prospect of increasing conservation gains by allowing eradication rather than fluctuating control of pest populations, thus reducing toxin use in the long term, and by allowing more pest species to be routinely targeted. This particularly meant adapting fences to prevent passage of a range of smaller mammals such as brushtail possums, rodents and mustelids.

In 1986 and 1987, a fence for excluding a range of such predators was trialled around a 10 ha exclosure at Burwood Bush, Te Anau, a takahe *Porphyrio hochstetteri* breeding centre, and at Twizel, around aviaries for rearing black stilt *Himantopus novaezelandiae* (Aviss and Roberts 1994). This fence was 1.3 m high with chain-mesh netting and electric wires facing out from the fence at top and bottom on fibreglass outriggers. As well, the mesh extended just into the ground then turned to lie parallel to the ground surface and beneath it for approximately 300 mm towards the exterior to deter burrowing animals (Aviss and Roberts 1994). This same fence design was used to enclose a constructed wetland at Twizel (Ruataniwha wetland) in 1993 to provide further protection for black stilt and a range of other ground-nesting wetland birds (Sanders et al. 2007) and in 1994, a similar electric fence was established around a 20 ha site at Kowhai Vale on Bank's Peninsula to protect a translocated population of eastern buff weka *Gallirallus australis hectori*. Although the programmes that used this general fence design were somewhat successful in maintaining and increasing the populations of these endangered birds, the fences were not effective in completely excluding all the target pests, and excluded some pests better than others (Sanders et al. 2007). Sanders et al. (2007) considered that the fence at Twizel was not tall enough, that electric fencing deterred but did not stop motivated pests and that waterways intersecting the fenceline were points of weakness.

Whether or not to use electric fencing has been an ongoing question in the development of fence designs for mammalian pests. Cowan and Rhodes (1992) conducted a trial to test whether an electric fence would be able to prevent brushtail possum movement in farmland. They concluded that it reduced possum movements across the fence by 60–80% but did not stop possums crossing the fence. They also found that maintaining the effectiveness of the electric fence over time required intensive management, and periods of low or no voltage were frequent because of the fence earthing from a variety of causes. Clapperton and Matthews (1996) also tested the interactions between brushtail possums and electric fences. They found that possums that had experienced electric fences still challenged those fences at later times and would climb over the fences when turned off. They also found that possums motivated by people driving them towards a fence would still climb over it even though they would receive shocks.

This experimentation clearly indicated that fences dependent on electric wires reduced but did not completely prevent passage of animals, and that fences based on exceeding the physical capability of pests would be a better focus for design development (Day and MacGibbon 2007). The search for fence designs that would completely exclude assemblages of mammalian pest species then led to a series of detailed experimental trials using test pens. The first of these was carried out starting in 1994 by a group in Wellington seeking to establish the Karori Wildlife Sanctuary (Zealandia) (Hitchmough 1994; Lynch 1995). The group observed the interactions between animals and fences of different designs within these pens to establish each species' physical capabilities in jumping, climbing, digging and passing through apertures of different sizes; and ultimately their ability to cross over the fences. These trials resulted in a fence design of approximately 2 m height (none of the animals tested could jump this high) with a curved top hood (to prevent climbing animals), a woven mesh wall with apertures small enough to prevent juvenile mice passing through (6×50 mm) and an underground skirt (to prevent burrowing animals). To provide comprehensive protection against mammalian pests, a fence of this design was erected around the Karori Wildlife Sanctuary (Zealandia) in Wellington in 1999 (Fig. 5.1).

A second set of trials was carried out independently starting in 1996 in the Waikato region initially to exclude herbivorous mammalian pests from restoration plantings but eventually to exclude all mammalian pests present in New Zealand (Day and MacGibbon 2007). These trials also used enclosures but, in this case, the enclosures were larger (approximately 12 m diameter) and entirely constructed of the fence design to be tested, with the barrier side of the fence facing inwards. Test animals were then introduced into the enclosure and their behaviour during escape attempts observed from a tower built beside the enclosure (Day and MacGibbon 2007). The researchers also undertook observations on the ability of mice to pass through mesh of different aperture size, confirming that even juvenile mice could not pass through mesh with an aperture of no more than 6 mm in one dimension (Day and MacGibbon 2007). These experiments led to the development of two pest-proof fence designs; one approximately 2 m tall that converged on the Karori design although with a different hood and welded mesh, and a second with a 1.3 m tall base fence and a 1.5–2.0 m tall section of horticultural "bird" netting mounted above the base fence on supple fibreglass rods (Day and MacGibbon 2007). In the latter design, the flexible upper portion is designed to sag back towards the ground on the outside of the fence if an animal climbs onto it. Both designs were effective in excluding all pest animals tested against them (Day and MacGibbon 2007). The first application of one of these fence designs was to enclose a 16 ha gully site at Warrenheip (Waikato Region) also in 1999.

The fences developed by this second group of researchers coincided with and facilitated an upsurge in community-led conservation projects within New Zealand by groups who saw the potential of this fencing technology in sustaining conservation gains made by the also newly acquired tools for pest eradication. A market for pest-proof fences therefore emerged and led to the formation of the Xcluder® Pest

Fig. 5.1 Karori Wildlife Sanctuary (Zealandia), Wellington, New Zealand enclosed by a pest-proof fence (photos: Bruce Burns)

Proof Fencing Company which has built this type of fencing for many such projects in New Zealand since 2000. Fences based on this design but adapted for local conditions have also been built by this company on Lord Howe Island, Hawaii, and Mauritius (Day and MacGibbon 2007). A second company, Pestproof Fences, was established in 2005 in competition with Xcluder®. Further discussion in this chapter will focus on the characteristics and impact of these pest-proof fence designs over the last decade in New Zealand.

Table 5.1 Twenty-eight areas at 24 sites in New Zealand cleared of mammalian pests and with mammalian pest-proof fences in December 2009

	Site	Island	Year completed	Area protected (ha)	Fence length (km)
1	Karori Wildlife Sanctuary (Zealandia)	North	1999	252	8.6
2	Warrenheip	North	1999	16	2.4
3	Ellen Elizabeth Preece Covenant	Pitt	2001	36	3.0
4	Rapanui Point, Taranaki	North	2002	1	0.4
5	Puketukutuku Peninsula, Lake Waikaremoana	North	2003	750	1.3
6	Southern Enclosure, Maungatautari	North	2004	65	3.5
7	Northern Enclosure, Maungatautari	North	2004	35	2.8
8	Tawharanui Open Sanctuary	North	2004	530	2.7
9	Riccarton Bush	South	2004	7.7	1.1
10	Tautari Wetland, Maungatautari	North	2005	3	0.8
11	Bushy Park	North	2005	98	4.7
12	Macraes Flat 1	South	2005	22	1.7
13	Nick's Head Predator Proof Enclosure	North	2005	35	0.6
14	Dancing Star Foundation Preserve	Stewart	2005	160	2.1
15	Maungatautari Ecological Island	North	2006	3,300	39.0
16	Garland Covenant, Maungatautari	North	2006	16	1.6
17	Whinray Ecological Charitable Trust	North	2006	1.4	0.5
18	Sweetwater Conservation Covenant	Chatham	2006	8	0.7
19	Macraes Flat 2	South	2007	11	1.3
20	Orokonui EcoSanctuary	South	2007	307	9.0
21	Cape Kidnappers and Ocean Beach Wildlife Preserve	North	2007	2,200	9.5
22	Pan Pac Kiwi Creche, Opouahi	North	2007	40	3.3
23	Rotokare Scenic Reserve Trust	North	2007	229	8.4
24	Glenfern Sanctuary	Great Barrier (Aotea)	2008	230	2.1
25	Kaipupu Point Sounds Wildlife Sanctuary	South	2008	40	0.6
26	Driving Creek Wildlife Sanctuary	North	2008	1.6	0.5
27	Sooty Shearwater Conservation Project	South	2008	0.5	0.3
28	Mokomoko Dryland Sanctuary	South	2009	0.3	0.2
	Totals			8395.5	112.7

Current State of Pest-Proof Fences in New Zealand

Between 1999, when an effective pest-proof fence design was developed, until the end of 2009, 28 conservation areas covering a total of 8,396 ha have been enclosed by 113 km of pest-proof fences and cleared of mammalian pests (Table 5.1). The 28 fenced areas occur at 24 discrete sites as 5 of the areas form a complex of enclosures adjacent to each other, based at Maungatautari (Table 5.1; Fig. 5.2). They cover a

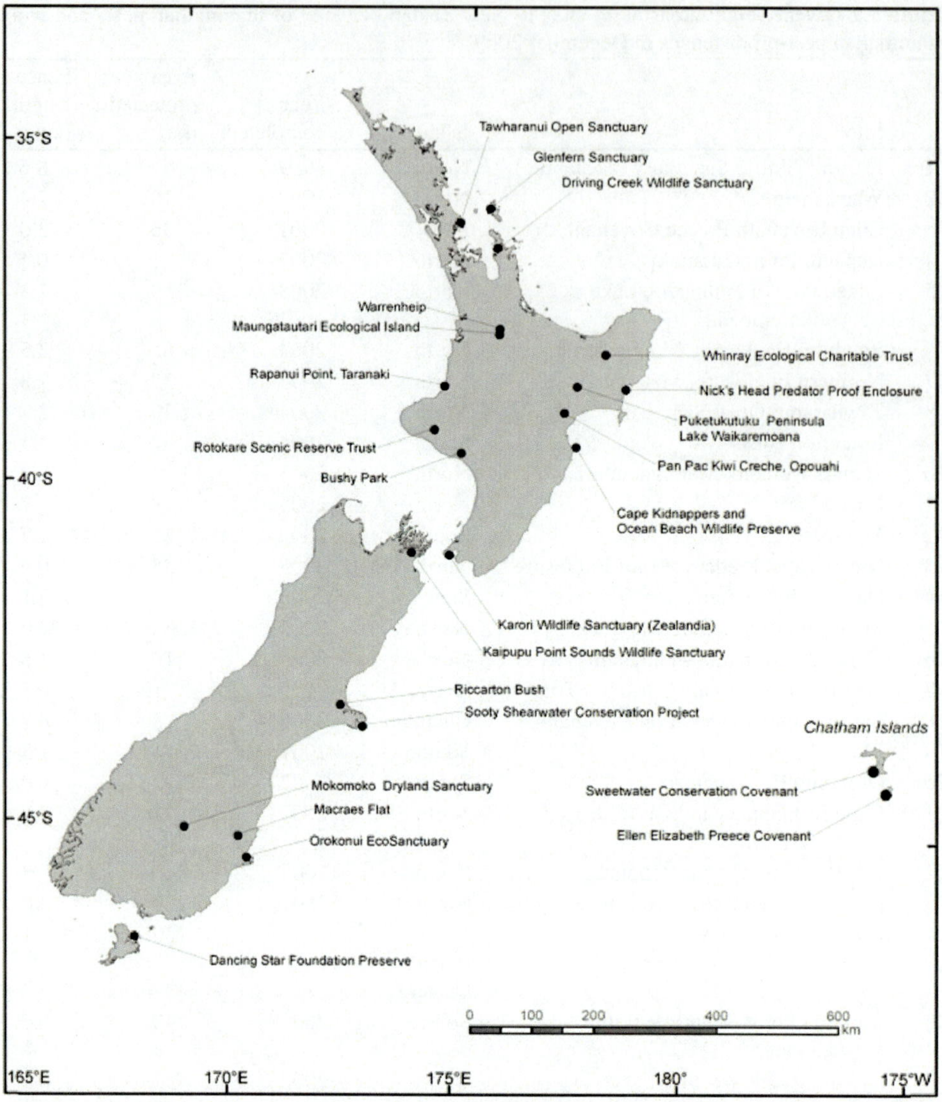

Fig. 5.2 Locations of pest-proof fence sites in New Zealand

wide range of sizes: 8 are <10 ha, 11 are between 10 and 100 ha, 7 are between 100 and 1,000 ha, and 2 are >1,000 ha (Cape Kidnappers and Ocean Beach Wildlife Preserve (2,200 ha) and the Maungatautari Ecological Island (Fig. 5.3; 3,400 ha)) (Table 5.1). Small pest-proof fenced enclosures of the same designs have also been used in zoo/wildlife park settings for threatened species protection or breeding programmes (e.g., Mt Bruce, Rainbow Springs). As well as completed fence projects, at least ten other community-led conservation projects in New Zealand (e.g., Shakespear Open Sanctuary, Brook Waimarama Sanctuary, Bream Head Trust, Hutton's Shearwater Charitable Trust) have current plans to build pest-proof fences around areas to be cleared of mammalian pests in the near future.

Fig. 5.3 Pest-proof fence at the Maungatautari Ecological Island (photo: Jillana Robertson)

Rather than being part of a coordinated national conservation strategy, the location of these pest-proof fenced sites has occurred through the independent decisions of local community groups and private individuals who have found the fence concept for restoring biodiversity inspiring and enabling. The areas they have chosen to fence have not generally been areas of high existing biodiversity, but areas of lesser biodiversity value often close to population centres and the presence of a motivated local human community. For example, the Karori Wildlife Sanctuary (Zealandia) is situated in the centre of the capital city Wellington, about 4 km from New Zealand's parliament building, and depends heavily on a volunteer labour force. The catchment it comprises was completely cleared of natural forest for farming purposes in the mid-nineteenth century and stands of the exotic tree radiata pine *Pinus radiata* were planted (Blick et al. 2008). Farming was then abandoned about 1900 and the catchment allowed to regenerate into secondary native scrub and forest though some stands of radiata pine still persist (Blick et al. 2008). The association of pest-proof fenced sites with human population is reflected in the higher number that occur on North Island (13 sites) than on the less populated South Island (seven sites) (Table 5.1).

Based on the existing sites, pest-proof fence projects in New Zealand have used one of three strategies to take advantage of natural landscape features in positioning fences: peninsulas, catchments, or fragments. In the peninsula approach, fences have been placed on the neck of a peninsula in order to protect a large area with a relatively short length of fence (e.g., Tawharanui Open Sanctuary, Cape Kidnappers and Ocean Beach Wildlife Preserve, Glenfern Sanctuary). These fences generally follow ridges across the peninsula so the fence does not cross any watercourses. New Zealand's convoluted coastline provides many opportunities for this type of fence position. The biggest disadvantage of this strategy is in constructing an effective barrier at the coastal interface, and preventing pests walking or swimming around the fence ends. Projects based on this strategy tend to control pests intensively within a buffer zone immediately outside the fence and around the coastal ends of the fence to minimize the risk of incursion at these points.

In the catchment approach, the pest-proof fence follows a ridgeline that encloses the entire top of a minor catchment (e.g., Karori Wildlife Sanctuary (Zealandia), Orokonui EcoSanctuary, Rotokare Scenic Reserve Trust). This strategy means that the fence need only cross one watercourse that flows out of the catchment, and is sited on slopes with the lowest gradient available. In these pest-proof fence designs, watercourses are generally accommodated by using a culvert with a weighted gate which will close off at times of low flow. The fence is then built over the top of the culvert. Such systems may provide opportunities for breaches in the barrier if gates are jammed open by debris, or floods undermine the structure. For this reason, minimizing the number of watercourses intersected by a fence may be an advantage as in this strategy. A potential disadvantage of this strategy is that fences may have tall forest on both sides of a fence. Adequate distances between tree branches inside and outside the fence need to be maintained at all times to prevent pest movement above the fence.

The third strategy for fenced areas has been to fence off an existing forest fragment surrounded by pasture (e.g., Maungatautari Ecological Island, Bushy Park). This strategy takes advantage of the many forest fragments present in rural landscapes within New Zealand (Ewers et al. 2006), and in being able to fence edges without trees on the outside therefore eliminating the risk of arboreal pest movement across the fence. A disadvantage may be the higher number of watercourses intersected by the fence, increasing the risks associated with these structures. The Maungatautari Ecological Island, for example, is a large forest fragment based on an isolated mountain surrounded by farmed pasture. The mountain has over 40 streams draining off its slopes, so that its surrounding fence contains a high number of watergates.

None of these landscape strategies for fence location has so far proved demonstrably more effective than the others in terms of security against pests. The peninsula strategy requires less fencing cost per hectare protected than the other strategies, but inherently requires greater pest management effort outside the fences to protect the exposed fence ends.

Contribution to Species Protection and Range Recovery

Several of the fence projects listed in Table 5.1 have been specifically established to protect existing populations of endangered species, or species uncommon on the main islands of New Zealand (Table 5.2). For some of these species, pest-proof fences are seen potentially as the last hope to prevent extinction. For example, within tussock grasslands at Macraes Flat in central Otago, two pest-proof fenced areas were established in 2005 and 2007 to conserve Otago skink *Oligosoma otagense* and grand skink *Oligosoma grande* which are classified as nationally critical (Norbury et al. 2007; Wilson et al. 2007). This was carried out as population modelling, based on data from 1996 to 2002, predicted high probabilities that both species would go extinct by 2010 if survival rates were not improved (Norbury et al. 2007). Predation by mammalian pests was identified as the most likely cause of decline (Tocher 2006). The populations protected by the pest-proof fences and other populations in areas intensively trapped for predators recovered significantly between 2005 and 2008 (Department of Conservation 2008). The success of the Otago and grand skink experience has led other researchers to strongly promote the use of pest-proof fenced areas for the conservation of other threatened skink species in New Zealand, particularly where other methods have failed (Hoare et al. 2007; Lettink et al. 2010).

Pest-proof fences have also been erected both to prevent pests invading an area containing an endangered species and to prevent that species dispersing to areas containing pests. A pest-proof fence was constructed across the Puketukutuku Peninsula on Lake Waikaremoana, Urewera National Park, in 2003 to protect a population of North Island brown kiwi *Apteryx mantelli*. Although designed to prevent pests entering the peninsula, the fence was also constructed as a means of preventing subadult kiwi from dispersing into less secure forest areas in their search to establish new territories. Unrestrained subadult kiwi average dispersal distances of more than 5 km from their natal range when seeking to establish territories (Basse and McLennan 2003).

New Zealand has an extremely diverse seabird community (Taylor 2000a) and large colonies of seabirds were once extremely common on all islands of the New Zealand archipelago (Worthy and Holdaway 2002; Wilson 2008). On the main islands, however, only a few colonies now remain, with many former colonies becoming extinct as a result of ongoing predation by mammalian pests (Taylor 2000a). Pest-proof fences are increasingly being recognized as an effective means of protecting remnant breeding seabird colonies, as a short fence enclosing a relatively small area on a cliff top is all that is necessary to secure colonies that occur at high densities. As examples, pest-proof fences now protect one of the last North Island colonies of grey-faced petrel *Pterodroma macroptera* at Rapanui Point in the Taranaki Region and one of the few remaining sooty shearwater *Puffinus griseus* colonies on South Island at Stony Bay on Bank's Peninsula. Both these species also have cultural importance as some chicks were traditionally harvested by local Maori (Taylor 2000b).

Table 5.2 Use of 24 pest-proof fenced sites for wildlife conservation in New Zealand as at December 2009

Species	Threat classification*	Resident populations fenced for protection[a]	Populations translocated since fencing[b]	Self-introduced populations since fencing[c]
Petroica traversi (black robin)	Nationally critical		1 (failed)	
Porphyrio hochstetteri (Takahe)	Nationally critical		1	
Pterodroma magentae (Taiko)	Nationally critical		1	
Oligosoma grande (Grand skink)	Nationally critical	2		
Oligosoma otagense (Otago skink)	Nationally critical	2	1	
Anas chlorotis (brown teal)	Nationally endangered		4	
Deinacrida mahoenui (Mahoenui giant weta)	Nationally endangered		1	
Gallirallus australis greyi (North Island weka)	Nationally endangered		1	
Leiopelma pakeka (Maud Island frog)	Nationally endangered		1	
Nestor meridionalis septentrionalis (North Island kaka)	Nationally endangered		2	1
Nestor meridionalis meridionalis (South Island kaka)	Nationally endangered		1	
Notiomystis cincta (Hihi)	Nationally endangered		2	
Philesturnus carunculatus carunculatus (South Island saddleback)	Nationally endangered		1	
Pterodroma axillaris (Chatham petrel)	Nationally endangered		2	
Charadrius obscurus aquilonius (Northern New Zealand dotterel)	Nationally vulnerable			1
Acanthisitta chloris chloris (South Island rifleman)	Declining		1	
Acanthisitta chloris granti (North Island rifleman)	Declining		1	
Apteryx mantelli (North Island brown kiwi)	Declining	1	7	
Apteryx haastii (Great spotted kiwi)	Declining		1	
Eudyptula minor iredalei (Northern little blue penguin)	Declining			1
Hemiphaga novaeseelandiae (Kereru)	Declining		1	
Cyanoramphus auriceps (Yellow-crowned kakariki)	Declining		1	
Galaxias argenteus (Giant kokopu)	Declining		1	
Naultinus elegans (Auckland green gecko)	Declining		2	
Naultinus gemmeus (Jeweled gecko)	Declining		1	

(continued)

Puffinus griseus (Sooty shearwater)	Declining	1		
Apteryx owenii (Little spotted kiwi)	Naturally uncommon		1	
Coenocorypha pusilla (Chatham Island snipe)	Naturally uncommon		1	
Deinacrida rugosa (Cook Strait giant weta)	Naturally uncommon		1	
Galaxias postvectis (Shortjaw kokopu)	Naturally uncommon		1	
Philesturnus carunculatus rufusater (North Island saddleback)	Naturally uncommon		2	
Sphenodon punctatus (Tuatara)	Naturally uncommon		1	
Anthornis melanura (Bellbird)	Not threatened		1	1
Aythya novaeseelandiae (New Zealand Scaup)	Not threatened		1	
Cyanoramphus novaezelandiae (Red-crowned kakariki)	Not threatened		1	
Galaxias fasciatus (Banded kokopu)	Not threatened		1	
Hemideina femorata (Canterbury tree weta)	Not threatened		1	
Hemideina thoracica (Auckland tree weta)	Not threatened		1	
Hoplodactylus granulatus (Forest gecko)	Not threatened		2	
Mohoua albicilla (Whitehead)	Not threatened		4	
Mohua novaeseelandiae (Brown creeper)	Not threatened		1	
Petroica australis longipes (North Island robin)	Not threatened		5	
Petroica macrocephala (North Island tomtit)	Not threatened		2	
Pterodroma macroptera (Grey-faced petrel)	Not threatened	1	1	2
Total		7	63	6

[*]Species ordered by threat status. Threat classification system is as per Townsend et al. (2008) and classification status as per Hitchmough et al. (2007)
[a]Number of wildlife populations pest-proof fences were established specifically to protect
[b]Number of wildlife populations translocated into pest-proof fenced sites
[c]Number of wildlife populations that have self-introduced into pest-proof fenced sites.

As well as protecting existing populations of endangered species, pest-proof fences have also provided secure sites for translocating populations of a large range of New Zealand species, and this is perhaps one of their greatest contributions to New Zealand conservation. Sixty-three translocations of 40 species have been made within the 24 pest-proof fenced sites established over the last 10 years (Table 5.2). Over the same time period in comparison, the IUCN Reintroduction Specialist Group reported 82 species translocations were made to pest-free off-shore islands in New Zealand (listed on http://rsg-oceania.squarespace.com/nz/ accessed on March 15, 2010), suggesting that pest-proof fenced areas are now playing a similar role in securing the future of threatened species.

Translocations to fenced sites have mostly been reintroductions of species previously known or considered highly probable to have been present in these areas. The translocation of the endangered takahe to Maungatautari, however, is a benign introduction of a South Island species to substitute for the North Island species (*Notornis mantelli*) which is now extinct (Wickes et al. 2009). Four of the 40 species translocated are ranked nationally critical, the category of species at highest risk of extinction (Hitchmough et al. 2007), 9 are ranked nationally endangered, 9 are declining, 6 are naturally uncommon and 12 are considered not threatened. The majority of these species are birds (27), although there have also been translocations of 5 reptiles, 4 invertebrates (all weta), 3 fish and 1 amphibian (Table 5.2).

Some of these translocations are particularly significant. The taiko *Pterodroma magentae* (nationally critical), a burrowing petrel, was thought to be extinct until rediscovered in 1978 (Aikman et al. 2001). Since then, only one breeding location on Chatham Island has been identified. Taiko chicks were translocated from this breeding location into the nearby fenced Sweetwater Conservation Covenant in 2007 and 2008 in an attempt to establish a second breeding colony (Miskelly et al. 2009). Similarly, Chatham petrel *Pterodroma axillaris* was reintroduced into the Ellen Elizabeth Preece Conservation Covenant on Pitt Island between 2002 and 2005. Six chicks have now returned to this site to breed. This is the first recorded breeding away from the only other known colony on Rangatira Island since the species' discovery in 1892 (Miskelly et al. 2009). Reintroductions at Karori Wildlife Sanctuary (Zealandia) have also contributed to the return of species to the main islands of New Zealand after long absences. The release of six species there (little spotted kiwi *Apteryx owenii*, Cook Strait giant weta *Deinacrida rugosa* (Watts et al. 2008), Maud Island frog (Lukis and Bell 2007), hihi *Notiomystis cincta* (Bell et al. 2006), North Island saddleback *Philesturnus carunculatus rufusater* (Parker 2008) and tuatara (Miller et al. 2009)) restored populations to North Island after they had been restricted to off-shore islands for many decades.

Although a large number of translocations have been carried out into fenced areas, one has not been successful to date and others may fail in the future. It is therefore uncertain whether all translocations will lead to new populations. Black robin *Petroica traversi*, another nationally critical species, is one of the most endangered bird species in the world and the subject of an intensive restoration effort (Merton 1992). As part of this, 20 black robins were reintroduced into the fenced Ellen Elizabeth Preece Conservation Covenant in February 2004. This reintroduction, however, failed after 3 years with the immaturity of the habitat and the continued presence of mice competing with the robins for invertebrates suggested as contributing factors.

Translocations into some pest-proof fenced areas have been specifically to temporarily hold animals and raise them for later supplementation of less protected populations. In particular, individuals of North Island brown kiwi and great spotted kiwi *Apteryx haastii* are raised as juveniles at small pest-proof fenced areas such as Warrenheip, Pan Pac Kiwi Creche, Whinray Ecological Charitable Trust and Riccarton Bush and then released as adults to supplement populations in larger tracts of forest. McLennan et al. (2004) found that young North Island brown kiwi suffered intense predation from stoats in their first 4 months of life, after which they had grown too large (>800 g) for stoats to kill. The vulnerability of juveniles has been identified as one of the main factors in population decline in wild kiwi populations (Holzapfel et al. 2008). Raising chicks to a "safe" weight in stoat-free sites such as fenced sanctuaries is now considered a crucial part of the recovery strategy for kiwi (Colbourne et al. 2005; Holzapfel et al. 2008).

Pest-proof fenced sites have also provided opportunities for natural dispersal and population founding events and six such events have been recorded (Table 5.2). Many of these are seabirds re-establishing on coasts, but perhaps the most significant of these has been the self-reintroduction of a population of bellbirds *Anthornis melanura* to the Tawharanui Open Sanctuary (Brunton et al. 2008). This species has been locally extinct on the mainland of this region for >100 years. A pest-proof fence was established and pest eradication achieved at Tawharanui in 2004, with bellbirds first recorded there in 2005. At that time, the presence of both male and female birds was confirmed and the song types recorded matched those of populations on Little Barrier Island, 23 km away (Brunton et al. 2008).

Finally, one of the unexpected byproducts of conservation management at pest-proof fenced sites has been the discovery of previously unknown populations of significant species through the enhanced activity and scrutiny at these sites. At Maungatautari Ecological Island, management and research activity has led to the serendipitous discovery of a population of the threatened Hochstetter's frog *Leiopelma hochstetteri* (threat classification = "naturally uncommon"; Baber et al. 2006), and a number of significant plants including a population of silver beech *Nothofagus menziesii* close to its northern limit (Brown et al. 2006).

Changes in Ecosystem Condition

Although not well documented yet, the removal of pest mammals from the ecosystems enclosed by pest-proof fences is leading to changes in the composition of residual communities, and the abundances of pest-sensitive populations as they are released from previous limits imposed by pest predation or herbivory. For vegetation in Karori Wildlife Sanctuary (Zealandia), seedling density in the forest understorey has increased markedly, and species previously known to be palatable to brushtail possums (e.g., kohekohe *Dysoxylum spectabile*, mahoe *Melicytus ramiflorus*, pate *Schefflera digitata* and kanono *Coprosma grandifolia*, Nugent et al. 2000, 2002) have strongly increased in importance (Blick et al. 2008). At Maungatautari, one of the more obvious vegetation responses has been the recruitment of kotukutuku *Fuchsia excorticata* and wineberry *Aristotelia serrata* around the forest edges and track margins after fence and track construction (BB, pers. obs.). Both these species are also preferred foods of brushtail possums, are declining throughout unprotected New Zealand forests (Nugent et al. 2000; Urlich and Brady 2005), and were in low abundance at Maungatautari prior to fence construction. Also, the percentages of common shrubs fruiting and flowering at Maungatautari have increased compared to equivalent unfenced areas (Ecoquest Education Foundation, personal communication).

For bird communities, limited initial changes have been documented 2 years after fence completion and pest eradication at Maungatautari based on call counts (Fitzgerald et al. 2009). Three native species were found to have increased in conspicuousness compared to an unfenced site with pest mammals (i.e., shining cuckoo *Chrysococcyx lucidus*, grey warbler *Gerygone igata* (host of the shining cuckoo), and tui *Prosthemadera novaeseelandiae*) and two species have decreased (i.e., blackbird *Turdus merula*, an exotic; and silvereye *Zosterops lateralis*, a native, self-introduced from Australia in the 1800s). The most dramatic difference was in the relative abundance of tui, with twice as many recorded in call counts at Maungatautari than the unfenced control (Fitzgerald et al. 2009). Tui are mobile birds, and surveys of landowners in 5 km bands around Maungatautari document 8–22% increases in the most tui seen at one time, up to 20 km away (JI, unpublished data). This also suggests that pest-proof fenced sites may act as source areas to enrich adjacent zones with mobile species.

Invertebrate communities have also shown changes in composition and abundances of key species. Significant increases in Auckland tree weta *Hemideina thoracica* and other weta (*Hemiandrus* spp. and *Gymnoplectron* spp.) were found 2 years after pest mammal eradication at Maungatautari (Watts et al. 2011). Watts (2007) also found a 300% increase in the abundance of ground-dwelling beetles at Maungatautari over the same time period with consistent though complex changes in composition. These are considered initial but not final changes to invertebrate communities and will probably continue to change as populations of insectivorous birds increase.

Negative Ecological Consequences of Fenced Sites: Real and Potential

Negative ecological consequences of pest-proof fences have not been obvious in New Zealand but this now needs critical assessment. Those that have been reported to date include (1) increased mortality of ducklings of forest edge nesting species (e.g., paradise shelduck *Tadorna variegata* when they are unable to follow parents to feed outside the fence (Sanders et al. 2007)), and (2) occasional bird deaths through collisions with fences (e.g., dead individuals of kereru *Hemiphaga novaeseelandiae* and long-tailed cuckoo *Eudynamys taitensis*; found at Maungatautari).

As the biotic communities enclosed by pest-proof fences develop in the absence of pest mammals, there are a number of potential negative ecological impacts that may also occur. Fencing may prevent dispersal of individuals from populations, as occurs for kiwi at Puketutuku Peninsula mentioned earlier, and lead to higher population densities within fences than might occur naturally. This overabundance may lead to overuse of resources and the potential for catastrophic decline (Hayward and Kerley 2009), although population limitation by intra- and inter-specific competition is probably a more natural or original state for New Zealand birds than top-down limitation by pest mammals (Innes et al. 2010). It is also often difficult to apply fencing at scales large enough to sustain genetically viable populations of some animals, due to inbreeding. For example, minimum North Island brown kiwi territory sizes are 2–3 ha (Holzapfel et al. 2008), so many of the existing pest-proof fenced areas in New Zealand are too small to support more than a few birds. The blocking of emigration and immigration to and from other populations may exacerbate inbreeding depression due to small founding populations (Hayward and Kerley 2009).

The eradication of pest mammals behind fences may also impact on species that now rely on these mammals for critical life history processes in the absence of their original mutualistic partners. For example, Dungan et al. (2002) found that brushtail possums were probably important as seed dispersers in seral vegetation in lowland Canterbury, and speculated on its possible modern role in dispersing large-seeded native species, particularly as large-gaped native birds are becoming increasingly uncommon. Also Lord (1991) attributed pollination and seed dispersal of the dioecious vine kiekie *Freycinetia banksii* to possums in areas in which its original pollinator short-tailed bat *Mystacina tuberculata* was locally extinct. Brushtail possum is one of the mammal species eradicated from within pest-proof fenced areas. In such areas, will the absence of both the indigenous and an adopted exotic pollinator and seed disperser now limit and eventually reduce large seeded native plant and kiekie populations?

Management Experience with Pest Exclusion and Pest-Proof Fences

All of the pest-proof fenced areas so far created in New Zealand have been operating for less than 10 years (Table 5.1), so experience with their management is still relatively immature. Here we discuss the initial pest eradication and maintenance of pest-free status within these areas, and ongoing maintenance requirements associated with the fences.

Pest eradication methods used in New Zealand have improved dramatically over recent years (Kelly and Sullivan 2010). For the pest-proof fenced projects considered here, the primary tool for initial eradication was aerial applications of cereal pellets containing the anticoagulant toxin brodifacoum applied in late winter/early spring when bait uptake by rodents is highest (Speedy et al. 2007). The aerial bait applications were accompanied by hand-spread baits immediately inside the fence to ensure complete coverage, followed by intensive trapping, and ground based bait stations where necessary (Goldwater 2007). As well as effects on pest mammal populations, the primary knockdown using brodifacoum inevitably resulted in some deaths of non-target wildlife. Hoare and Hare (2006) record 16 native New Zealand bird species that have suffered mortality as a result of exposure to brodifacoum, and Dowding et al. (2006) specifically noted the deaths of eight shorebirds (three pied stilts *Himantopus himantopus*, four New Zealand dotterel *Charadrius obscurus* and one spur-winged plover *Vanellus miles novaehollandiae*) following a broadifacoum operation at the pest-proof fenced Tawharanui Open Sanctuary.

Although the initial bait applications in pest-proof fenced areas removed the majority of pest mammals, small numbers of survivors often remained (Speedy et al. 2007). At Maungatautari, small numbers of surviving ship rats, mice and hedgehogs were found up to 20 months after the initial eradication operations within the southern and northern enclosures (Speedy et al. 2007). These were eventually removed with persistent trapping and by deploying extra baits in bait stations. At Tawharanui, the most persistent pest mammals have been mice, hedgehogs and rabbits that were still present 2 years after the initial knockdown (Goldwater 2007).

Despite the demonstrated efficacy of the fences used on these projects in trials, ongoing but infrequent reinvasions have been recorded. Some of these have been the result of damage to the fence incurred by tree falls or flood scours after storm events, or by human error such as gates being left open, vehicles being left close to the fence from which pests jumped over the fence, or vehicle or other damage to the fence (Day and MacGibbon 2007). At Maungatautari, tree falls on the fence occurred at a rate of 10 over a 3 year period (Day and MacGibbon 2007). Video surveillance of 20 simulated fence breaches at Maungatautari indicated that pest mammals were common directly outside the fence and constantly challenging it (Connolly et al. 2009). This study estimated a 99 and 85% probability of reinvasion at a fence breach by at least one mammal within 24 h in summer and winter respectively, with mice and ship rats being the most likely invaders (Connolly et al. 2009). Rapid response to a fence breach is therefore critical to prevent reinvasion, and managers of pest-proof fenced sites have developed systems to ensure this, e.g., development of a surveillance wire on top of fences that when tripped will immediately alert staff to breach events (Day and MacGibbon 2007).

Although most reinvasion events into pest-proof fenced sites have been quickly mitigated once detected, reinvasion (or repopulation) by mice has resulted in persistent populations of this species in some sites which are proving very difficult to eliminate. At Karori Wildlife Sanctuary (Zealandia), mice were thought to have been removed completely during the initial eradication but reinvaded within 2 years, indicating that this initial design was probably not completely mouse-proof. Removal of this population has not been achieved yet though they are maintained at low densities. At Tawharanui Open Sanctuary which is on a peninsula, mice were probably never completely eradicated and were also able to reinvade after the initial eradication attempt by moving around the space between the fence ends and the sea (Goldwater 2007). Here, mouse populations increased rapidly in the absence of mammalian predators and competitors (a type of mesopredator release predicted by the models of Tompkins and Veltman 2006) to attain densities higher than any previously recorded in New Zealand (157 mice per ha, Goldwater 2007) and are still present. The renewed elimination or control of mice, and understanding the impact on biodiversity of mouse populations within pest-proof fenced sites, are now key ongoing research questions for managers. Without targeted research, it is also difficult to tell if residual individual mice are survivors (or their progeny) or invaders; better monitoring tools and strategies are needed for this species.

The threat of reinvasion by pest mammals into pest-proof fenced sites ensures that monitoring to detect such events is a key activity and cost. One of the main tools for detecting incursions within these areas is the footprint tracking tunnel (King and Edgar 1977; Brown et al. 1996) deployed in grids ranging in intensity from 25×25 to 100×100 m (Speedy et al. 2007). This method has proved generally effective in detecting the presence of several key pest mammals (mice, ship rats, stoats, hedgehogs and possums) and may provide additional benefits by monitoring recovery of some valued indigenous species, such as insects (Watts et al. 2011). The use of trained rodent search dogs *Canis familiaris* to detect and locate pest individuals at low densities is increasingly being recognized as an alternative and effective technique (Gsell et al. 2010).

Other management specifically associated with pest-proof fences includes the need to maintain both the condition of the fences and vegetation clearance around them. Although pest-proof fences are generally constructed of durable materials with a long life expectancy (Day and MacGibbon 2007) even in coastal environments, e.g., marine grade stainless steel mesh, the fences require regular and frequent inspections to ensure all parts of the fence retain their pest-proof capability. The 47 km-long Maungatautari pest-proof fence is inspected at a general level weekly, and more intensively monthly (T. Rolley, Maungatautari Ecological Island Trust, personal communication). Fences that bisect forested areas also require regular inspection to ensure that the gap between canopies on either side of the fence remains wide enough (approximately 4 m) to prevent pests jumping between them. As well, grass and weeds growing around the base of the fence require regular control to prevent them compromising fence effectiveness. It is a common misconception that fences provide an "erect and forget" strategy for pest exclusion. In reality, effective fence designs provide an "erect and check forever more" tool that is always under pressure from potential invaders (Connolly et al. 2009). Fences therefore require a high level of vigilance and maintenance to ensure prolonged success.

Social and Economic Considerations

The upsurge in pest-proof fenced sites in New Zealand is as much a social phenomenon as a conservation initiative. Rather than being government-led, the agencies responsible for establishing these fenced areas are mostly community-led trusts or private individuals (Campbell-Hunt et al. 2010). Building fences has not generally been the result of a coordinated conservation programme but independent decisions by a number of communities. Campbell-Hunt et al. (2010) suggest that the emergence of these groups has been triggered by a shared perception of ecological loss within the community together with the motivation to take action provided by the novel availability of pest-proof fences. These fences have provided an attractive solution as they are not perceived as requiring the ongoing application of toxins that alternative approaches demand, and are also tangible and familiar.

The long-term commitment required by community groups to this restoration is also surprising because of the large capital and ongoing costs associated with pest-proof fences. Current costs for this type of fence in New Zealand are approximately US$150,000/km (Campbell-Hunt 2008a) and significant ongoing operational costs to maintain and monitor the fences and develop the reserve are required. Although funds for the establishment of these reserves and construction of the fences have been gained from charitable trusts, private donations and one-off government grants (Campbell-Hunt 2008b), sources of money to ensure the long-term financial sustainability of these pest-proof fenced sites are more difficult to find, and the future of many of these projects is uncertain. Many of these trusts view ecotourism as a partial source of revenue for ongoing operating costs for these sites, and have either started or are preparing to charge visitors (Campbell-Hunt 2008a). Karori Wildlife Sanctuary (Zealandia) is the site with the longest experience at ecotourism, now with approximately 60,000 visitors annually providing 30% of its annual income (Campbell-Hunt 2008a). Nevertheless, the financial viability of these projects without more underlying support is currently being tested.

A major social benefit from community-led pest-proof fenced projects that has not been fully valued is the increase in both social capital that has occurred as a result of these endeavours and in the education and reconnection of community group members with the natural world. Local communities have found a common goal that has provided a unifying focus. The increase in understanding and appreciation for the natural world by community members associated with these groups is often most evident during translocations of valued species (Parker 2008).

The high cost of pest-proof fences has led to an ongoing debate on the cost-effectiveness of such protection compared to sustained pest removal over time on mainland sites (although such alternative techniques have not been able to duplicate the near-zero pest densities achieved in fenced areas). Clapperton and Day (2001) compared the cost-effectiveness of pest-proof fencing when compared against the cumulative cost of conventional pest control based on repeated use of toxins and traps. Their scenarios suggest that pest-proof fencing becomes cost effective between 4 and 9 years compared with conventional control depending on the size of the reserve and assumptions on the cost of fence maintenance and monitoring. Experience with pest-proof fences since this analysis suggest that the costs of fence maintenance, continued control of residual or reinvading pests, and ongoing pest surveillance in fenced sites may be higher than assumed in these models, so a reanalysis of the cost-effectiveness of different strategies is urgently needed.

The challenge to find lower cost solutions to conservation management in New Zealand still remains.

Comparisons are also made between pest-proof fenced sites and offshore islands in terms of determining effective investment strategies for the conservation of threatened species. Offshore islands probably have a higher level of security following eradication than fenced sites in that their borders are not continually challenged by pest species (Connolly et al. 2009). However, invasions by pests capable of swimming to near shore islands or off boats occur regularly (Russell et al. 2008; Bellingham et al. 2010), so pest reinvasion is a threat to both types of reserve and continuous surveillance to detect invasions is necessary. Offshore islands are characterized by oceanic and lowland environments and do not cover the full range of climates and soils represented on the New Zealand mainland (Meurk and Blaschke 1990). Therefore, pest-proof fenced sites may provide opportunities to secure populations of threatened species in environments not found on offshore islands. Fenced sites can act as sources for the movement of mobile species into the surrounding landscape more so than offshore islands. Also fenced sites are more accessible to the public than offshore islands so can bring valued biodiversity into the life experiences of more people. Lastly, the establishment and maintenance of pest-proof fences incurs a high cost on these sites, but offshore islands incur ongoing costs associated with difficulties of access, e.g., remoteness and often poor landing sites. Pest-proof fenced sites and offshore islands therefore provide different advantages and disadvantages to the conservation of threatened species, and should be viewed as different options rather than as strict alternatives.

Conclusions

There has been a major surge in development of conservation fencing in the last decade in New Zealand, and projects adopting this tool have already achieved many important conservation outcomes in extending the ranges of threatened species and in ecosystem restoration. They must now be considered an important component of any national conservation strategy. After only a decade of development, management experience with conservation fences must still be considered immature, and adaptive approaches will be necessary to realize their potential. Conservation fencing in New Zealand has required high investment but this has occurred with a high degree of community involvement and ownership with attendant social benefits. Fencing has certainly enabled rare and valued biodiversity to be much more accessible to the public and is fulfilling an important advocacy role. Although the economic sustainability of projects adopting this technology is not yet clear, there are high hopes for their success. In the long term, the high cost of fences means they are probably stop-gap measures until other more cost-effective conservation solutions are found. The ultimate (and currently unrealistic) measure of conservation success will be when the threat is removed outside the fence, and the fence itself can be taken down.

Acknowledgements We thank members of the Sanctuaries of New Zealand network (www.sanctuariesnz.org) from many different projects for their assistance in writing this chapter and for their inspiring and unwavering commitment to New Zealand conservation. We also thank Matt Hayward, Michael Somers, and Graham Kerley for organizing the original symposium on which this chapter is based. This chapter was partly funded by the Foundation for Research, Science and Technology in New Zealand through the Sustaining and Restoring Biodiversity Outcome Based Investment (OBI) (Contract C09X0503). We thank Craig Briggs for drafting Fig. 5.2, and Phil Seddon and two anonymous referees for their helpful comments to improve the manuscript. Finally, we dedicate this chapter to the late Diane Campbell-Hunt who understood the potential of fenced sanctuaries and provided quiet and rational leadership to many.

References

Aikman, H., Davis, A., Miskelly, C., O'Conner, S. & Taylor, G.A. (2001) Chatham Island taiko recovery plan. Threatened species recovery plan 36. Department of Conservation, Wellington.

Atkinson, I.A.E. (2001) Introduced mammals and models for restoration. *Biological Conservation*, **99**, 81–96.

Aviss, M. & Roberts, A. (1994) Pest fences: notes and comments. Threatened Species Occasional Publication No. 5, Department of Conservation, Wellington.

Baber, M., Moulton, H., Smuts-Kennedy, C., Gemmell, N. & Crossland, M. (2006) Discovery and spatial assessment of a Hochstetter's frog (*Leiopelma hochstetteri*) population found in Maungatautari Scenic Reserve, New Zealand. *New Zealand Journal of Zoology*, **33**, 147–156.

Basse, B. & McLennan, J.A. (2003) Protected areas for kiwi in mainland forests of New Zealand: how large should they be? *New Zealand Journal of Ecology*, **27**, 95–105.

Bell, B.D., Empson, R. & Booth, M. (2006) Returning the hihi or stitchbird to the New Zealand mainland: Sightings and movements after two translocation to the Karori Wildlife Sanctuary near Wellington. *Journal of Ornithology*, **147: Suppl. 1**, 136.

Bellingham, P.J., Towns, D.R., Cameron, E.K., Davis, J.J., Wardle, D.A., Wilmshurst, J.M. & Mulda, C.P.H. (2010) New Zealand island restoration: seabirds, predators, and the importance of history. *New Zealand Journal of Ecology*, **34**, 115–136.

Blick, R., Bartholomew, R., Burrell, T. & Burns, K.C. (2008) Successional dynamics after pest eradication in the Karori Wildlife Sanctuary. *New Zealand Natural Sciences*, **33**, 3–14.

Brook, C.S. (1939) Value of electric fencing: new possibilities for closer and more economic subdivision. *New Zealand Journal of Agriculture*, **58(6)**, 499–504.

Brown, D. (1994) Transfer of Hamilton's frog, *Leiopelma hamiltoni,* to a newly created habitat on Stephens Island, New Zealand. *New Zealand Journal of Zoology*, **21**, 425–430.

Brown, K.P., Moller, H., Innes, J. & Alterio, N. (1996) Calibration of tunnel tracking rates to estimate relative abundance of ship rats (*Rattus rattus*) and mice (*Mus musculus*) in a New Zealand forest. *New Zealand Journal of Ecology*, **20**, 271–275.

Brown, P., Burns, B., Clarkson, B. & Robertson, J. (2006) First record of silver beech (*Nothofagus menziesii*) on Maungatautari, central Waikato. *New Zealand Botanical Society Newsletter*, **84**, 16–18.

Brunton, D.H., Evans, B.A. & Ji, W. (2008) Assessing natural dispersal of New Zealand bellbirds using song type and song playbacks. *New Zealand Journal of Ecology*, **32**, 147–154.

Campbell-Hunt, D. (2002) Developing a sanctuary – the Karori experience. Victoria Link, Wellington.

Campbell-Hunt, D.M. (2008a) Ecotourism and sustainability in community driven ecological restoration; case studies from New Zealand. *Sustainable Tourism III* (eds. C.A. Brebbia & F.D. Pineda). Ecology and the Environment volume 115, pp. 231–240. WIT Press, Southampton.

Campbell-Hunt, D.M. (2008b) Community-driven ecological restoration in New Zealand. *Pacific News*, **30**, 12–15.

Campbell-Hunt, D.M., Freeman, C. & Dickinson, K.J.M. (2010) Community-based entrepreneurship and wildlife sanctuaries; case studies from New Zealand. *International Journal of Innovation and Regional Development*, **2**, 4–21.

Clapperton, B.K. & Day, T.D. (2001) Cost-effectiveness of exclusion fencing for stoat and pest control compared with conventional control. DOC Science Internal Series 14, Department of Conservation, Wellington.

Clapperton, B.K. & Matthews, L.R. (1996) Trials of electric fencing for restricting the movements of common brushtail possums, *Trichosurus vulpecula* Kerr. *Wildlife Research*, **23**, 571–579.

Colbourne, R.M., Bassett, S., Billing, B., McCormick, H., McLennan, J.A., Nelson, A. & Robertson, H.A. (2005) The development of Operation Nest Egg as a tool in the conservation management of kiwi. *Science for Conservation 259*. Department of Conservation, Wellington. 24 p.

Connolly, T.A., Day, T.D. & King, C.M. (2009) Estimating the potential for reinvasion by mammalian pests through pest-exclusion fencing. *Wildlife Research*, **36**, 410–421.

Cowan, P.E. & Rhodes, D.S. (1992) Restricting the movements of brushtail possums, *Trichosurus vulpecula*, on farmland with electric fencing. *Wildlife Research*, **19**, 47–58.

Craig, J., Anderson, S., Clout, M.N., Creese, B., Mitchell, N., Ogden, J., Roberts, M. & Ussher, G. (2000) Conservation issues in New Zealand. *Annual Review of Ecology and Systematics*, **31**, 61–78.

Day, T. & MacGibbon, R. (2007) Multiple-species exclusion fencing and technology for mainland sites. *Managing Vertebrate Invasive Species: Proceedings of an International Symposium* (eds. G.W. Witmer, W.C. Pitt & K.A. Fagerstone), pp. 418 – 433. USDA/APHIS/WS, National Wildlife Research Centre, Fort Collins.

Department of Conservation (2008) Annual report for the year ended 30 June 2008. Department of Conservation, Wellington.

Dodd, M.B. & Power, I.L. (2007) Recovery of tawa-dominated forest fragments in the Rotorua Basin, New Zealand, after cessation of livestock grazing. *Ecological Management and Restoration*, **8**, 208–217.

Dowding, J.E., Lovegrove, T.G., Ritchie, J., Kast, S.N. & Puckett, M. (2006) Mortality of northern New Zealand dotterels (*Charadrius obscurus aquilonius*) following an aerial poisoning operation. *Notornis*, **53**, 235–239.

Dungan, R.J., O'Cain, M.J., Lopez, M.L. & Norton, D.A. (2002) Contribution by possums to seed rain and subsequent seed germination in successional vegetation, Canterbury, New Zealand. *New Zealand Journal of Ecology*, **26**, 121–128.

Ewers, R.M., Kliskey, A.D., Walker, S., Rutledge, D., Harding, J.S. & Didham, R.K. (2006) Past and future trajectories of forest loss in New Zealand. *Biological Conservation*, **133**, 312–325.

Fitzgerald, N., Innes, J. & Forrester, G. (2009) *Changes in bird counts at Maungatautari after pest mammal eradication: results to 2008*. Landcare Research Contract Report 0910/044, Landcare Research, Hamilton, New Zealand.

Forsyth, D.M., Wilmhurst, J.M., Allen, R.B. & Coomes, D.A. (2010) Impacts of introduced deer and extinct moa on New Zealand ecosystems. *New Zealand Journal of Ecology*, **34**, 48–65.

Gibbs, G. (2006) *Ghosts of Gondwana: The History of Life in New Zealand*. Craig Potton Publishing, Nelson.

Gillies, C.A., Leach, M.R., Coad, N.B., Theobald, S.W., Campbell, J., Herbert, T., Graham, P.J. & Pierce, R.J. (2003) Six years of intensive pest mammal control at Trounson Kauri Park, a Department of Conservation "mainland island", June 1996–July 2002. *New Zealand Journal of Zoology*, **30**, 399–420.

Goldwater, N. (2007) *Ecology of house mice within the Tawharanui Open Sanctuary*. MSc thesis, University of Auckland.

Gsell, A., Innes, J., de Monchy, P. & Brunton, D. (2010) The success of using trained dogs to locate sparse rodents in pest-free sanctuaries. *Wildlife Research*, **37**, 39–46.

Hayward, M.W. & Kerley, G.I.H. (2009) Fencing for conservation: restriction of evolutionary potential or a riposte to threatening processes. *Biological Conservation*, **142**, 1–13.

Hitchmough, R. (1994) Feasibility study. *Pest fences: notes and comments* (eds. M. Aviss & A. Roberts), pp. 31–35. Threatened Species Occasional Publication No. 5, Department of Conservation, Wellington.

Hitchmough, R., Bull, L. & Cromarty, P. (2007) New Zealand Threat Classification lists – 2005. Department of Conservation, Wellington.

Hoare, J.M. & Hare, K.M. (2006) The impact of brodifacoum on non-target wildlife: gaps in knowledge. *New Zealand Journal of Ecology*, **30**, 157–167.

Hoare, J.M., Adams, L.K., Bull, L.S. & Towns, D.R. (2007) Attempting to manage complex predator-prey interactions fails to avert imminent extinction of a threatened New Zealand skink population. *Journal of Wildlife Management*, **71**, 1576–1584.

Holzapfel, S., Robertson, H.A., McLennan, J.A., Sporle, W., Hackwell, K. & Impey, M. (2008) Kiwi (*Apteryx* spp.) recovery plan: 2008–2018. *Threatened Species Recovery Plan 60.* Department of Conservation, Wellington.

Innes, J., Warburton, B., Williams, D., Speed, H. & Bradfield, P. (1995) Large-scale poisoning of ship rats (*Rattus rattus*) in indigenous forests of the North Island, New Zealand. *New Zealand Journal of Ecology*, **19**, 5–17.

Innes, J., Hay, R., Flux, I., Bradfield, H. & Jansen, P. (1999) Successful recovery of North Island kokako *Callaeas cinerea wilsoni* populations by adaptive management. *Biological Conservation*, **87**, 201–221.

Innes, J., Kelly, D., Overton, J.M. & Gillies, C. (2010) Predation and other factors currently limiting New Zealand forest birds – a review. *New Zealand Journal of Ecology*, **34**, 86–114.

Jansen, P.W. (1993) Eradication of Norway rats and rabbits from Moutohora (Whale) Island, Bay of Plenty. *Ecological Management*, **1**, 10–15.

Kelly, D. & Sullivan, J. (2010) Life histories, dispersal, invasions, and global change: progress and prospects in New Zealand ecology, 1989–2029. *New Zealand Journal of Ecology*, **34**, 207–217.

King, C.M (ed.) (2005) *The handbook of New Zealand mammals.* Oxford University Press, Melbourne.

King, C.M. & Edgar, R.J. (1977) Techniques for trapping and tracking stoats (*Mustela erminea*); a review, and a new system. *New Zealand Journal of Zoology*, **4**, 193–212.

Lee, W.G., Wood, J.R. & Rogers, G.M. (2010) Legacy of avian-dominated plant-herbivore systems in New Zealand. *New Zealand Journal of Ecology*, **34**, 48–65.

Lettink, M., Norbury, G., Cree, A., Seddon, P.J., Duncan, R.P. & Schwarz, C.J. (2010) Removal of introduced predators, but not artificial refuge supplementation, increases skink survival in a coastal duneland. *Biological Conservation*, **143**, 72–77.

Lord, J.M. (1991) Pollination and seed dispersal in *Freycinetia baueriana,* a dioecious liane that has lost its bat pollinator. *New Zealand Journal of Botany*, **29**, 83–86.

Lukis, K. & Bell, B.D. (2007) Restoring native frogs to a mainland island reserve: translocation and adaptive management of the Maud Island frog (*Leiopelma pakeka*) in the presence of house mice (*Mus musculus*) at Karori Wildlife Sanctuary. *New Zealand Journal of Zoology*, **34**, 267.

Lynch, J. (1995) Back to the future. Karori – from reservoir to wildlife sanctuary. *Forest and Bird*, **275**, 12–19.

McLennan, J.A., Dew, L., Miles, J., Gillingham, N. & Waiwai, R. (2004) Size matters: predation risk and juvenile growth in North Island brown kiwi (*Apteryx mantelli*). *New Zealand Journal of Ecology*, **28**, 241–250.

Merton, D. (1992) The legacy of "Old Blue". *New Zealand Journal of Ecology*, **16**, 65–68.

Meurk, C.D. & Blaschke, P.M. (1990) How representative can restored islands really be? An analysis of climo-edaphic environments in New Zealand. *Ecological Restoration of New Zealand Islands* (eds. D.R. Towns, C.H. Daugherty & I.A.E. Atkinson), pp. 52–72. Conservation Sciences Publication No. 2. Department of Conservation, Wellington.

Miller, K.A., Nelson, N.J., Smith, H.G. & Moore, J.A. (2009) How do reproductive skew and founder group size affect genetic diversity in reintroduced populations? *Molecular Ecology*, **18**, 3792–3802.

Miskelly, C.M., Taylor, G.A., Gummer, H. & Williams, R. (2009) Translocations of eight species of burrow-nesting seabirds (genera Pterodroma, Pelecanoides, Pachyptila and Puffinus: Family Procellariidae). *Biological Conservation*, **142**, 1965–1980.

Murphy, E. & Bradfield, P. (1992) Change in diet of stoats following poisoning of rats in a New Zealand forest. *New Zealand Journal of Ecology*, **16**, 137–140.

Norbury, G., Reardon, J. & McKinlay, B. (2007) *Grand and Otago skink recovery plan 2006–2016*. Working draft. Dunedin, Department of Conservation.

Nugent, G., Sweetapple P., Coleman J. & Suisted, P. (2000) Possum feeding patterns: dietary tactics of a reluctant folivore. *The brushtail possum: biology, impact and management of an introduced marsupial* (ed T. L. Montague), pp. 10–23. Manaaki Whenua Press, Lincoln.

Nugent, G., Whitford, J., Innes, J. & Prime, K. (2002) Rapid recovery of kohekohe (*Dysoxylum spectabile*) following possum control. *New Zealand Journal of Ecology*, **26**, 73–79.

Parker, K. (2008) Translocations: providing outcomes for wildlife, resource managers, scientists, and the human community. *Restoration Ecology*, **16**, 204–209.

Parkes, J. & Murphy, E. (2003) Management of introduced mammals in New Zealand. *New Zealand Journal of Zoology*, **30**, 335–359.

Parrish, R., Sherley, G. & Aviss, M. (1995) Giant land snail recovery plan *Placostylus* spp., *Paryphanta* spp. Threatened Species Recovery Plan Series No. 13. Department of Conservation, Wellington.

Porteous, T. (1993) *Native forest restoration: a practical guide for landowners.* Queen Elizabeth II National Trust, Wellington.

Pryde, P.R. (1997) Creating offshore island sanctuaries for endangered species: The New Zealand experience. *Natural Areas Journal*, **17**, 248–254.

Russell, J.C., Towns, D.R. & Clout, M.N. (2008) Review of rat invasion biology: implication for island biosecurity. Science for Conservation 286. Department of Conservation, Wellington.

Sanders, M., Brown, K.P. & Keedwell, R.J. (2007) Testing the effects of a predator-exclusion fence on predator abundance and wetland bird breeding success at Ruataniwha wetlands, Twizel. Department of Conservation, Christchurch.

Saunders, A. & Norton, D.A. (2001) Ecological restoration at mainland islands in New Zealand. *Biological Conservation*, **99**, 109–119.

Simberloff, D. (1990) Reconstructing the ambiguous: can island ecosystems be restored? *Ecological Restoration of New Zealand Islands* (eds. D.R. Towns, C.H. Daugherty & I.A.E. Atkinson), pp. 37–51. Conservation Sciences Publication No. 2. Department of Conservation, Wellington.

Smale, M.C., Ross, C.W. & Arnold, G.C. (2005) Vegetation recovery in rural kahikatea (*Dacrycarpus dacrydioides*) forest fragments in the Waikato region, New Zealand, following retirement from grazing. *New Zealand Journal of Ecology*, **29**, 261–269.

Speedy, C., Day, T. & Innes, J. (2007) Pest eradication technology – the critical partner to pest exclusion technology: the Maungatautari experience. *Managing Vertebrate Invasive Species: Proceedings of an International Symposium* (eds. G.W. Witmer, W.C. Pitt & K.A. Fagerstone), pp. 115 – 126. USDA/APHIS/WS, National Wildlife Research Centre, Fort Collins.

Sweetapple, P.J. & Nugent, G. (2007) Ship rat demography and diet following possum control in a mixed podocarp-hardwood forest. *New Zealand Journal of Ecology*, **31**, 186–201.

Taylor, G.A. (2000a) *Action plan for seabird conservation in New Zealand. Part A, Threatened Seabirds*. Department of Conservation, Wellington, Threatened Species Occasional Publication No. 16. 233 p.

Taylor, G.A. (2000b) *Action plan for seabird conservation in New Zealand. Part B, Non-threatened Seabirds*. Department of Conservation, Wellington, Threatened Species Occasional Publication No. 17. 201 p.

Taylor, R.H. & Thomas, B.W. (1993) Rats eradicated from rugged Breaksea Island (170 ha), Fiordland, New Zealand. *Biological Conservation*, **65**, 191–198.

Tennyson, A.J.D. (2010) The origin and history of New Zealand's terrestrial vertebrates. *New Zealand Journal of Ecology*, **34**, 6–27.

Timmins, S.M. (2002) Impact of cattle on conservation land licensed for grazing in South Westland, New Zealand. *New Zealand Journal of Ecology*, **26**, 107–120.

Tocher, M. (2006) Survival of grand and Otago skinks following predator control. *Journal of Wildlife Management*, **70**, 31–42.

Tompkins, D.M. & Veltman, C.J. (2006) Unexpected consequences of vertebrate pest control: predictions from a four-species community model. *Ecological Applications*, **16**, 1050–1061.

Towns, D.R. & Broome, K.G. (2003) From small Maria to massive Campbell: forty years of rat eradications from New Zealand islands. *New Zealand Journal of Zoology*, **30**, 377–398.

Townsend, A.J., de Lange, P.J., Duffy, C.A.J., Miskelly, C.M., Molloy, J. & Norton, D.A. (2008) New Zealand threat classification system manual. Department of Conservation, Wellington.

Urlich, S.C. & Brady, P.J. (2005) Benefits of aerial 1080 possum control to tree fuchsia in the Tararua Range, Wellington. *New Zealand Journal of Ecology*, **29**, 299–309.

Watts, C.H. (2007) Beetle community responses to mammal eradication in the southern exclosure on Maungatautari. Landcare Research Contract Report LC0607/170, Landcare Research, New Zealand, Hamilton.

Watts, C.H., Stringer, I., Sherley, G., Gibbs, G. & Green, C. (2008) History of weta (Orthoptera: Anostostomatidae) translocation in New Zealand: lessons learned, islands as sanctuaries and the future. *Journal of Insect Conservation*, **12**, 359–370.

Watts, C.H., Armstrong, D.P., Innes, J. & Thornburrow, D. (2011) Dramatic increases in weta (Orthoptera) following mammal eradication on Maungatautari – evidence from pitfalls and tracking tunnels. *New Zealand Journal of Ecology,* **35(3),** 261–272

Wickes, C., Crouchley, D. & Maxwell, J. (2009) Takahe (*Porphyrio hochstetteri*) recovery plan 2007–2012 . Threatened Species Recovery Plan 61. Department of Conservation, Wellington.

Wilson, D.J., Mulvey, R.L. & Clark, R.D. (2007) Sampling skinks and geckos in artificial cover objects in a dry mixed grassland–shrubland with mammalian predator control. *New Zealand Journal of Ecology*, **31**, 169–185.

Wilson, K-J. (ed.) (2008) The state of New Zealand's birds 2008. Conservation of birds on the mainland. Ornithological Society of New Zealand, Nelson.

Worthy, T. & Holdaway, R. (2002) *The lost world of the moa: prehistoric life of New Zealand.* Canterbury University Press, Christchurch.

Chapter 6
Fencing for Purpose: A Case Study of Elephants in South Africa

Rob Slotow

Introduction

Conservation management requires planning and then implementation in order to effectively achieve objectives. Historically, major interventions were undertaken without a complete understanding (or regard) of the potential consequences, for example, fencing the western border of the Kruger National Park (KNP) negatively affects herbivore populations (Whyte 1985 cited in Grant et al. 2008; Grant et al. 2008; see also Loarie et al. 2009). Further, because the fences separated the animals from traditional water-sources (Grant et al. 2008), that intervention resulted in the necessity to create artificial water-holes, which in turn had unforeseen consequences (e.g. population decline of rare plains species – Harrington et al. 1999). Fencing for conservation management is a strong potential tool (Hayward and Kerley 2009), but we need to understand better our ability to nuance its implementation in order to maximise benefits while reducing costs.

Hayward and Kerley (2009) provide a comprehensive review of fencing and conservation, and highlight the complexity of creating and maintaining barriers in conservation land. South African conservation managers have been leaders in developing fencing as a conservation management tool (e.g. van Dyk 1997; Hayward et al. 2007; Grant et al. 2008; Gusset et al. 2008). One of the major reasons for fencing is to prevent human–wildlife conflict by preventing dangerous animals, such as megaherbivores or large carnivores, from entering human communities (Grant et al. 2008; Hayward and Kerley 2009).

R. Slotow (✉)
Amarula Elephant Research Programme, School of Biological and Conservation Sciences, Westville Campus, University of KwaZulu-Natal, Private Bag X54001, Durban 4000, South Africa
e-mail: slotow@ukzn.ac.za

M.J. Somers and M.W. Hayward (eds.), *Fencing for Conservation: Restriction of Evolutionary Potential or a Riposte to Threatening Processes?*,
DOI 10.1007/978-1-4614-0902-1_6, © Springer Science+Business Media, LLC 2012

While there are clearly benefits from fencing, there are also a range of costs (Hayward and Kerley 2009). Fencing may be essential for a particular conservation objective or purpose, such as mitigating human–wildlife conflict (e.g. Grant et al. 2008; Hayward and Kerley 2009), but there are ways in which the associated costs can be reduced (e.g. through economies of scale, Lindsey et al. 2009). In this paper, I interrogate and explore some of the innovations in fencing for conservation which allow reduction of some of the costs associated with fencing, using African elephants (*Loxodonta africana*) in South Africa as a case study.

Elephant Management with Fences in South Africa

The history of elephant management in South Africa was recently reviewed (Carruthers et al. 2008), and here I focus on the detail of reintroductions that occurred subsequent to 1980 (see Garai et al. 2004; Slotow et al. 2005), as this provides the main context for management through fencing. Naturally occurring populations of elephants occur at KNP, Addo Elephant National Park, Tembe Elephant Park (Carruthers et al. 2008), while elephants have been reintroduced to over 80 state-owned, communally owned and privately owned reserves (Garai et al. 2004; Carruthers et al. 2008).

There are many different types of fences that have been used for wildlife management, and a range of studies testing their effectiveness or appropriateness (e.g. Moseby and Read 2006; Vercauteren et al. 2006). Fencing has been used to specifically control the movement of elephants since the erection of the first "Armstrong" fence in Addo Elephant National Park which consisted of a strong, unelectrified barrier (Grant et al. 2008). Since 1980/1981, when elephants were introduced to Pilanesberg National Park (Anderson 1994), small reserves (exceptions were Ithala and iSimangaliso (see below)) in South Africa have been required to have an electrified complete perimeter fence of a minimum standard (see Grant et al. 2008 for fence information). The effectiveness of electric fences such as these, and other barriers in restricting elephants have been extensively studied (see Grant et al. 2008 for review).

The vast majority of reserves that have reintroduced elephants use fences 2.4 m high, and with either three or four live electric strands (Fig. 6.1). Having a fence capable of keeping elephants in does not mean that it is successful in keeping people out. Reserves have to patrol the border fence for two purposes, firstly reserve security against intruders, and secondly, to check whether the fence requires maintenance. A survey was conducted of all small reserves with elephants in 2001, and there was wide variation in how reserves dealt with these issues (Fig. 6.2). Guards in most state reserves were armed, whereas 40% of private reserves had unarmed guards, and about 40% of private reserves employed <5 guards (Fig. 6.2a). There was also variation in the frequency of fence patrolling, with almost all private reserves patrolling on a daily basis, i.e. a specific point on the fence was passed daily (Fig. 6.2b). Ten out of 57 (17.5%) reserves surveyed reported fence damage with no breakout, indicating that the fence does work despite elephants testing it, but

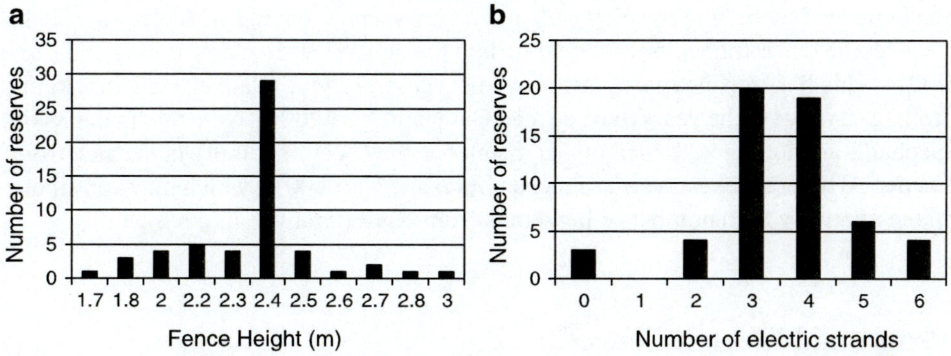

Fig. 6.1 Fencing structure in reserves which introduced elephants. (**a**) Maximum height of fence (**b**) Number of live electric strands (most have an associated earth wire). Data are from a questionnaire survey of reserves conducted by the Elephant Owners and Managers Association by Dick Carr and collated and analysed by myself (see Slotow et al. 2005 for details of methods)

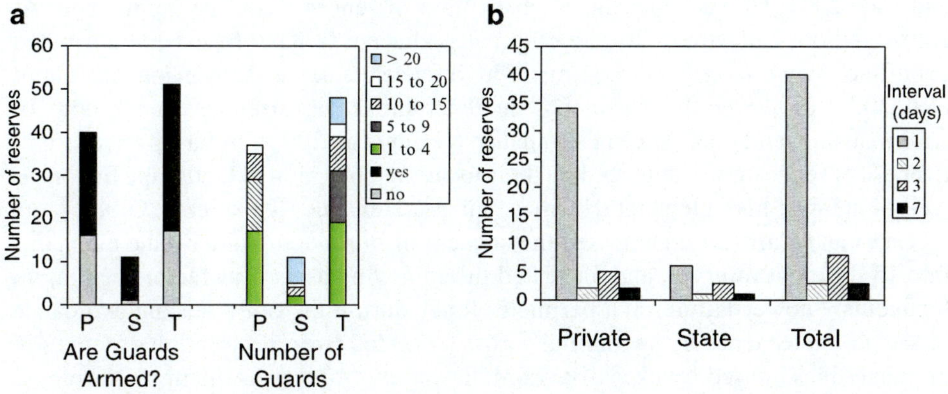

Fig. 6.2 Security associated with fenced small reserves. (**a**) Security guards indicating if they are armed or not, and the number employed by the reserve. Legend reflects number of guards. *P*=Privately owned, *S*=State owned, *T*=Total regardless of ownership. (**b**) Frequency of patrolling of the border fenceline where numbers reflect the interval between patrols in days. Figure indicates whether the guards are armed or not, and secondly, the number of guards employed by the reserve. Data are from a questionnaire survey of reserves conducted by the Elephant Owners and Managers Association by Dick Carr and collated and analysed by myself (see Slotow et al. 2005 for details of methods)

emphasizes the need for continual patrolling to detect damage. However, breakouts do occur, with five out of 15 reserves that introduced older adult male elephants reporting breakouts (Garai and Carr 2001).

Large mammals are capable of learning about fences, and developing strategies to overcome the barrier. This is particularly true for electrified fences which become inoperative (Garai and Carr 2001). The process for introducing elephants into small reserves in South Africa includes a short period in a small (about 1 ha) holding facility (boma) (Garai and Carr 2001). The boma is electrified in the same manner as the boundary fence, and the elephants learn to associate fencing with electric shock, i.e. become fence-trained prior to release into the general reserve area (Garai and Carr 2001). This has, for the most part, proved successful, with relatively few breakouts from reserves over the years (but see read Garai and Carr 2004). Note that in the KNP, elephants are not fence-trained in this manner as they occur naturally in the area from the period before the reserve was fenced (Garai and Carr 2001), which may contribute to the relatively high number of breakouts from KNP (Grant et al. 2008).

People and Electric Fences

A fence is only effective as a barrier if it is completely operational. Entry and egress require gates, and these have to be properly closed (Vercauteren et al. 2006), and if the fence is electrified, the integrity of the electrification of the gate needs to be maintained. Elephants have broken out of at least two reserves through gates (Garai and Carr 2001). In addition, the electrification of fences has to be maintained. At least five Foot and Mouth disease outbreaks adjacent to KNP have been attributed to buffalo (*Syncerus caffer*) escaping through fence breaks made by elephants (Grant et al. 2008). However, the reason why the fence-breaks occurred was poor quality of workmanship, and poor fence maintenance (Grant et al. 2008). In East Africa, effectiveness of fences was influenced by their location in relation to landscape factors as well as areas of high elephant densities, and maintenance (Kioko et al. 2008).

Garai and Carr (2001) assessed the success of introducing older male elephants into 15 different reserves, and identified the following aspects as factors leading to breakouts: "power failure on a perimeter fence during the rainy season; avoidance of the perimeter fence by swimming across a flooded river; destruction of fences in drainage lines caused by excessive rainfall; inability of management to patrol and maintain the perimeter fence during an excessively wet season; and lack of electrification of gates in the perimeter fence". Fence maintenance was clearly a major issue, despite the fact that many of these reserves regularly patrol their fences (Slotow personal observation).

There are a range of natural factors that can affect fencing, particularly electric fencing. Fences have to cross water-courses which flood in summer thunderstorms and deposit debris which shorts out the fence. A solution to this has been the use of sacrificial fences across drainage lines with continual problems, whereby a cheap, easily replaced section is constructed across the drainage line, and the whole section is simple replaced when compromised (personal observation). Large mammals pose a major problem in that carnivores use fences as hunting aids (van Dyk and Slotow 2003), often resulting in breakage or compromise of the electrics (personal observation). Unless quickly repaired, these can become weak-points through which elephants can break-out.

Fence maintenance is not only about natural factors, as the main problems with the KNP fence is theft and vandalism providing opportunity for animals to then break out (Grant et al. 2008).

In terms of damage to infrastructure within the reserve, managers have responded in three ways. Firstly, doing nothing, and absorbing the costs. This is particularly in cases where resources are not available, and as yet the costs have not escalated to the extent that intervention is deemed necessary. However, in most cases damage tends to escalate as repeat offences tend to build up into problem individuals (Slotow et al. 2008). In such cases, management has no option other than to remove the offending individual (who becomes known through repeat events (Slotow et al. 2008)). In most cases, these tend to be adult males (various managers personal communication; personal observation). The third management intervention is to fence out key infrastructure, which includes accommodation areas (camps), and water storage tanks of various forms (elephants damage tanks or break piping to get to water). Most often these are solar-powered, electrified fences which work effectively (if maintained).

Unfenced Boundaries in Small Reserves

Unfenced boundaries are common in reserves outside of South Africa, and currently exist along international borders within Transfrontier parks. There are two reserves within South Africa that have elephants and that currently have unfenced boundaries, Ithala Game Reserve and iSimangaliso Wetland Park (iSWP, previously Greater St. Lucia Wetland Park). I will discuss each of these in turn as case studies. There have also been two occurrences where elephants moved out of reserves through gaps in the fence where large rivers exit reserves. Firstly, shortly after ten large male bulls were introduced from KNP to Hluhluwe-iMfolozi Park (HiP), one of these males moved out the reserve when crossing the Hluhluwe River. Apparently, he walked along the fence, entered the river, and then exited the river on the other side on the wrong side of the fence. He was captured and returned to HiP (Slotow unpublished data). In the second incident, two males moved out of Songimvelo Game Reserve in Mpumulanga, and into Malolotja Nature Reserve, Swaziland, by crossing the Komati River. These two animals remained in Malolotja, and thus pose a concern to managers there (Mtui and Owen-Smith 2006; Norman Owen Smith personal communication 2009).

Ithala Game Reserve introduced elephants between 1990 and 1993, young orphans from the KNP culls (see Slotow et al. 2005). The reserve is bordered in the north by the Phongola River, and that river frontage is not fenced. The land on the north of the river belongs to a number of communities and a small mining company. It is very sparsely populated, with little crop farming, and is mainly untransformed natural vegetation suitable for elephants. There is a steep ridge forming the northern edge of the river valley, which may prevent movement out of the valley bottom. More dense human habitation starts on top of that ridge. If any situation in South Africa was suitable for having an unfenced boundary, this would be it. As expected, elephants have started making excursions out of the reserve (the following is based on Ezemvelo KZN Wildlife (EKZNW) management reports collated by Taryn Gilson), with a brief movement out in 1992, the next by males in 1995, followed by the herd in 1998. In 1999, the whole herd moved out 3–4 times per year, and by 2003, they went out >10 times, staying out for up to 2 weeks. Since 2005, when satellite collars were fitted to the female groups, the furthest north that they have ventured is 2 km from the river. Excursions were mainly in the wet summer, when resources are more widely available. In general, the staff responded to excursions by chasing the animals back, either using a helicopter or on foot using load noises including gunshots. Three males were shot on return to the reserve in 2005 in a hope of discouraging future excursions (disturbance culling, see Grant et al. 2008), but this has been unsuccessful.

The river provides an important resource, so there will be fundamental problems with fencing the river out of the reserve as this will remove access to that resource, particularly for the endangered black rhinoceros (*Diceros bicornis*). In addition, the river has major aesthetic value for tourists, which will be compromised by a fence. Reserve management, therefore, faces a conundrum, and is currently working with the concept of virtual fences (see below).

iSimangaliso Wetland Park introduced elephants in 2001, and faces a problem in the southern part of the park around Lake St. Lucia itself. The land within the reserve does not completely surround the lake, which makes it possible for elephants to move into the lake, and then out onto land that is not contained by fencing. This land is either occupied by communities, or private land-owners practicing various land-uses. In addition, the eastern border of the lake is made up by the Indian Ocean, making it theoretically possible for elephants to walk around the fence on the beach. The lake is shallow enough for elephants to easily cross, and large areas dry out during drought years, meaning that the reserve has to erect temporary fencing along those areas when the lake dries up, and remove the fences when the lake floods. Shortly after introduction into the reserve, two young males walked north along the lake shore, and walked around the fence on the northern border of the lake (Slotow unpublished data). They had to be recaptured, and were returned to Hluhluwe-iMfolozi Park from which they were originally translocated from (Slotow unpublished data).

A further problem faced by iSWP is that two key reserve boundaries are made up of rivers. The northern and eastern border of Mkuze Game Reserve is along the Mkhuze River, which is accessed by the communities on the north/east banks for their own use, as well as for their livestock and agriculture. Historically, the river was not fenced, which led to people and their livestock moving freely into the reserve, and animals leaving the reserve. The proclamation of the Greater St. Lucia World Heritage Site, and the plan to introduce large carnivores, as well as a major raid by most of the elephants into an orchard on the east bank, led to a decision to fence that boundary. This resulted in the river being fenced out of the reserve, requiring innovative planning to provide artificial water for the animals. The second major water-source that was fenced out of the reserve was Nyalazi River, on the west bank of Lake St. Lucia. This river has a community living on the western shore, and is the most substantial and consistent fresh-water source in the region (the lake becomes hypersalinic during drought years), necessitating artificial provisioning of water. The ecological effects of fencing-out these rivers have not been studied. In Mapungupwe National Park, which has its own elephant population, SANParks have erected a fence along the Limpopo River border with the Tuli Block, primarily to prevent additional elephants moving into the reserve from the larger Tuli population. This fence reduces the aesthetic of the river frontage, and also prevents other animals from access to the river (Norman Owen Smith, personal communication).

Enclosures vs. Exclosures

Fencing of elephants is generally considered to be a problem of enclosing them inside a reserve in order to prevent them escaping. However, fences are also important for excluding elephants from a specific area within the reserve, such as camps, staff housing, infrastructure or even key natural resources of high conservation value, including threatened plant species (reviewed in Grant et al. 2008). The purpose of the fence needs to be clearly understood, and an exclosure and enclosure may have very different purposes. For example, it may be desirable to preclude only elephants from entering a particular area, whereas all species may be prevented from leaving the reserve. Further, fences may also be established to prevent humans from entering a protected area. In addition to direct management benefits, exclosures also provide a key resource for scientific understanding of the effects of elephants on the ecosystem (Grant et al. 2008).

Ecological Effects of Fences

Fences constrain the movement of elephant, and in a relatively small reserve, increasing elephant densities (Slotow et al. 2005; Mackey et al. 2006) may negatively affect ecological processes or biodiversity (see Grant et al. 2008 for review). We do need to separate the concern over elephant concentration into small areas on biodiversity (Kerley et al. 2008, but see Landman et al. 2008) from the effect of the fence as a barrier per se. Fences affect elephant movement, and in extensive areas may cause them to bunch up against the fence (Loarie et al. 2009). However, we have recently shown for Pilanesberg National Park, female elephant use the fence more for movement between foraging areas, presumably with low foraging impact near the fence, but that the edge effect of fencing may result in higher vegetation impacts deeper in the reserve (Vanak et al. 2010).

Permeable Fences

Fencing may reduce gene flow from natural populations outside of protected areas and also has localised effects on biodiversity through focussing biotic processes (Hayward and Kerley 2009). If a fence needs to exist, then it should be as permeable as possible, preventing only movement of the target species. This would allow all non-target species to move freely, avoiding some of the negative consequences of fencing (see Hayward and Kerley 2009 for costs associated with fencing).

One of the first such fences in Africa was developed by Natal Parks Board (now EKZNW) to restrict the movement of rhinoceros. A singe strong cable was strung about 30 cm above the ground. This prevented rhinoceros from crossing the fence as they could not lift their legs over it. This allowed rhinoceros to be restricted into specific areas of a reserve, or to prevent them from moving onto a public road through a reserve and thus posing a threat to motorists (and vice versa) such as at Weenen Game Reserve, KwaZulu-Natal (personal observation).

This concept was applied by EKZNW to the conservation management of elephants at iSWP. In this case, an electrified strand was placed about 2 m above the ground with an associated earth strand. This prevented elephants that were introduced to the park from moving out into the surrounding community areas. This proved very effective in that no elephants broke-out of the reserve through this fence (Slotow unpublished data). The fence did allow other species to move freely, and has now been modified to a complete standard electrified fence to allow the introduction of a wider range of species (personal observation).

Besides this fence having the ecological advantage of allowing movement of other species, it is also much cheaper than a complete fence, and requires less maintenance as the electrics or fence itself are not broken by other species (van Dyk and Slotow 2003; Hayward et al. 2007).

The concept of single electric-strand fencing to control the movement of elephants has become more widely applied, for example some areas within the Associated Private Nature Reserves (APNR) adjacent to KNP use such fencing to exclude elephants from camps (personal observation), and Phinda Private Game Reserve (Lagendijk et al. 2011) and Tembe Elephant Park have used it to exclude elephants from key threatened Sand Forest areas. Managers are learning through experience to modify the fences to be more effective. For example, in the APNR, for the single electric strand exclusion fence, the electric wire is led down the support poles to prevent the elephants from pulling out the poles, and thereby shorting the fence (Slotow personal observation). At Phinda, the height of the single electric strand has been lowered slightly to prevent young elephants moving under the fence (whereupon the mothers become agitated and break through the fence (Tarynne Dickerson, pers. comm.)).

Virtual Fences

Modern technology provides the opportunity to view fencing from another perspective. In the context of elephants, fencing is either to protect people from risk, or to prevent elephants from entering a particular area for an extended time. A large portion of the risk associated with elephants, i.e. when people and elephants come into contact, is the surprise factor, both in terms of location and timing (Slotow unpublished data). If we look a the Ithala example above, the elephants are not spending a lot of time out of the reserve, and even when outside the reserve, they are not necessarily entering areas of high human activity. It is relatively unlikely that they will encounter people, i.e. it is a low-risk situation. However, there may be times or circumstances when the risk of an incident escalates greatly. In that case, managers would need to intervene, which may simply mean warning, for example, the staff of the mine across the river. GPS collars can provide real-time data on elephant locations. The management of Ithala have set up a process, whereby each of the elephant herds is collared with a GPS collar, and they have identified "hot-spots" of risk outside the reserve. As soon as the elephants enter a buffer area around those hot-spots, the management of the staff respond in some way. This means that they can potentially effectively manage the risk without incurring the huge ecological and financial costs of fencing-off the river.

iSimangaliso Wetland Park has a different situation, in that there are two elements of risk, of the elephants leaving the reserve, but also of risks of human–elephant conflict within the reserve. Since iSWP has excellent GSM cell-phone coverage, it is possible to use real-time feedbacks on their location relative to specific parts of the landscape. Computer technology (www.yrless.co.za) allows an ARCVIEW shape-file containing the border of the reserve to be loaded, and any time the elephant collar moves across that border, a notification is immediately sent to pre-specified cell-phone numbers giving the location of the elephant, the time and a local geographic reference point.

Within iSWP itself, there are two kinds of risk, firstly to staff and visitors within the camps which are not fenced. Secondly, to contractors who are working on the reserve, for example in harvesting from forestry plantations within the reserve. The same principle is applied in these instances, were each camp has a virtual border loaded into the computer system, and each time an elephant enters or leaves that zone an SMS is sent to the pre-determined staff member, for example the camp manager. The camp manager can then make an appraisal of the situation, and whether any intervention is necessary, such as moving people to safety. In terms of the forestry workers, the zone can be moved around, i.e. the shape-file updated when they move into a new block, and the foreman can be notified of any elephants within the zone. In the same way as Ithala above is managing the risk, iSWP has a mechanism whereby it can manage the risk of human-elephant conflict using virtual fences.

Some of the problems that have been encountered with the system are that (1) The border needs to be moved outside the actual border by about 100 m to allow for GPS error, which otherwise gives false alarms when elephants walk alongside the fence. (2) SMS numbers for alerts need to be updated when there is staff turnover, otherwise the correct staff member is not notified. (3) Contractors do not really use the system as they do not see the elephants much, i.e. they do not associate a major risk with elephants.

Discussion

Fencing of reserves was primarily aimed at preventing elephants from leaving reserves into the surrounding communities and farmland. Despite the breakouts mentioned above, fences have been largely effective at preventing human–elephant conflict outside reserves (Slotow et al. 2008; Twine and Magome 2008), other than the Greater KNP complex (which includes the adjacent private reserves) (Grant et al. 2008). For managers of smaller reserves, the main concern tends to be human–elephant conflict within the reserve, with 72% of human fatalities by elephants between 2002 and 2007 occurring within protected areas (Twine and Magome 2008), including aggression towards people and other species, and damage to infrastructure (Slotow et al. 2008). Although it is possible to fence out accommodation areas, all fatal interactions have occurred in the general veld (Slotow unpublished data), making such conflict impossible to manage through conventional fencing. More widespread use of virtual fences may be an alternative worth pursuing.

Fencing for ecological exclosures in order to remove the effect of elephants has been relatively successful (Addo: Lombard et al. 2001; Phinda: Lagendijk et al. 2011). Such exclosures, and those erected for scientific study can also produce useful insights into the ecological effects of elephants (Grant et al. 2008).

Badly aligned fences can have major consequences for biodiversity, for example by interrupting key movement patterns (Hayward and Kerley 2009), or when elephant bunch up against them in the wet season (Loarie et al. 2009). This can result in major mortality of key species, or indirectly by shifting the ecological influence of biotic factors such as grazing or browsing (see examples in Hayward and Kerley 2009). The key issue of fencing out of rivers, and the ecological and aesthetic consequences of this, needs study.

Hayward and Kerley (2009) coin the phrase "metaphorical" fences to describe alternative approaches to barriers such as using *Capsicum* repellents (Osborn and Rasmussen 1995). Other alternatives being tested also include the use of bee (*Apis mellifera*) hives (King et al. 2009), or even the sound of agitated bees (King et al. 2007; see also Grant et al. for discussion of alternatives to fencing). An important result to emerge from this overview is the potential for permeable fencing which achieves a specific purpose, but reduces both the financial and ecological costs. It is important that proper studies be undertaken to refine different methods (e.g. Moseby and Read 2006) for permeability. It is clear that permeable fences are not relevant only for controlling larger species, for example in Australia it is the smaller species such as feral cat (*Felis catus*) that need to be constrained, with larger species such as kangaroos being able to jump over the fences; sophisticated designs could facilitate this (see Moseby and Read 2006). I take the conceptualisation of alternative methods a step further, explaining the use of virtual fences using remote-sensing technology as alternatives both within reserves to avoid fencing camps, or outside reserves to avoid fencing perimeters. The uses of such alternatives need to be better understood and tested.

Fences are not infallible, particularly given the maintenance necessary for electrification (particularly when human vandalism and theft are rising), and the sophisticated ability of elephants to learn how to overcome fences (Grant et al. 2008). The key issue for using fences to contain elephants is not necessarily the absolute strength of the fence, but rather the integrity of the electrical system. It may be possible to reduce vandalism and theft, through working with the community, and to reduce human shorting of the fence for transit, for example by illegal immigrants through KNP, by placing a second permeable electric fence just within the boundary fence, which will allow people to move freely under it while maintaining a barrier to elephants. To improve consistency, remote sensing technology could be placed at key points on the boundary to alert managers when power goes down, at sacrificial river-crossings, for instance, especially in reserves that do not patrol the fenceline daily.

Gates are weak points, and some form of barrier that does not require human intervention at points of entry/egress is ideal (see Vercauteren et al. 2006 for some examples), and electrified grids across the road work to contain African wildlife, including elephants (personal observation at a range of reserves). Fencing major water-bodies is difficult, and similar problems to those indicated for elephant are evident in electrified road-side fences used to control moose (*Alces alces*), where moose enter the road area through gaps in the fence-line such as at lakes (Leblond et al. 2007). Fences in such situations require management flexibility, for example erection of temporary fences that can be removed when flooding occurs, or localised use of repellents (e.g. *Capsicum*, Osborn and Rasmussen 1995). Virtual fences may also prove useful in these situations.

There was a lot of variation among reserves as to the effort invested in patrolling for maintenance and security, which is a high, ongoing cost to fencing. While effort could relate to the local circumstances, it may prove valuable to develop effective and "best practice" in the industry through sharing information among stakeholders.

Private land owners tend to be strongly independent, but there is value in combining together to form larger conservancies (Lindsey et al. 2009). In such conservancies, the internal fences between properties are removed, and all members contribute to the perimeter fence of the overall conservancy. This leads to economies of scale regarding fencing (see Lindsey et al. 2009) as the overall distance of fence relative to the area of land declines with increasing size (Vercauteren et al. 2006). In addition, maintenance cost per individual owner would decline in a conservancy through economies of scale (a single patrolling programme rather than each separate farmer having their own patrolling programme). Finally, within a conservancy, the number of corners within the fenceline and river-crossings decline, resulting in a reduction in weak-points and expense: corners and end points contribute >80% of the material costs of fences (Vercauteren et al. 2006). Importantly, cadastral boundaries are often formed by rivers, which are key ecological resources, and joining land parcels will result in the inclusion of rivers within the fenced area. There are obviously ecological and economic incentives for forming conservancies (see Lindsey et al. 2009), and from a fencing perspective, it makes sense for land-owners to join together in conservancies.

While fencing to prevent human–elephant conflict has become required in South Africa (DEAT 2008), it is not widely used elsewhere, and other Governments are going to need to consider fencing along hard boundaries where transformed human community land borders natural areas, such as along the western border of Serengeti National Park, Tanzania. Governments have tended to be slow to react to changing scenarios that potentially require fencing (Hayward and Kerley 2009), but the private sector is initiating fencing of some reserves in East Africa (Kioko et al. 2008).

Hayward and Kerley (2009) conclude that the costs of fencing far outweigh the benefits, but their approach was simply a listing of benefits next to costs. The issues that they list in their Table 2 should not be equally weighted, and I believe that, depending on the circumstances, the benefits far outweigh the costs. For example, private game reserves in South Africa would not be allowed to introduce key tourism species if they do not upgrade their boundary fence to a minimum regulated requirement for particular species such as elephant (DEAT 2008). If they did not fence, then these reserves would not exist, and the benefits that they have brought to conservation (e.g. Gusset et al. 2008, but see Hunter et al. 2007; Slotow and Hunter 2009) would not persist. Hayward and Kerley (2009) conclude that fencing to mitigate Human–wildlife conflict is likely to persist, and in this chapter, I have demonstrated how careful planning can mitigate some of the costs (including ecological) while still achieving conservation objectives. Importantly, the costs associated with fencing can be reduced through economies of scale with increasing area, and consolidation of land-parcels through conservancies (e.g. Lindsey et al. 2009) and other partnerships such as Transfrontier parks (van Aarde and Jackson 2007). The key purpose of a fence needs to be defined, so that the most effective solution, permeability, can be implemented.

Acknowledgements Data from the Elephant Owners and Managers Association survey in 2001 were provided to me for analysis and interpretation, and I thank and acknowledge Dick Carr who conducted the survey, and Dick Carr and Marion Garai who initiated and planned the survey. Thanks to the management staff of Ezemvelo KZN Wildlife for interesting discussions around the management of Ithala Game Reserve and iSimangaliso Wetland Park, particularly Rob Blok and Tony Conway. Many of the ideas and results presented here are shared, and developed in partnership with the management teams. Funding was provided by University of KwaZulu-Natal, Distell (PTY) Ltd to the Amarula Elephant Research Programme, and the National Research Foundation Grant FA2006032300024. Thanks to Norman Owen-Smith and Michael Somers for comments on the manuscript.

References

Anderson, J.L. (1994) The introduction of elephant into medium-sized conservation areas. *Pachyderm*, **18**, 35–38.

Carruthers, J., Boshoff, A., Slotow, R., Biggs, H.C., Avery, G. & Matthews, W. (2008) The elephant in South Africa: History and distribution. *Assessment of South African Elephant Management*. (eds. R.J. Scholes & K.G. Mennell), pp 23–83.Witwatersrand University Press, Johannesburg.

DEAT (2008) National Norms and Standards for the management of elephants in South Africa. Department of Environment and Tourism. *Government Gazette*, 8 May 2008.

Garai, M.E. & Carr, R.D. (2001) Unsuccessful introductions of adult elephant bulls to confined areas in South Africa. *Pachyderm*, **31**, 52–57.

Garai, M.E., Slotow, R., Reilly, B. & Carr, R.D. (2004) History and success of elephant re-introductions to small fenced reserves in South Africa. *Pachyderm*, **37**, 28–36.

Grant, C.C., Bengis, R., Balfour, D., Peel, M., Davies-Mostert, H., Killian, H., Little, R., Smit, I., Garai, M., Henley, M., Anthony, B. & Hartley, P. (2008) Controlling the distribution of elephants. *Assessment of South African Elephant Management*. (eds. R.J. Scholes & K.G. Mennell), pp 329–369.Witwatersrand University Press, Johannesburg.

Gusset, M., Ryan, S.J., Hofmeyr, M., van Dyk, G., Davies-Mostert, H.T., Graf, J., Owen, C., Szykman, M., Macdonald, D., Monfort, S., Wildt, D., Maddock, A., Mills, M.G.L., Slotow, R. & Somers, M. (2008) Efforts going to the dogs? Evaluating attempts to re-introduce endangered wild dogs in South Africa. *Journal of Applied Ecology,* **45**, 100–108.

Harrington, R., Owen-Smith, N., Viljoen, P.C., Mason, D.R. & Funston, P.J. (1999) Establishing the causes of the roan antelope decline in the Kruger National Park, South Africa. *Biological Conservation*, **90**, 69–78.

Hayward, M.W. & Kerley, G.I.H. (2009) Fencing for conservation: Restriction of evolutionary potential or a riposte to threatening processes? *Biological Conservation*, **142**, 1–13.

Hayward, M.W., Adendorff, J., O'Brien, J., Sholto-Douglas, A., Bissett, C., Moolman, L.C., Bean, P., Fogarty, A., Howarth, D., Slater R. & Kerley G.I.H. (2007) Practical considerations for the reintroduction of large, terrestrial, mammalian predators based on reintroductions to South Africa's Eastern Cape Province. *The Open Conservation Biology Journal*, **1**, 1–11 doi 10.2174/1874-8392/07.

Hunter, L., Skinner, J.D., Pretorius, K., Carlisle, L.C., Rickleton, M., Walker, C. & Slotow, R. (2007) Reintroducing lions *Panthera leo* to northern Kwazulu-Natal, South Africa: short-term biological and technical successes but equivocal long-term conservation. *Oryx*, **41**, 196–204.

Kerley, G.I.H., Landman, M., Kruger, L., Owen-Smith, N., Balfour, D., de Boer, W.F., Gaylard, A., Lindsay, K. & Slotow, R. 2008. Effects of elephant on ecosystems and biodiversity. In: RJ Scholes and KG Mennell (eds) Assessment of South African Elephant Management. Witwatersrand University Press, Johannesburg. pp 146–205.

King, L.E., Douglas-Hamilton, I. & Vollrath, F. (2007) African elephants run from the sound of disturbed bees. *Current Biology*, **17**, R832–R833.

King, L.E., Lawrence, A., Douglas-Hamilton, I. & Vollrath, F. (2009) Beehive fence deters crop-raiding elephants. *African Journal of Ecology*, **47**, 131–137.

Kioko, J., Muruthi, P., Omondi, P. & Chiyo, P.I. (2008) The performance of electric fences as elephant barriers in Amboseli, Kenya. *South African Journal of Wildlife Research*, **38**, 52–58.

Lagendijk, D.D.G., Mackey, R.L., Page, B.R., Slotow, R. 2011. The Effects of herbivory by a mega- and mesoherbivore on tree recruitment in Sand Forest, South Africa. *PLoS ONE* 6(3): e17983. doi:10.1371/journal.pone.0017983.

Landman, M., Kerley, G.I.H. & Schoeman, D.S. (2008) Relevance of elephant herbivory as a threat to important plants in the Addo Elephant National Park, South Africa. *Journal of Zoology*, **274**, 51–58.

Leblond, M., Dussault, C., Ouellet, J.-P., Poulin, M., Courtois, R.H. & Fortin, J. (2007) Electric Fencing as a Measure to Reduce Moose–Vehicle Collisions. *Journal of Wildlife Management*, **71**, 1695–1703.

Lindsey, P.A., Romañach, S.S. & Davies-Mostert, H.T. (2009) The importance of conservancies for enhancing the value of game ranch land for large mammal conservation in southern Africa. *Journal of Zoology*, **277**, 99–105.

Loarie, S.R., Van Aarde, R.J. & Pimm, S.L. (2009) Fences and artificial water affect African savannah elephant movement patterns. *Biological Conservation*, **142**, 3086–3098.

Lombard, A.T., Johnson, C.F., Cowling, R.M. & Pressey, R.L. (2001) Protecting plants from elephants: botanic reserve scenario within the Addo Elephant National Park, South Africa. *Biological Conservation*, **102**, 191–203.

Mackey, R.L., Page, B.R., Duffy, K.L. & Slotow, R. (2006) Modeling elephant population growth in small, fenced, South African reserves. *South African Journal of Wildlife Research*, **36**, 33–43.

Moseby, K.E. & Read, J.L. (2006) The efficacy of feral cat, fox and rabbit exclusion fence designs for threatened species protection. *Biological Conservation*, **127**, 429–437.

Mtui, D. & Owen-Smith, N. (2006) Impact of elephants (*Loxodonta africana*) on woody plants in Malolotja Nature Reserve, Swaziland. *African Journal of Ecology*, **44**, 407–409.

Osborn, F.V. & Rasmussen, L.E.L. (1995) Evidence for the effectiveness of an oleo-resin capsicum aerosol as a repellent against wild elephants in Zimbabwe. *Pachyderm*, **2**, 55–64.

Slotow, R. & Hunter, L.T.B. 2009. Reintroduction decisions taken at the incorrect social scale devalue their conservation contribution: African lion in South Africa. *The Reintroduction of*

Top-order Predators. (eds. M.W. Hayward and M.J. Somers), pp 43–71. Wiler-Blackwell Publishing, Oxford.

Slotow, R., Garai, M.E., Reilly, B., Page, B. & Carr, R.D. (2005) Population ecology of elephants re-introduced to small fenced reserves in South Africa. *South African Journal of Wildlife Research*, **35**, 23–32.

Slotow, R., Whyte, I., Hofmeyr, M., Kerley, G.H.I., Conway, T. & Scholes, R.J. (2008) Lethal management of elephant. *Assessment of South African Elephant Management*. (eds. R.J. Scholes and K.G. Mennell), pp 370–405.Witwatersrand University Press, Johannesburg.

Twine, W. & Magome, H. (2008) Interactions between elephants and people. *Assessment of South African Elephant Management*. (eds. R.J. Scholes and K.G. Mennell), pp 206–240. Witwatersrand University Press, Johannesburg.

van Aarde, R.J. & Jackson, T.P. (2007) Megaparks for metapopulations: addressing the causes of locally high elephant numbers in southern Africa. *Biological Conservation*, **134**, 289–297.

Vanak, A.T., Thaker, M. & Slotow, R. (2010) Do fences create an edge-effect on the movement patterns of a highly mobile mega-herbivore? *Biological Conservation* 143: 2631–2637.

van Dyk, G. (1997) Reintroduction techniques for lion (*Panthera leo*). *Lions and Leopards as Game Ranch Animals*. (ed Penzhorn, B.L.), pp. 82–91. SAWVA, Onderstepoort, South Africa.

van Dyk, G & Slotow, R. (2003) The effects of fences and lions on the ecology of African wild dogs reintroduced to Pilanesberg National Park, South Africa. *African Zoology*, **38**, 79–94.

Vercauteren, K.C., Lavelle, M.J. & Hyngstrom, S. (2006) Fences and deer-damage management: A review of designs and efficacy. *Wildlife Society Bulletin*, **34**, 191–200.

Whyte, I.J. (1985) *The present ecological status of the blue wildebeest (*Connochaetes taurinus taurinus *Berchell, 1823) in the Central District of the Kruger National Park*. M.Sc. Thesis, University of Natal, Pietermaritzburg.

Chapter 7
An Adaptive Monitoring Programme for Studying Impacts Along the Western Boundary Fence of Kruger National Park, South Africa

Ken Ferguson, Laura Adam, and Ferran Jori

Introduction

In this chapter, we introduce our participatory and adaptive *Fence Incident Surveillance System* (FISS), which monitors wild mammal permeability to fencing and the damage to these structures that a number of species of African large mammals can cause along a 90 km pilot study section of the western boundary fence of Kruger National Park (KNP).

The entire 400 km western boundary of the KNP (the southern boundary adds a further 350 km of fencing) is demarcated by a veterinary fence, primarily designed to contain the foot-and-mouth disease virus within the park (Bengis et al. 2003; Jori et al. 2011). The western fence varies in structural types and different sections can be exposed to different degrees and causes of damage (Ferguson 2009a, b; Jori et al. 2009). The resultant large mammal fence permeability patterns that we present here represent a vital pre-requisite to an understanding of the underlying processes and the potential mitigation of the impact of such cross-boundary animal movement.

Historically, the sustainable monitoring and management of fences in Southern Africa has been primarily perceived as an issue of disease control, with often the more direct human–wildlife conflict aspects and the threats posed by these structures to conservation goals being relegated to a secondary management issue (Taylor and Martin 1987; Hayward et al. 2008; Lamarque et al. 2008).

K. Ferguson (✉)
Institute of Biodiversity, Animal Health and Comparative Medicine, University of Glasgow, G12 8QQ, Glasgow, Scotland
e-mail: kenneth.ferguson@glasgow.ac.uk

L. Adam
Mammal Research Institute, Department of Zoology and Entomology, University of Pretoria, Pretoria 0002, South Africa

F. Jori
UPR AGIRs, CIRAD, Montpellier, France, Research Fellow at the Mammal Research Institute, Department of Zoology and Entomology, University of Pretoria, Pretoria 0002, South Africa

M.J. Somers and M.W. Hayward (eds.), *Fencing for Conservation: Restriction of Evolutionary Potential or a Riposte to Threatening Processes?*,
DOI 10.1007/978-1-4614-0902-1_7,

Fencing can "constrain and contain" in symbolic, physical, ecological and economic senses (Spierenburg and Wels 2006), and fences represent an important structure that moderates the biological exchange between the "natural" and "human-modified" domains (Boone and Hobbs 2004; Kock 2005; Lindenmayer and Fischer 2006). The practical issues surrounding fencing should be viewed in the broader light of human–wildlife conflict theory and practice (Lamarque et al. 2008).

Containing wildlife by means of fencing can only serve a potential combination of four purposes: firstly, to reduce human–wildlife conflict by reducing contact between the two; secondly, to reduce disease transmission risk between wild and domestic animals; thirdly, to increase the security of a protected area and fourthly, on occasion, to demarcate an international boundary (Hayward and Kerley 2009; Newmark 2008).

In South Africa, fences could therefore be seen to be part of a well-worked out containment strategy within a multiple zoning framework (e.g. fenced and disease risk zones) endorsed by international bodies such as the World Animal Health Organization (OIE) that governs the import/export of livestock products (Bengis et al. 2003; Jori et al. 2009).

In terms of the threats posed by fences to large mammal conservation goals, these can be divided into two categories: direct and indirect. Factors such as direct mortality, the entrapment of a relict population by fencing and the dissection of land that reduces the carrying capacity of the source area (effectively creating a source-sink paradigm; Lindenmayer and Fischer 2006) can each affect large mammal populations (Williamson and Mbano 1988; Whyte and Joubert 1988; Mbaiwa and Mbaiwa 2006).

Indirect factors related to fencing that may induce a reduction in population size, such as changes in reproductive parameters in either the relict or source population, remain obscure. Certainly, the manifold and subtle effects of fencing on wild animal populations are only now being slowly determined (Ben-Shahar 1993; Martin 2005).

Therefore, fences "impose themselves" (by enhancing fragmentation via edge effects and creating biological and human impact gradients) on landscapes, which often due to cost, run with unremitting linearity (Albertson 1998). Lindenmayer and Fischer (2006) discuss in detail the creation and impact of natural and artificial edge effects. Edge effects are major drivers of change in many fragmented landscapes, but they are often highly variable in space and time.

Alternatives to fencing that allow some degree of merger between these "natural" and human-modified domains, such as the development of buffer zones, wildlife corridors and the zoning of different forms of wildlife usage, require much more effort and more planning (e.g. corridor development; Chetkiewicz et al. 2006; Newmark 2008) than simply erecting fencing and will inevitably fail unless animal health experts can be convinced that diseases of economic importance to the nation can still be effectively contained within this new partially or wholly "unfenced" dispensation (Bengis et al. 2003).

However, fencing can also have conservation benefits (Hayward et al. 2008), for example by physically protecting rare species such as rhino or protecting wild herbivore species from grazing competition with livestock by excluding the latter (Dunham et al. 2003). It is the achievement of a balance between "responsible fencing" (Bengis et al. 2003) and the use of fences as a "blunt" instrument that is often the ignition point for "fences conflict" between conservationists and the animal production agro-industry (Albertson 1998).

A Brief History of KNP Fences

For nearly half of its existence, the KNP has been bounded to a greater or lesser degree by varying lengths and types of fence. Starting in the early 1960s, the Government of South Africa, following several outbreaks of foot-and-mouth disease in 1958, decreed that the risk of foot-and-mouth disease spreading to neighbouring livestock production areas that lay to the west of the KNP had to be curbed (Bengis et al. 2003; Joubert 2007).

On completion of the western fence in 1961, much of the park became effectively isolated from the rest of the progressively more human-dominated, greater Kruger ecosystem (du Toit 2003). Smuts (1982) commented that the "effective size of the park is gradually shrinking". Pollard et al. (2003) point out that the onset of the isolation of the Kruger Park pre-dates fences; the isolation itself was largely due to mechanisms such as human population increases (released by a decline in malaria outbreaks and the eradication of the tsetse fly) in the Lowveld from the 1930s to the 1950s and also to the exclusivist policies of the park management at that time.

The cessation of the migrations of large herbivores at the KNP western interface may have been partially due to the imposition of fencing (Whyte and Joubert 1988; Ben-Shahar 1993). However, the removal of fencing can rapidly lead to a resumption of regular wildlife movements as has been witnessed by recent elephant movement into Limpopo National Park from the Kruger Park (D. Pienaar personal communication; 2009). Joubert (2007) commenting from an intimate knowledge of KNP states that "at the time of demarcation of the boundaries, important considerations such as animal migrations were not taken into account". And yet it is clear from the historic records that the zebra and wildebeest migration routes westwards out of the park were largely mapped prior to fence construction (see Pienaar's 1960 map reproduced in Joubert 2007) and the fact that a water provision programme near the western border pre-dates fencing suggests that the fence was "on balance" considered by the then park managers to be a positive contribution to the preservation of the KNP (Joubert 2007).

The western boundary fence received an upgrade to electrical status that was completed in 1999 at a cost of twelve million Rand. The next year floods devastated the electric fence and caused seven million Rands worth of damage. Increased permeability to large mammals in parts of the western fence caused by the floods and elephant damage in particular meant that the electric fence was no longer a viable option.

Has the KNP western boundary fence performed to expectation with regard to the containment of disease? Between the years 1983 and 1999, no foot-and-mouth disease outbreaks were recorded. Between the years 2000 and 2009, six outbreaks have been recorded in KNP (Grant et al. 2007; Jori et al. 2009).

The current managers of the western Kruger fence (Department of Agriculture, Forestry and Fisheries (DAFF)) have a clear mandate to protect both the park and the surrounding areas from epizootic impacts (Bengis et al. 2003). The containment of other forms of human–wildlife conflict is not within the DAFF remit unless these issues are pertinent to disease control.

In fencing terms, the KNP has come full circle, from being an "open" management system at inception to becoming a completely "closed" management system from the mid-1970s to one which is again opening up (due to transfrontier conservation; Hanks 2000) in new and expansive ways (Joubert 2007).

Materials and Methods

Types of Fence

Our research fence line spans approximately 90 km in length (from the Luvuvhu River in the north of KNP to the Olifants River further South) along the northern-western KNP boundary fence. The topography along the fence line is generally flat and the dominant vegetation type is Mopane (*Colophospermum mopane*) shrub and bushveld. Habitat "inlets" of a diverse array of other vegetation types (various Sandveld habitats) occasionally intrude along the fence.

Along this perimeter, we encountered five types of fence which have been erected at various times. Table 7.1 shows fence incidents statistics per type of fence and Fig. 7.1 shows temporal permeability data collected over the duration of the field study. Figure 7.2 shows the position of the various types of fences and the fence intensity density data.

(a) *New "I" Beam* (specifications: 2.4 m high; wooden droppers; 4 × 13 mm steel cable; barbed wire strands; steel uprights implanted at 50 m intervals). This is the preferred DAFF fencing structure and it is now considered by them to be the universal standard for sealing off elephant populations from livestock-dominated landscapes. This fence type composes 42% of the total fence length in the study area. The fence is designed to withstand sustained assault by elephants such that this species cannot facilitate buffalo excursions (and the resultant spread of disease). It is still unclear if elephants have the capacity to successfully challenge this fence (two isolated cases seem to suggest that they can; K. Ferguson personal observation; D. Keet personal communication; 2010). The "I" Beam fence is being erected in priority areas prior to the planned complete coverage along the western boundary. It is being constructed at the relatively slow rate of 25 km per annum (D. Keet personal communication; 2009). The "I" Beam should be relatively maintenance free. It is not and was not intended to be carnivore proof.
(b) *2.4 m Steel Fence* (specifications as for *New "I" Beam* except with metal droppers and hollow steel uprights). In some sections, many of the metal droppers have not been bent out of shape (K. Ferguson personal observation; 2009; usually by animals or fence workers during repairs) which tends to suggest a low incidence of fence challenges. The reason for this remains unknown, but it may be due to the inherent strength of the fence, lower densities of bull elephants in this area or landscape differences (e.g. the presence of a large barren sodic pan adjacent to much of this fence section).

Table 7.1 FISS per species and per type of fence between November 2008 and July 2009

Type of fence	Proportion of the fence type on the total fence line	Total fence incidents/km	Fence incidents/km of type of fence by species					
			Unknown	*Buffalo*	*Elephant*	*Human*	*Hyena*	*Lion*
1.8 m rail post fence	19.8	8.9	0.1	0.2	6.7	0.1	1.2	0.7
2.4 m electric fence	8.8	3.3	0	0.6	1.3	0.8	0	0.4
2.4 m new I-Beam fence	41.8	3.3	0.2	0.1	0.5	0.4	1.7	0.4
2.4 m old fence	24.2	17	0.1	0.2	13.2	0.5	1.4	1.5
2.4 m steel fence	5.5	2.8	0	0	0.4	0.4	0.8	1.2
Total FISS $n=727$	?	Buffalo	Elephant	Human	Hyena	Lion		
Percentage (%)	1.5	2.3	64.1	4.7	16.8	10.6		

?=unknown species

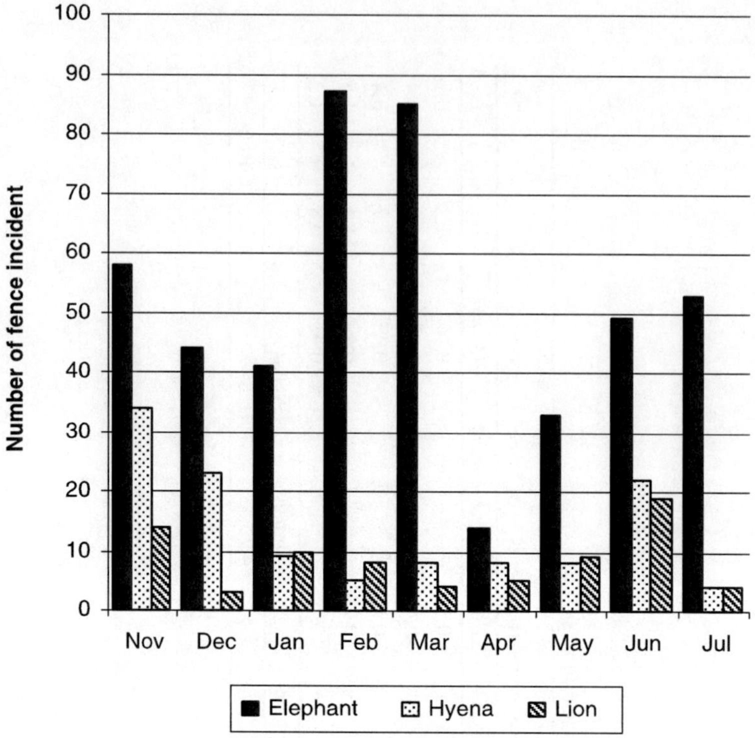

Fig. 7.1 FISS per species (elephant, lion and hyena only) per month between November 2008 and July 2009

(c) *1.8 m Rail/Cable Fence* (specifications as for *New "I" Beam* except for height). Parts of this fence show severe signs of sustained elephant damage and it is currently being heightened to 2.4 m by means of added cables and girders. This represents one of the highest impact zones for elephants and carnivores which epicentres at the joining of this fence and the "old inactive" electric fence.
(d) *2.4 m "Old inactive" electric fence* (specifications: steel posts embedded in concrete, metal droppers, barbed wire strands and conductor strands only). This was the KNP's major fence electric upgrade of the 1990s. After the floods of 2000 and the continuous theft of electric components (solar panels, batteries, wire) it fell into disrepair. Along most of its length it has suffered sustained damage by elephants and it is no longer an effective barrier to the movement of most species. Most recorded excursions occur in these sections.
(e) *2.4 m Electric fence* (specifications as for *"Old inactive" electric fence*, but the electric current is active). One small section is maintained as an electric fence (by the use of solar panels and batteries, but since 2001 thirty-six solar panels have been stolen in the study area (D. Keet personal communication; 2009). This fence records very few excursions and usually only when the electric current fails (usually over a week-end when a skeleton fence maintenance staff remains in the

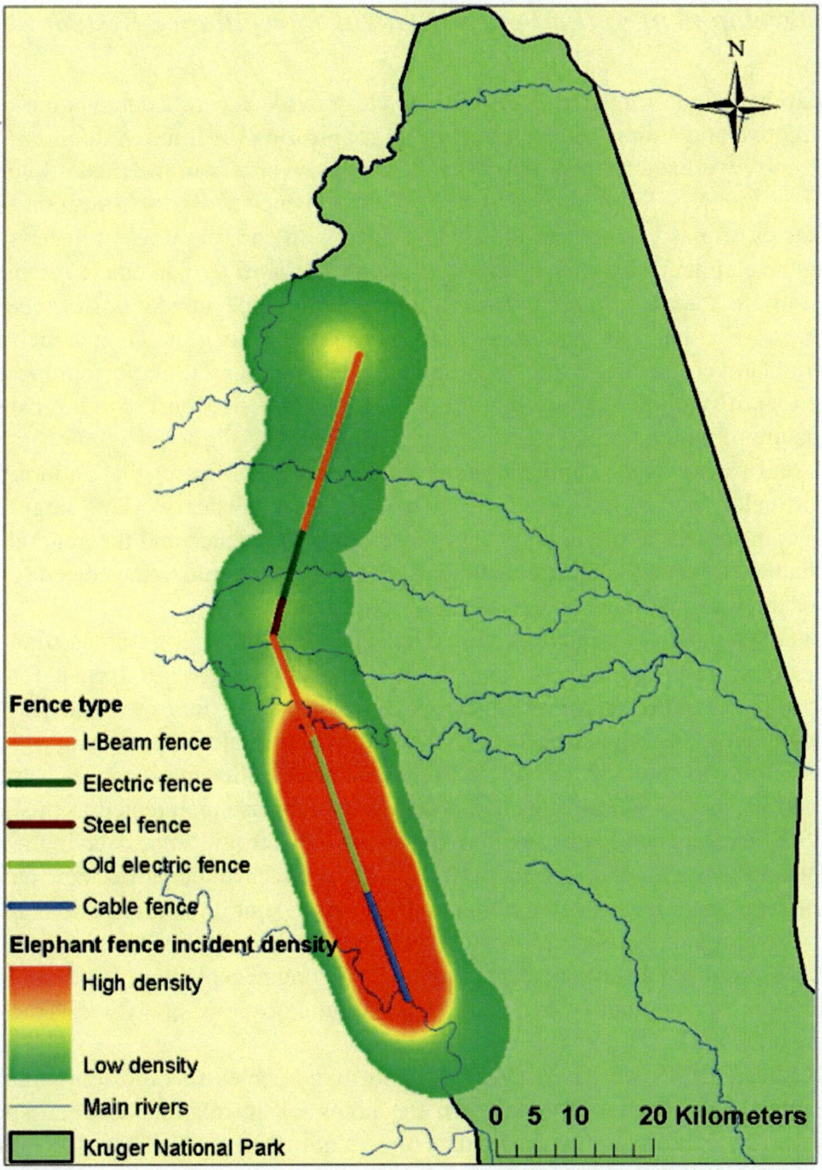

Fig. 7.2 The position of the various types of fences used in the study

park). It is remarkable how quickly elephants can determine a decrease in voltage (within hours; K. Ferguson personal observation; 2009) of the electric fence. The fence requires constant maintenance and is costly in terms of guarding the solar panels. It is an effective anti-predator fence, except across small drainage lines. We agree with Thouless and Sakwa (1995), Bonnington et al. (2009) and Graham et al. (2009b) that electric fencing is the most effective anti-elephant barrier, if it is properly maintained and appropriately sited.

The Development of the Fence Incident Surveillance System

Our adaptive monitoring strategy is intended to provide accurate data via our novel, cost-effective and simple fence permeability monitoring/profiling system.

Currently we have trained over 40 DAFF fence workers in the methodology of the Fence Incident Surveillance System (FISS). Fence workers are requested to attach strips of plastic tape to the fence inscribed with one (or more) of five symbols that denote elephant (*Loxodonta africana*), lion (*Panthera leo*), hyena (*Crocuta crocuta*), buffalo (*Syncercus caffer)* and human theft fence incidents (respectively: X=lion; •=Elephant; ▲=buffalo; □=hyena; and ╫=fence theft), and each symbol is indicative of whether the recorded species has crossed and/or damaged the fence. A sixth symbol, a question mark (?), allows the fence worker to indicate that he is unsure of which species has challenged the fence. Additional symbols (↑=out of park and ↓=into park) denote whether the fence worker deems that an individual of a particular species has exited or entered the park. Fence workers largely use spoor, scrape marks at the base of the fence, hair on the fence and the general state of the damaged fence to determine which species is considered to have been responsible for the fence damage and/or a species "excursion" event.

Pocket PC (personal computer) trained fence workers then geo-reference their own data on to a pre-programmed (using Cyber-Tracker™ Software) format for later downloading on to the laptop spreadsheet of a mobile senior fence worker. The latter then downloads this spreadsheet on to the PC of the State Veterinarian with GIS responsibilities in Skukuza (KNP DAFF/HQ). Permeability maps for each type of fence and its species interactions and a dedicated database are created at this point. The FISS, therefore, facilitates an unbroken chain of electronic fence data. In the event of a Pocket PC malfunctioning, the fence workers can leave the barrier tape hanging from the fence for later geo-referencing (the "lifespan" legibility of ink on the tape can be up to several months; K. Ferguson personal observation; 2009). Training protocols have been produced in the local language – XiTsonga (explaining instructions for operating the tape/symbol system and the use of the Pocket PC and Cyber-Tracker™ software).

ARCView™ GIS was used to record and map species-specific fence incident permeability and to attempt to discern the potential correlation between species with regard to patterns of fence-related movement. Our profile maps accumulate data over time allowing the managers of the fence to detect recurrent patterns of fence damage and to rectify their fencing deterrent strategies as necessary.

Results

Fence Profiling in KNP: Results per Species per Type of Fence

Within our study area, incident statistics per type of fence are presented in Table 7.1 and temporal FISS data collected over the duration of the field study are illustrated in Fig. 7.1. We concluded that none of the fence types are 100% effective.

The highest numbers of fence incidents were recorded in the older fences where electric power was very irregular or nonexistent or the structure of the fence was weak (2.4 m old fence and 1.8 m rail post fence), ranging between 9 and 17 incidents/km. Incidents were much lower in those sections of the fence where electricity was maintained and also in the I-Beam fence (dropping to 3.3 and 2.8 incidents/km).

Elephant was by far the species most often recorded causing damage and/or crossing through the fence with a rate of 6.4 incidents/km of fence. It was followed by carnivores, with 16.8 and 10.6 incidents recorded/km for hyena and lion, respectively.

Buffaloes were by far, from those monitored, the species that caused the lowest number of incidents with only 2.3 incidents recorded/km of fence during the study period.

There was a clear increase in the number of fence incidents caused by elephants in the months of February and March. The other species monitored did not show a specific temporal trend of fence incidents (see Fig. 7.1).

Discussion

Elephants and Fencing Impacts

Elephants represent the major challenge for fence management along the KNP western boundary. This "keystone" fence breaker can directly cause conflict with local people and indirectly lead to massive national economic losses by facilitating the release of diseased buffaloes (Grant et al. 2007).

Our study area encompasses a "bull enclave" (historically, it is bull elephants that have home ranges along the western boundary; Smit et al. 2007b) and the only cow/calf herd detected during the 10 months of fieldwork was photographed at night, adjacent to the fence, by means of a camera trap. Habitual fence offenders are, therefore, likely to be almost always bull elephants. The "sub-population" of elephants whose home ranges adjoin the fence is relatively small (data from ear-notch identification files), but this group is also likely to be continually replaced after lethal incidents (approximately ten bull elephant deaths reported during the study period) caused by humans.

The reasons for our study area being a predominantly "bull area" are not clear. It may be in part due to the physiological limits placed on cow/calf groups in this location. The interface also represents an anthropogenic danger to elephants where humans are an ever present threat to these animals; bulls may be more "risk-prone" in this regard (Hoare 1999). It is also possible that bull home ranges and movements vary widely due to social pressures induced by hierarchy and musth (Wittemyer et al. 2008). The processes which motivate elephants to cross the fence are therefore likely to be multi-factorial and seasonal. Osborn and Parker (2003) contend that in the case of crop raiding, elephant bulls exit protected areas when the quality of the available forage declines below the level of the quality of the crop species grown outside of the area.

In the Kruger case, it has been long suspected that a seasonal peak in elephant excursions is caused by the Marula tree (*Sclerocarya birrea* subspp. *caffra*) fruiting season (Grant et al. 2007) and this implies that the number of fruiting mature female trees has attained a higher density outside of the park (due mainly to elephant damage of stock within the park and perhaps also to differential fire regimes across the boundary than within, preventing seedling recruitment (Helm et al. 2009).

Our research from line transect data (inside and outside of the boundary) has shown that a tree density differential does exist and that the February–March peak of elephant excursions is likely to be driven by the availability of this seasonal resource (Adam et al. unpublished data). This result suggests that elephant-induced mortality of marula trees inside the park acts as a seasonal driver of "excursions". Presumably, as the park's elephant population has increased so has the mortality of marula trees within the park to such a degree that the marula population could face local extinction within KNP (Helm et al. 2009). Our data suggest that a partially protected tree population may exist adjacent and external to the park boundary. This marula tree population size difference suggests either that fences in the past have worked and they have only recently become highly permeable (due to the park-wide floods in the year 2000) to elephants or that an increase in elephant numbers has led to a segment of the "risk-prone" bull population seeking out the last of these resources beyond the fence.

These observations lead to an interesting management scenario. A potential way of mitigating the effects of this driver of elephant excursions would be for local people to pre-harvest the fruit and to store and ripen it for later use (F.V. Osborn personal communication; 2010). Another alternative is to deter fence breakages by the selective displacement culling of identifiable "habitual" elephant fence breakers or to deploy non-lethal mitigation methods (e.g. direct protection of the fence by chilli pepper or indirect protection by aversion therapy, e.g. "shooting" elephants with "pepper balls").

A second hypothesised "excursion peak" in winter has yet to be defined by our data, but it has been suggested that the resource sought after in this case seems to be surface water (R. Bengis personal communication; 2008). Grant et al. (2007) showed that in the winter of 2005 up to 35 elephant fence breaks/day occurred along a 12 km stretch of the western fence, where presumably the elephants were breaking out to drink in one of the few remaining pools situated in the Klein Letaba River.

The closure of boreholes within the park over the last several decades may have also led to recent increases of fence challenges by elephants (and possibly lions) in order to access external (to the park) water sources (Grant et al. 2007).

Density-independent permeability factors such as the floods of 2000 were largely claimed to have been responsible for the disabling of the recently upgraded electric fence, and thereafter it could not be maintained faster than elephants destroyed it, and humans stole components from it. Limited data for past years fence breakages suggest that the variation in the number of breaks between the years 2001 and 2004 (Grant et al. 2007) may be as significant as the intra-annual average variation in fence breakages, thus contesting a simple "more elephants equal more fence breaks" hypothesis. A localised density-dependent effect in the number of breakages is more likely to be due to the greater abundance of the fruiting marula trees found outside of the park which may encourage the highly seasonal grouping of bull elephants.

The most likely alternative to a direct density-dependence effect is to postulate an increase in habitual fence breakage as a result of one or more of the following factors (i) by more bulls being available and younger ones being "taught" to break by older bulls (ii) and/or by younger bulls avoiding an increased number of musth males and/or an increased number of musth males searching more widely (iii) and/or by bulls seeking other essential resources, e.g. forage and water (iv) by dominant herds forcing sub-dominant herds to the marginal periphery, often out of a protected area (Wittemyer et al. 2008).

General rules governing the movement and landscape selection by African elephants are slowly being revealed largely due to the development of more sophisticated telemetry devices. Chase and Griffin (2009), Loarie et al. (2009) and Thomas et al. (2008) have presented general evidence that park and veterinary fences in southern Africa (in addition to civil war and other human perturbations) have had a major impact on the traditional movement patterns of these populations. Particular concern in shown by these authors to the excising of wet season ranges caused by the erection of these barriers. In East Africa, the use of fencing to aid the rapid privatisation of rangeland has had a major impact on elephant and wildlife distribution whilst proving ineffective at controlling human–wildlife conflict and indeed engendering a greater degree of human–human conflict (Okello and D'Amour 2008).

More specifically, the rules that define short- to long-term elephant movement across fences (and therefore the positive selection by this species) in either direction, i.e. from or to heavily "human-dominated" and "human-sparse" landscapes are likely to be primarily based on seasonal and rainfall parameters and food resources. Surface water availability often presents the best predictor of seasonal range use (Redfern et al. 2003; Smit et al. 2007a; Harris et al. 2008). Landscape topography (e.g. avoidance of hills; Wall et al. 2006), corridor linkages (Douglas-Hamilton et al. 2005), direct and indirect conflict with humans (Hoare 1999; Sitati et al. 2003; Dublin and Hoare 2004; Knickerbocker and Waithaka 2005; Graham et al. 2009a), sexual dimorphism in feeding behaviour (Stokke and du Toit 2002) and even the eco-physiology of temperature thermoregulation (Kinahan et al. 2007) are all further potential drivers of movement in and out of a fenced, protected area.

Carnivore Excursions

Lion depredations of livestock are a major cause of local community dissatisfaction with the park (K. Ferguson personal observation; 2010; S. Midzi personal communication; 2010) in terms of human–wildlife conflict. Lions regularly exit the park via the fence and kill livestock (mainly cattle). In June 2009, a lioness was found dead near the boundary of the park. She was photographed by camera trap exiting through the same scrape under the fence that she had used the previous night (on this occasion she killed 16 goats, apparently to teach her elderly cubs how to hunt). Post-mortem results show that she was poisoned by the agro-pesticide carbofuradan, KNP's first recorded case of the use of this agro-pesticide chemical for this purpose.

This event illustrates that lions tend to be site-faithful when exiting the park. Preference seems to be to exit via rivers, but they can also dig through any part of any fence (including the electric fence, though this is likely to be when the current is inactive). The only hyena-proof fence at present is the electric fence. Hyenas also prefer regular exit routes, and there is some evidence that these routes overlap with lion excursion patterns. We conjecture that hyenas follow lions across the fence to feed on kills made by lions.

Seasonal peaks of carnivore movement across fences may also be related to water distribution. Social factors are likely to confound the picture. Dispersing males may be more willing to risk fence excursions than established territorial males (D. Keet, personal communication; 2009).

We also endorse Anthony's (2006) recommendation that as far as human–lion conflict is concerned there is a pressing need to harmonise the response between the provincial and park authorities to the lethal control of this species. A compensation scheme which positively rewards good husbandry techniques (particularly, efficient nocturnal kraaling) whilst at the same time financially remunerating verified livestock losses should be considered by SANParks as a less wasteful alternative to direct killing.

Large Ungulates and Fencing Impacts

Our buffalo excursion rate was recorded as the second lowest incident rate. Buffaloes do not challenge fences directly unless they are being stampeded by predators (F. Jori personal observation; 2009) and otherwise they react to all fences in a passive manner. The low level of buffalo incidents (elephants are 27 times more likely to exit the park than buffaloes) may indicate that the fence workers did not easily record buffalo spoor in the vicinity of a fence break. This may be due to the heavy presence of cattle spoor on the outside fence track and the large number of buffalo dung piles that accumulate along the inner fence track (evidence from dung pile transect data). Buffalo herds regularly use the fence track and a "parallel path" created by wildlife which runs along the fence track (up to 100 m within the park boundary fence). The hypothesis that buffaloes track elephant fence breaks has not been substantiated, although overlap in the density of fence incidents between these two species does exist. The importance of bachelor herds/solo "dagga boys" vs. the mixed herd's use of the interface needs to be examined. Buffaloes are often chased back into the park by helicopter and only euthanised if they prove too troublesome to direct back inwards. Solitary males seem to make extensive home range use of areas near fencing and on occasion spend large periods of time outside of the park (K. Ferguson personal observation; 2009). Whilst it is generally accepted that habitual fence excursions by elephants are common, the movements of buffaloes across fences have so far received little attention. How "habitual" are the buffaloes that are crossing the boundary and whether there is a sexual dimorphism or demographic variability in the profile of "escapees" remain to be studied.

Species Interactions and Overlaps at Fence Breaks

Fence permeability data and maps can be used to detect general overall trends in fence excursions between species. It is clear from the maps we have produced that all species benefit from the damage caused by elephants, especially in terms of this species facilitating the excursions of other large mammals through the dilapidated "old" sections of fencing (although the rate of excursions of "diggers" such as lion and hyena are not limited by a low number of elephant breakages).

However, we must add the caveat that the data interpretation in this regard is limited because elephant breaks can extend for up to 50 m in length whilst a lion "scrape" may be less than a meter in width (GPS accuracy is between 6 and 10 m, further complicating direct referencing of inter-species facilitation of movement). Therefore, we did not find a statistically proven correlation between elephant breaks and the movements of some species (e.g. buffaloes).

We have also recorded that the current fence contractor is using unsuitable (it can be used to make heavy-duty snares) 13 mm cable in the construction of the new "I" beam fence. It remains a future possibility that human theft of cable from the new "I" beam fence may weaken the structure and allow elephants and buffaloes to cross at the same point along the fence with the former facilitating the release of the latter.

Limitations and Biases of the FISS Method

Spatial permeability to fences by this guild of large mammals is determined by multiple factors, the most important of which are species abundance and distribution. Species-specific behavioural mechanisms for circumventing these barriers will be predicated by the type of fence (and its state of maintenance), the landscape that the fence is embedded in (including access to water points) and the biology of the species in the surrounding matrix (e.g. density, social structure). Temporal permeability will be affected by the above factors but also by the abundance and location of the resources that "drive" each species to cross this hostile barrier. Our analysis of these multiple factors and the processes which underpin these animal "excursions" is currently focussing on readily observable patterns of fence damage as this is the easiest and most sustainable way of building up a database of events. Further research, including telemetry of key species, will be needed to unravel the complex sequence of events that have emerged in terms of fence excursion rates and timing.

The discernment of FISS patterns for each species is not without its limitations. Human observer bias is likely and whilst the fence workers can estimate the spatial and temporal frequency of species excursions from spoor, they cannot estimate the number of animals involved in each incident (unlike other studies that use spoor density indices, e.g. Houser et al. 2009). However, broad agreement between FISS patterns and a second questionnaire-based fence permeability method (Jori et al. 2011) suggests that that interpretations of spoor reading may not be a significant source of bias. The majority of fence workers experienced few problems related to the use of the Pocket PC and these were overcome with time and extra training.

Fences and Human Neighbours

It has been estimated that at least two million people reside within 50 km of the KNP western boundary fence (Pollard et al. 2003). The demise of the tsetse fly allowed an expansion of agriculture and increased human densities which often exceed 300 persons/km for some communal areas (Pollard et al. 2003).

The mixture of private and subsistence farming adjacent to the western boundary of the park has meant that park fencing is viewed by these bordering constituents in different ways. In the past, protection of commercial livestock from disease would have been seen as a greater priority than protecting indigenous people from direct conflict with wildlife (Wels 2000).

The dual role of the fence, as a barrier to disease and a barrier to more direct human–wildlife conflict, has been the single most important reason for the lack of a coherent strategy by the park authorities in terms of monitoring and mitigating the fence impacts on people and wildlife (e.g. "fences" are rarely mentioned in park management plans; Ferguson 2009a, b).

Anthony (2006) has highlighted the disparity in equity between the park and its indigenous neighbours. Although there has been no direct attitudinal survey of the local people to find out their views on the Kruger fence, we can extrapolate to a degree from the findings of other studies of the effects of animal movements across Kruger fences on the perception of local people.

Anthony (2006) has suggested that the "fences and fines" approach to Kruger park management will be increasingly more problematic to enforce. Local populations often only interact with the park authorities after human–wildlife conflict incidents or when they are caught within the park whilst illegally harvesting resources. The fence engenders in some villagers a feeling of security against wild animal depredations, although this attitude could change quite rapidly (see Knickerbocker and Waithaka 2005). To others, the fence is undoubtedly viewed as a symbol of oppression – or even a "white signature" scarring the land and acting as a metaphor of appropriation (Wels 2000; Spierenburg and Wels 2006; but see Wolmer 2007). The theft of fence components or the vandalism of the fence may be in part due to resentment towards the park and partly a means of subsistence or making a profit. The irony is that the creation of a greater permeability flow of dangerous animals due to humans weakening the fence structure should be noted.

Lagendijk and Gusset (2008) found that 52% of respondents who lived along the border of KNP owned livestock and 81% had lost animals to wild carnivores, the implication here being that none of the fences are predator-proof and currently no plans exist to upgrade any KNP fences to this standard.

Conclusion

The FISS methodology that we have tested highlights the need for a continuation of fence permeability monitoring and we suggest that participatory monitoring and the communal management of fences should also be encouraged (Ferguson 2009a, b; Lindenmayer and Likens 2009).

The "edge" (buffer or boundary) of any protected area represents the "soft underbelly" of conservation efforts (Woodroffe and Ginsberg 1998). In the Kruger case, the fence acts as a mediator between two different organisational systems (protected areas and communal lands; Newmark 2008) and two management regimes with different objectives (DAFF and SANParks). For the adjoining human communities, the fence has both symbolic and practical implications (Spierenburg and Wels 2006): it demarcates mobility in relation to the boundary and in certain cases requires to be negotiated either physically or in terms of financial loss if epizootic disease or dangerous animals permeate through.

For animal species that require access to resources outside the park (the seasonal movements of some species, in the KNP case study, may also represent an innate westerly drift through "corridors past") the fence represents a "gateway" into a more dangerous anthropogenic setting. Many of the species observed at the fence will make strenuous efforts (including self-harm) to gain rapid access back into the park. There is much to learn about fence interface interactions and the various factors that regulate species movement in terms of the "movement resistance" across the landscape created by fencing. These lessons cannot be learned without the long-term application of a robust fence monitoring system (Lindenmayer and Likens 2009).

In the future, we foresee that the concept of "*SMART*" *(specific, measurable, achievable, realistic and targeted)* fencing will more often be applied. We define this as fencing that attempts to be more sensitive to the "grain" of ecological landscapes and in "tune" with ecological processes (such as migration, wherever possible). For example, should the opportunity arise (and cost permitting) for the placement of fencing in direct relation to iso-height contours or that encompasses external water points rather than excluding these variables by means of a straight fenceline, then this may be a more preferable outcome for a protected area.

We speculate that in any given year, within the southern African sub-region, the amount of kilometre of fence that will be erected will be much greater than the lengths of fence removed due to the expansion of transfrontier or game conservancy conservation policies. Therefore, it is important that the conflict that ensues around fencing is seen as part of the general human–wildlife conflict.

Furthermore, intensifying rates of climate change may have unforeseen consequences for animal movement in relation to fencing and protected area boundaries (Ogutu and Owen-Smith 2003).

The "One Health" (Osofsky et al. 2008) approach to human–wildlife interface dynamics is a promising start that seeks to place fencing in a proper context, and perhaps the use of fences as a "blunt" instrument will give way to more wildlife friendly modes of animal (both wildlife and livestock) movement control.

It is our hope that our standardised method of collecting permeability data can be harmonised across the sub-region and perhaps lead to an information-driven regional position encompassed within a *Pan Southern African Fences Charter* that will seek to lay down guidelines for the judicious erection and/or the removal of these unnatural barriers. It is not possible to ignore the role that fences play in the agricultural sector and they will continue to play a major role in protected area and animal disease management.

Acknowledgements We would like to thank Drs. Roy Bengis and Dewald Keet of the State Veterinary Services for their consistent support. We also extend our thanks to Dr. Rina Grant of SANParks. Mr. Patrick Baloyi and Mr. Robert Mshlanga (of DAFF) aided us in many ways; we are in your debt. This research was funded by USFWS Grant Afe-0401 to Professor Elissa Cameron and WWF-SA Grant 1490 to Mr. Ken Ferguson and Dr. Rina Grant as part of a designated SANParks Special Project. This chapter is dedicated to the memory of the late Simon Thirgood.

References

Albertson, A. (1998) Northern Botswana Veterinary Fences: Critical Ecological Impacts. The Wild Foundation/Kalahari Conservation Society. Botswana. pp. 1–59.

Anthony, B. (2006) *A view from the other side of the fence: Tsonga communities and the KNP, South Africa*. PhD Thesis, Central European University, Budapest.

Ben-Shahar, R. (1993) Does fencing reduce the carrying capacity for populations of large herbivores? *Journal of Tropical Ecology,* **9**, 249–253.

Bengis, R.G., Grant, R., de Vos, V. (2003) Wildlife diseases and veterinary controls: a savanna ecosystem perspective. In du Toit, J.T., Rogers, K.H. & Biggs, H.C., (eds). *The Kruger experience.* Island Press. Washington. pp. 349–369.

Bonnington, C., Grainger, M., Dangerfield, S. & Fanning. E. (2009) The influence of electric fences on large mammal movements in the Kilombero Valley, Tanzania. *African Journal of Ecology*, **46**, 138–145.

Boone, R.B. & Hobbs, N.T. (2004) Lines around fragments: effects of fencing on large herbivores. African *Journal of Range & Forage Science,* **21**, 147–158.

Chase, M.J. & Griffin, C. (2009) Elephants caught in the middle: Impacts of war, fences and people on elephant distribution and abundance in the Caprivi Strip, Namibia. *African Journal of Ecology*, **47**, 223–233.

Chetkiewicz, C-L., St. Clair, B.C.C., Boyce, M.S. (2006) Corridors for Conservation: Integrating Pattern and Process. Annual Review of Ecology, *Evolution and Systematics*, **37**, 317–342.

Douglas-Hamilton, I., Krink, T. & Vollrath, F. (2005) Movements and corridors of African elephants in relation to protected areas. *Naturwissenschaften*, **92**, 158–163.

Dublin, H. & Hoare, R. (2004) Searching for solutions: The evolution of an integrated approach to understanding and mitigating human-elephant conflict in Africa. *Human Dimensions of Wildlife*, **9**, 271–278.

Dunham, K.M., Robertson, E.F. & Swanepoel, C.M. (2003) Population decline of tsessebe antelope (*Damaliscus lunatus lunatus*) on a mixed cattle and wildlife ranch in Zimbabwe. *Biological Conservation*, **113**, 111–124.

du Toit, J.T., Rogers, K.H. & Biggs, H.C. Eds. (2003) The Kruger Experience: Ecology and Management of Savanna Heterogeneity. Island Press, Washington.

Ferguson, K.J. (2009a) The spatial dynamics of wildlife populations across and along the north-western Kruger National Park boundary fence. Final Report: WWF-ZA Project 1490.

Ferguson, K.J. (2009b) Longitudinal Archiving and Concurrent Capture of Fence Egress Data in the Kruger National Park. Final Performance Report USFWS Project : AFE-0401-8-G752.

Graham, M.D., Douglas-Hamilton, I., Adams, W.M. & Lee, P.C. (2009a). The movement of African elephants in a human-dominated land-use mosaic. *Animal Conservation*, **12**, 445–455.

Graham, M.D., Gichohi, N., Kamau, F., Aike, G., Craig, B, Douglas-Hamilton, I. & Adams, W.M. (2009b) The Use of Electrified Fences to Reduce Human Elephant Conflict: A Case Study of the Ol Pejeta Conservancy, Laikipia District, Kenya, Working Paper 1, Laikipia Elephant Project, Nanyuki, Kenya.

Grant, R., Bengis, R., Balfour, D. & Peel, M. (2007) Controlling the distribution of elephants. In: Mennell, K. & Scholes, R. (eds). *The 2007 Scientific Assessment of Elephant Management in South Africa.* Witwatersrand University Press, Johannesburg.

Hanks J. (2000) The role of transfrontier conservation areas in southern Africa in the conservation of mammalian biodiversity. In: Entwistle A. Dunstone N. (eds). *Priorities for the conservation of mammalian diversity: has the panda had its day?.* Cambridge University Press, Cambridge, UK.

Harris, G.M., Russel, G.J., van Aarde, R.J. & Pimm, S.L. (2008) Rules of habitat use by elephants *Loxodonta africana* in southern Africa: insights for regional management. *Oryx*, **42**, 66–75.

Hayward, M.W., Hayward, G.J., Druce, D. & Kerley, G.I.H. (2008) Do fences constrain predator movements on an evolutionary scale? Home range, food intake and movement patterns of large predators reintroduced to Addo Elephant National Park, South Africa. *Biodiversity and Conservation*, **18**, 887–904.

Hayward, M. & Kerley, G. (2009) Fencing for conservation: restriction of evolutionary potential or a riposte to threatening processes? *Biological Conservation*, **142**, 1–13.

Helm, C.V., Witkowski, E.T.F., Kruger, L., Hofmeyr, M. & Owen-Smith, N. (2009) Mortality and utilisation of *Sclerocarya birrea* subsp. *caffra* between 2001 and 2008 in the Kruger National Park, South Africa. *South African Journal of Botany*, **75**, 475–484.

Hoare, R.E. (1999) Determinants of human-elephant conflict in a land-use mosaic. *Journal of Applied Ecology*, **36**, 689–700.

Houser, A-M., Somers, M.J. & Boast, K.L. (2009) Spoor density as a measure of true density of a known population of free-ranging wild cheetah in Botswana. *Journal of Zoology*, **270**, 108–115.

Joubert, S. (2007*) The Kruger National Park – A History*. High-Branching, Johannesburg.

Jori, F., Vosloo, W., Du Plessis, B., Bengis, R. Brahmbatt, D., Gummow, B. & Thomson, G.R. (2009) A qualitative risk assessment of factors contributing to foot and mouth disease outbreaks on the Western Boundary of Kruger National Park. *Revue Scientifique et Technique Office Internationale Epizooties,* **28**, 917–931.

Jori, F., Brahmbhatt, D., Fosgate, G., Thompson PN, Budke, C., Ward, M., Ferguson and Gummow, B. (2011) A questionnaire-based evaluation of the veterinary cordon fence separating wildlife and livestock along the Kruger National Park, South Africa. *Preventive Veterinary Medecine*, **100**, 210–220 doi:10.1016/j.prevetmed.2011.03.015.

Kinahan, A.A., Pimm, S.L. & van Aarde, R.J. (2007) Ambient temperature as a determinant of landscape use in the savanna elephant, *Loxodonta africana*. *Journal of Thermal Biology*, **32**, 47–58.

Knickerbocker, T.J & Waithaka, J. (2005) People and Elephants in Shimba Hills, Kenya. In: Woodrooffe, R., Thirgood, S. & Rabinowitz, A., (eds.) *People and Wildlife: Conflict or Coexistence?.* Cambridge University Press. Cambridge.

Kock, R. (2005) What is this infamous "wildlife/livestock disease interface?" A review of current knowledge for the African continent In: Osofsky, S.A. (ed.), Conservation and Development Interventions at the Wildlife/Livestock Interface: Implications for Wildlife, Livestock and Human Health. pp. 1–13. Occasional paper Number 30 of the IUCN Species Survival Commission. Gland.

Lagendijk, G.D. & Gusset, M. (2008) Human–Carnivore Coexistence on Communal Land Bordering the Greater Kruger Area, South Africa. *Environmental Management*, **42**, 971–976.

Lamarque, F., Anderson, J., Chardonnet, P., Fergusson, R., Lagrange, M., Osei-Owusu, Y., Bakker, L., Belemsobgo, U., Beytell, B., Boulet, H., Soto, B. & Tabi Tako-Eta, P. (2008). Human-wildlife conflict in Africa: an overview of causes, consequences and management strategies. FAO report, available from http://www.wildlifeconservation.org/en/content/download/1034/4568/file/HumanWildlifeConflict2008.pdf.

Lindenmeyer, D.B. & Fisher, J. (2006) Habitat Fragmentation and Landscape Change: An Ecological and Conservation Synthesis. *Island Press. Washington*.

Lindenmayer, D.B. & Likens, G.E. (2009) Adaptive monitoring: a new paradigm for long-term research and monitoring. *Trends in Ecology & Evolution*, **24**, 482–6.

Loarie, S.R, Van Aarde, R.J & Pimm, S.L. (2009) Fences and artificial water affect African savannah elephant movement patterns. *Biological Conservation*, **142**, 3086–3098.

Martin, R.B. (2005) The influence of veterinary control fences on certain wild large mammal species in the Caprivi strip, Namibia. In: Osofsky, S.A. (ed.), Conservation and Development Interventions at the Wildlife/Livestock Interface: Implications for Wildlife, Livestock and Human Health. pp. 27–39. Occasional Paper IUCN SSC, Gland, Switzerland/Cambridge, UK.

Mbaiwa, J.E & Mbaiwa, O.I. (2006) The effects of veterinary fences on wildlife populations in Okavango Delta, Botswana. *International Journal of Wilderness*, **12**, 17–41.

Newmark, W.D. (2008) Isolation in African protected areas. *Frontiers in Ecology and Environment*, **6**, 231–328.

Ogutu, J.O. & Owen-Smith, N. (2003) ENSO, rainfall and temperature influences on extreme population declines among African savanna ungulates. *Ecology Letters*, **6**, 412–419.

Okello, M.M. & D'Amour, D.E. (2008) Agricultural expansion within Kimana : electric fences and implications for natural resource conservation around Amboseli National Park, Kenya. *Journal of Arid Environments*, **72**, 2179–2192.

Osborn, F.V. & Parker, G.E. (2003). An integrated approach for reducing the conflict between elephants and people: a review of current research. *Oryx*, **37**, 80–84.

Osofsky, S.A., Cumming, D.H.M. & Kock, M.D. (2008) Transboundary management of natural resources and the importance of a 'One Health' approach: perspectives on Southern Africa. *State of the Wild 2008–2009 A Global Portrait of Wildlife, Wildlands, and Oceans*. In: K.H.R. Fearn (ed). Island Press, Washington, D.C. pp. 89–98.

Pollard, S., Shackleton, C. & Carruthers, J. (2003) Beyond the fence: people and the lowveld landscape (eds. du Toit, J.T., Rogers, K.H., Biggs, H.C.), pp. 422–446. Island Press, Washington.

Redfern, J.V., Grant, R., Biggs, H. & Getz, W.M. (2003) Surface-water constraints on herbivore foraging in the Kruger National Park, South Africa. *Ecology*, **84**, 2092–2107.

Sitati, N.W., Walpole, M.J., Smith, R.J. & Leader-Williams, N. (2003) Predicting spatial aspects of human elephant conflict. *Journal of Applied Ecology*, **40**, 667–677.

Smit, I.P.J., Grant, C.C. & Devereux, B.J. (2007a) Do artificial waterholes influence the way herbivores use the landscape? Herbivore distribution patterns around rivers and artificial surface water sources in a large African savanna park. *Biological Conservation*, **136**, 85–99.

Smit, I.P.J., Grant, C.C. & Whyte, I.J. (2007b) Elephants and water provision: what are the management links? *Diversity and Distributions*, **13**, 666–669.

Smuts, G.L. (1982) *Lion.* Macmillan South Africa Publishers, Johannesburg.

Spierenburg, M. & Wels, H. (2006) Securing space. *Space and Culture*, **9**, 294–312.

Stokke, S. & du Toit, J.T. (2002) Sexual segregation in habitat use by elephants in Chobe National Park, Botswana. *African Journal of Ecology*, **40**, 360–371.

Taylor, R.T. & Martin, R.B. (1987) Effects of veterinary fences on wildlife conservation in Zimbabwe. *Environmental Management*, **11**, 327–334.

Thomas, B., Holland, J.D. & Minot, E.O. (2008) Elephant (*Loxodonta africana*) home ranges in Sabi Sand Reserve and Kruger National Park: a five-year satellite tracking study. *PLoS ONE*, **3**, e3902.

Thouless, C.R. & Sakwa, J. (1995) Shocking elephants, fences and crop raiders in Laikipia District, Kenya. *Biological Conservation*, **72**, 99–107.

Wall, J., Douglas-Hamilton, I.D.H. & Vollrath, F. (2006) Elephants avoid costly mountaineering. *Current Biology*, **16**, 527–529.

Wels, H. (2000) *Fighting over fences: organisational co-operation and reciprocal exchange between the Savé Valley Conservancy and its neighbouring communities*, PhD thesis, Free University, Amsterdam.

Whyte, I.J. & Joubert, S.C.J. (1988) Blue wildebeest population trends in the Kruger National Park and the effects of fencing. *South African Journal of Wildlife Research*, **18**, 78–87.

Williamson, D. & Mbano, B. (1988). Wildebeest mortality during 1983 at Lake Xau, Botswana. *African Journal of Ecology*, **26**, 341–344.

Wittemyer, G., Polansky, L., Douglas-Hamilton, I.D.H & Getz, W.M. (2008) Disentangling the effects of forage, social rank, and risk on movement autocorrelation of elephants using Fourier and wavelet analyses. *Proceedings of the National Academy of Sciences*, **105**, 191–198.

Wolmer, W. (2007) *From wilderness to farm invasions: Conservation and development in Zimbabwe's South-east Lowveld.* James Currey Publishers. London.

Woodroffe, R. & Ginsberg, J.R. (1998) Edge effects and the extinction of populations inside protected areas. *Science*, **280**, 2126–2128.

Chapter 8
Does the Vastness of the Serengeti Limit Human–Wildlife Conflicts?

Marion L. East, Julius W. Nyahongo, Katja V. Goller, and Heribert Hofer

Introduction

As wildlife-rich habitats shrink in size and become isolated within areas degraded by human activities, the importance of ensuring that wildlife can move between species-rich areas has increased, leading to greater emphasis on landscape planning and the creation, maintenance and quality of wildlife corridors, including those that cross national boundaries (Newmark 1996; Smith et al. 1998; Puth and Wilson 2001; Hofer et al. 2004; Chetkiewicz et al. 2006). Fences that prevent the movement of wildlife are at odds with this new paradigm, but may provide conservation benefits to specific areas or species (Hayward and Kerley 2009). Fences, as a form of demarcation of land, have an ancient history, and the extent to which they have been used in agricultural landscapes is influenced by cultural traditions (Rackham 2003). Tradition, and construction and maintenance costs are probably the key factors that have influenced the extent to which fences have been used as a wildlife management tool in different countries. This may explain why in southern Africa, conservation areas are often fenced, whereas in eastern Africa few conservation areas have been fenced.

The Serengeti National Park (SNP) in northern Tanzania (Fig. 8.1) covers an area of 14,763 km^2. It is surrounded by a network of protected areas that together encompass the majority of the Serengeti ecosystem that spans the international border between Tanzania and Kenya. This ecosystem (also termed the Serengeti-Mara ecosystem or the Serengeti ecological unit) is defined as the approximately 25,500 km^2 covered by migratory herbivores during their annual migration (Schaller 1972;

M.L. East (✉) • K.V. Goller • H. Hofer
Leibniz Institute for Zoo and Wildlife Research, Alfred-Kowalke-Street 17, Berlin 10315, Germany
e-mail: east@izw-berlin.de

J.W. Nyahongo
School of Natural Sciences and Mathematics, University of Dodoma, Dodoma, Tanzania

M.J. Somers and M.W. Hayward (eds.), *Fencing for Conservation: Restriction of Evolutionary Potential or a Riposte to Threatening Processes?*,
DOI 10.1007/978-1-4614-0902-1_8,

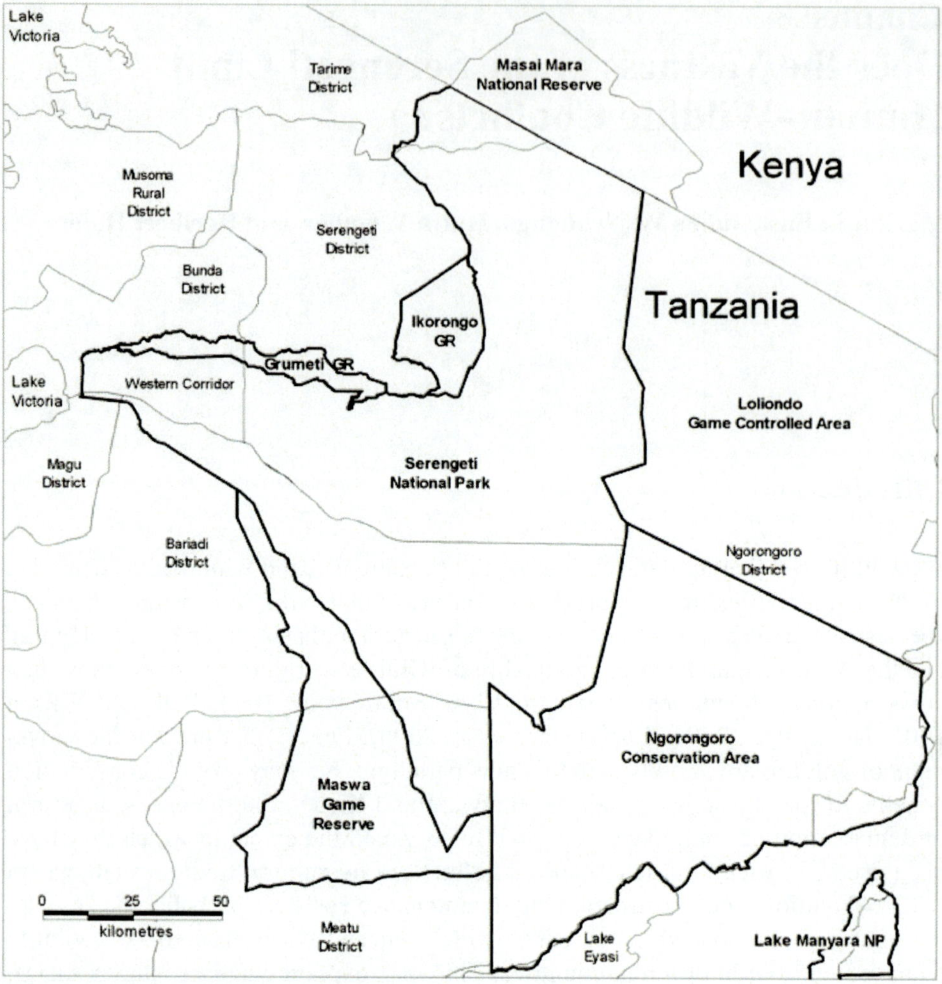

Fig. 8.1 The Serengeti National Park and surrounding protected areas (Ikorongo Game Reserve, Grumeti Game Reserve, Maswa Game Reserve, Ngorongoro Conservation Area, Loliondo Game Controlled Area) plus boundaries of administrative districts. The Western Corridor is the area of the Serengeti National Park that extends westwards towards Lake Victoria

Sinclair 1979, 1995). The vast migratory herds drive many key ecosystem processes, including plant community dynamics (McNaughton 1990; McNaughton and Banyikwa 1995) and predator–prey interactions (e.g. Hanby and Bygott 1979; Hofer and East 1993a–c). Given the disastrous effect of fences on herbivore migrations elsewhere in Africa (e.g. Williamson and Williamson 1984; Spinage 1992) there can be little doubt that fencing the SNP would be entirely detrimental to the annual herbivore migration, and thus the entire Serengeti ecosystem.

In this chapter we consider whether the vast size of SNP and associated protected areas is sufficient to minimise sources of human–wildlife conflict between the SNP and local people, and whether non-physical barriers reduce such conflict. We start by briefly describing the protected areas that surround the SNP. We then consider three key forms of human–wildlife conflict: predation of livestock by wild carnivores, the illegal offtake of bushmeat by local hunters and the transmission of infectious pathogens between domestic animals and wildlife. We discuss predation by wild carnivores in terms of the economic cost of livestock losses to farmers, consider how the costs and benefits of hunting determine the spatial distribution of hunting activities and review the use of mass vaccination campaigns to induce non-physical barriers of immunity against the spread of three pathogens, two *Morbilliviruses* that are relatively new to Africa (rinderpest and canine distemper virus [CDV]) and one virus with a more ancient history in Africa (rabies).

The Protected Areas and Surrounding Communities

The SNP is administered by the Tanzania National Parks Authority. Photographic tourism is permitted in the SNP but hunting, human settlements, crop cultivation, livestock grazing and domestic animals are not permitted. Along its western edge are three game reserves (GRs), the Ikorongo GR, Grumeti GR and Maswa GR (Fig. 8.1), that form buffer-zones between the SNP and mostly subsistence farming communities that grow crops and rear livestock (Loibooki et al. 2002). Communities to the west of the SNP practice bushmeat hunting, for home consumption and to generate income through the sale of dried meat (Hofer et al. 1996; Loibooki et al. 2002). The Wildlife Division of the Ministry for Natural Resources, Environment and Tourism administer GRs and the Loliondo Game Controlled Area (LGCA) to the east of the SNP. Only licensed hunting is permitted in GRs and the LGCA. Crop cultivation, livestock grazing and human settlements are not permitted in GRs. The Ngorongoro Conservation Area (NCA) and LGCA are mainly inhabited by low density pastoralist communities that rarely hunt bushmeat (Campbell and Hofer 1995). The NCA is a multiple land use area administered by the NCA Authority. To the north, across the international border is the Masai Mara National Reserve (MMNR) in Kenya (Fig. 8.1). We will refer to the SNP and its associated protected areas as the protected area (PA) of the Serengeti ecosystem.

In the SNP, apparent anthropogenic effects are limited and natural ecosystem processes mostly predominate. Beyond the PA, particularly in the west, human activities and those of domestic animals and livestock predominate. Recently, the term Serengeti ecological region has been applied by studies in areas well beyond the western limit of the Serengeti ecosystem (e.g. Cleaveland et al. 2000; Lembo et al. 2008; Kaare et al. 2009) – an important point to bear in mind.

Livestock Predation

Farmers worldwide view wild carnivores as a threat to their livestock and as a result many carnivore species have been exterminated from much of their historic range (Nowell and Jackson 1996; Mills and Hofer 1998; Treves and Karanth 2003). Here

Table 8.1 Reported losses of cattle per household per year (mean ± SE) and their value in US-$ equivalents

	Animals/household/year			US$/household/year		
Cause of loss	Serengeti district	Bunda district	Magu district	Serengeti district	Bunda district	Magu district
Strayed	0.4±0.1	1.6±0.3	0.5±0.1	20	80	25
Predation	0.9±0.3	1.9±0.5	0.6±0.1	45	95	30
Theft	1.3±0.4	2.6±0.4	1.1±0.2	65	130	55
Disease	2.8±0.5	7.3±1.4	1.4±0.3	140	365	98
Total losses	5.4	13.4	3.6	275	670	280

The average price of one cow in 2002 was approximately US$50

we examine human–predator conflict in terms of livestock predation reported by farmers living close to the western boundary of the PA. We recognise that the number of livestock losses claimed by farmers may not be strictly accurate, but consider estimates of this kind to reveal the level of conflict farmers perceive they have with wild carnivores and provide a rough index of livestock predation.

Livestock Predation in Areas West of the Protected Area

Between 2001 and 2002 the economic costs of different causes of livestock loss were surveyed in 21 villages close to the western boundary of the PA (Nyahongo 2004). A total of 210 livestock owners were interviewed, representing ten farmers from seven randomly selected villages in each of three administrative districts (Serengeti, Bunda and Magu Districts, Fig. 8.1). Respondents tallied their livestock losses during the preceding 5 years and allocated losses to four categories: predation by wild carnivores, disease, theft and animals that strayed when herds were moved to and from grazing or watering areas. The estimated economic cost of livestock losses in these categories (Tables 8.1 and 8.2) was based on the approximate value of an adult cow of US$50 and of an adult goat or sheep of US$12 (Nyahongo 2004) even though the price received at sale would depend on size and condition (Loibooki et al. 2002).

In all three districts, the largest reported livestock loss (cattle, Table 8.1; sheep and goats combined, Table 8.2) was caused by disease (US$202, 409 and 110 per farmer, per year in Serengeti, Bunda and Magu districts respectively, Tables 8.1 and 8.2). These losses were several times higher than losses from other causes, including losses caused by predation (US$52, 119 and 36 per farmer per year in the three districts, respectively, Tables 8.1 and 8.2). Furthermore, the likelihood of a carnivore attack on livestock was low, given that only 20% of those surveyed reported one or more predator attacks per annum. Farmers in all three districts reported a similar relative importance for the four categories of loss for cattle (disease >> theft > predation > strayed) and the same was true for losses of goats and

Table 8.2 Reported losses of sheep and goats per household per year (mean ± SE) and their value in US-$ equivalents

	Animals/household/year			US$/household/year		
Cause of loss	Serengeti district	Bunda district	Magu district	Serengeti district	Bunda district	Magu district
Strayed	0.4±0.1	1.5±0.3	0.2±0.1	4.8	18.0	2.4
Predation	0.6±0.1	2.0±0.5	0.5±0.1	7.2	24.0	6.0
Theft	0.4±0.1	1.5±0.3	0.2±0.1	4.8	18.0	2.4
Disease	5.2±1.0	3.7±0.7	1.0±0.1	62.4	44.4	12.0
Total losses	6.6	8.7	1.9	79.2	104.4	22.8

The average price per sheep or goat in 2002 was approximately US$12

sheep (disease >> predation > theft = strayed). Losses caused by predation may have been overestimated because they are based on the average sale price of an adult animal of reasonable size and condition, whereas some animals taken by predators might have been in poor condition or might have died before reaching adulthood. There was considerable variation in the number of livestock owned by farmers, with some owning few or no livestock and others owning large herds (e.g. 490 cattle plus 270 sheep). Ownership of a large cattle herd is an important source of income and prestige, and a resource used to acquire wives through the presentation of between 2 and 16 animals to the bride's family (Loibooki et al. 2002). For a farmer with few livestock, the loss of any livestock to predators can represent a considerable economic loss. For farmers with an average sized herd (with approximately 18 cattle and 9 sheep/goats) the value of such a herd, based on the price of adult animals, would be approximately US$1,008 and the maximum loss of livestock to predators per year (reported by farmers in Bunda District, Tables 8.1 and 8.2) of approximately 2 cows and 2 sheep or goats would represent roughly US$124 per year, whereas the minimum loss (reported in Magu District) of approximately 1 cow and 0.5 sheep or goats would represent US$56 per year. A study of livestock losses in 2004 (Holmern et al. 2007) estimated that only 27% of the 481 households surveyed claimed to have lost livestock to predators. For these households, the estimated annual value of losses to predators was US$98 per household and similar to those estimated in 2001/2002. There are no government schemes to compensate farmers for livestock killed by wild carnivores. However, some pastoralists living to the east of the SNP are compensated for livestock killed by African wild dogs *Lycaon pictus* through a scheme run by local communities and funded by tourist companies.

Predator attacks west of the PA occurred predominantly (98%) at night on livestock in protective enclosures. Most attacks were on goats or sheep, and involved small groups of 1–3 spotted hyenas *Crocuta crocuta* (83% of 143 claimed witnessed attacks). Attacks by lions *Panthera leo* and leopards *Panthera pardus* were rarely witnessed; all six observed lion attacks were on cattle. Predation of cattle caused greater resentment among farmers than predation of goats or sheep, probably because of the larger economic loss involved (Nyahongo 2004).

The density of livestock outside the western boundary of the PA was far higher (38.0 ± 4.1 animals km^{-2}) than the density of wild herbivores (2.8 ± 0.01 animals km^{-2}, Nyahongo et al. 2005). Given such a high livestock density, why don't wild carnivores kill livestock more often? Possibly because the density of wild herbivores inside the SNP is equivalent to or higher than livestock densities outside the PA, and hunting livestock protected in enclosures and guarded by domestic dogs and people may be risky (Nyahongo et al. 2005). The spotted hyena, the carnivore most often observed attacking livestock, chiefly preys on brindled wildebeest *Connochaetes taurinus*, plains zebra *Equus burchelli* and Thomson's gazelle *Gazella thomsoni* (Hofer and East 1993a, b; Höner et al. 2005). These herbivore species are predominantly migratory in the Serengeti ecosystem except for a small resident population of wildebeest in the western corridor of the SNP (Fig. 8.1). Migratory movements of these prey species are mostly restricted to the PA and result in large temporal fluctuations in prey density in spotted hyena clan territories inside the SNP (Hofer and East 1993a, b). As a result, the total population of approximately 5,300 spotted hyenas in the SNP (Hofer and East 1995) regularly travel long distances from their "home" territory to forage in areas in the PA that contain large concentrations of migratory herbivores (Hofer and East 1993b, c). It is probably more beneficial and less risky to forage in distant areas with high densities of preferred migratory prey species than risk attacking livestock in villages.

Although farmers generally think that livestock predation is caused by carnivores from the SNP (Loibooki et al. 2002), how likely is this? Spotted hyena predation on livestock does not decline with distance from the PA boundary, whereas predation by large cats does (Holmern et al. 2007). This suggests that attacks on livestock by large cats are probably caused by animals from the SNP or PA, whereas most spotted hyena attacks involve members of resident clans outside the PA. The suggestion that wild carnivores in protected areas preferentially hunt natural prey rather than livestock is consistent with the finding that the number of carnivores killed in retaliation for livestock predation in areas around Tarangire and Manyara National Parks in Tanzania, was small in villages close to these parks and increased in villages at greater distances from these parks (Kissui 2008).

Livestock Predation in Other Areas of East Africa

In areas surrounding Tarangire and Manyara National Parks in Tanzania, the largest cause of livestock loss was also attributed to disease; predation accounted for minimal losses (Kissui 2008). On commercial farmland in Laikipia District, northern Kenya, in an area without any formally protected reserves, predation levels by wild carnivores (lion, leopard, cheetah *Acinonyx jubatus*, spotted hyena, striped hyena *Hyaena hyaena* and African wild dog) were similar to those west of the SNP and although tolerance of wildlife was high, the incidence of retaliatory killing of wild carnivores was related to the number of livestock thought to have been killed by wild carnivores (Ogada et al. 2003).

Conclusions

Although farmers living close to the western boundary of the Serengeti PA suffered economic losses caused by livestock predation, these losses were minor compared to those caused by disease. The belief that all livestock predation is caused by wild carnivores from the SNP or PA is not consistent with current evidence and should be tested rather than assumed. Currently, the spotted hyena population in the SNP chiefly prey on migratory species throughout the year even though this requires regular long-distance foraging trips to areas with high densities of migratory herbivores. For this reason it is likely that livestock losses outside the PA will remain low, provided high densities of the preferred prey of spotted hyenas are preserved inside the PA. The economic losses suffered by farmers from livestock predation would best be ameliorated by processes that reduced the current high loss of livestock to disease such as improved veterinary services and education on animal husbandry. At a local level, community initiatives to compensate farmers for livestock killed by predators can also be a useful tool to decrease predator-human conflict. Several studies of livestock predation by wild carnivores in East Africa suggest that more robust "carnivore-proof" enclosures would decrease predation (e.g. Ogada et al. 2003; Nyahongo 2004), but the material costs (when fencing materials are purchased) and opportunity cost (time spent collecting fencing material from a distant commercial supplier or from local natural resources, plus the time spent constructing and maintaining enclosures rather than on other activities) appear a sufficient disincentive (at the current low level of predation) to greater investment in overnight enclosures.

As already mentioned, wildlife-proof fences would damage ecosystem processes in the PA should they be employed as a management tool to decrease human–wildlife conflict. Furthermore, the extremely high cost of constructing, maintaining and patrolling a wildlife-proof perimeter fence around an area the size of the SNP (or the entire PA) would exceed, by many orders of magnitude, the cost of crop damage or livestock predation caused by wildlife. As fencing materials are a useful resource, any section of fence that was not regularly patrolled would probably be dismantled by locals and used elsewhere or sold for profit. Well administered compensation schemes for local farmers, based on a quantification of losses caused to wildlife and/or the provision of improved agricultural and livestock education facilities and extension services would most likely be far less expensive than fencing and more effective at decreasing human–wildlife conflict.

Bushmeat Hunting

Prior to the establishment of the current boundaries of the SNP and PA, bushmeat hunting was practiced by communities living in the west of the Serengeti ecosystem (Turner 1987). The establishment of the SNP in the 1950s outlawed hunting inside the park and permitted hunting only under licence in the adjacent GRs, yet (illegal)

Fig. 8.2 A spotted hyena in the Serengeti National Park with a wire snare through its mouth

bushmeat hunting continued in the west of the PA, to obtain meat for home consumption and to generate income through the sale of dried meat and hides (Arcese et al. 1995; Campbell and Hofer 1995; Hofer et al. 1996; Mduma et al. 1998; Loibooki et al. 2002; Nyahongo 2004). The current level of illegal bushmeat hunting is considerable and has resulted in the local extinction of resident herbivores in some areas of the SNP (Arcese et al. 1995; Campbell and Hofer 1995). Hunters mainly use inexpensive methods such as wire snares (Fig. 8.2), self-made traps and poisoned arrows (Turner 1987; Arcese et al. 1995; Nyahongo et al. 2005). The vast majority of hunters walk from their villages to hunting areas inside the PA. Some hunters establish camps in concealed areas so that they can operate for an extended period inside PA and employ porters to carry accumulated dried meat out of the PA (Arcese et al. 1995; Hofer et al. 1996).

Distance Barriers and Costs of Illegal Bushmeat Hunting

Several spatial models have explored the pattern and extent of illegal hunting in the PA and these models can then be tested using empirical data (Campbell and Hofer 1995; Hofer et al. 1996, 2000). These models assumed that illegal hunters weigh the costs and benefits of hunting in different areas as a function of a number of spatially explicit variables such as their travel distance to and from a hunting area, the proximity of hunting areas to ranger posts, habitat quality and the likelihood of detection by ranger patrols. These models were conservative in that they calculated the benefit obtained from illegal hunting on the basis of the distribution of relatively low density populations of resident herbivores and ignored the potential benefit-enhancing, temporary, yet unpredictable presence of large concentrations of migratory herbivores.

In a recent model (Hofer et al. 2000), the costs of hunting included the logistic effort of travelling by foot to hunting areas, the penalty incurred if arrested, capital invested in hunting equipment and the opportunity cost of hunting. The benefit was the income obtained from the sale of bushmeat from resident species. Results from the model suggested travel distance and investment in hunting equipment were key costs and that the cost of travel was sufficient to prevent extensive illegal bushmeat hunting by local communities deep inside the SNP (Hofer et al. 2000). This is consistent with other studies of bushmeat hunting that have found travel distance to be a significant barrier to illegal hunting and a significant factor determining the spatial distribution of illegal hunting (Nyahongo et al. 2005; Ling and Milner-Gulland 2008). Furthermore, local people have migrated from distant villages to settle in villages closer to the boundary of the PA, resulting in an unusually high population growth in these villages during the 1980s (Hofer et al. 1996). One likely reason for moving close to the PA boundary would be to minimise travel distance and increase the economic benefit obtained from bushmeat hunting (Hofer et al. 1996).

Opportunity costs are probably not important for bushmeat hunters because available alternative sources of income provided low economic returns (Hofer et al. 2000). It is worth noting that the returns from bushmeat hunting in terms of meat for home consumption, to trade for other commodities or to generate income (Loibooki 1997; Section "Economic Benefits of Illegal Bushmeat Hunting") are likely to be greater than the returns gained from small herds of livestock, and this may explain why few livestock owners invest heavily in terms of time or money in night-holding areas for livestock, when in any given year the probability of a predator attack is low (see Section "Livestock Predation"; Holmern et al. 2007).

The results of Hofer et al. (2000) also indicated that the cost of arrest was not an important factor for hunters because their chance of arrest was very low. Certainly, most bushmeat hunters use inconspicuous wildlife capture techniques that are hard to detect such as wire snares, and are chiefly active at night (Arcese et al. 1995; Hofer et al. 1996; Loibooki et al. 2002; Nyahongo 2004). Hunters can also reduce their chance of arrest by operating in areas with dense vegetation and complex topography that are rarely patrolled by rangers (Campbell and Hofer 1995). A combination of these factors may explain why high hunting effort can occur close to some ranger posts in the PA (Nyahongo et al. 2005).

Despite high livestock densities close to the PA boundary, livestock is not often taken inside the PA, possibly because livestock owners, in contrast to bushmeat hunters, consider the chance of detection and financial penalties (fines or confiscation of livestock) too high in relation to the benefit gained from illegally acquired forage and water for their stock (Nyahongo et al. 2005). Furthermore, owners of medium to large livestock herds are relatively wealthy members of local communities and rarely engage in illegal bushmeat hunting (Loibooki et al. 2002).

Economic Benefits of Illegal Bushmeat Hunting

Expected returns from the sale of dried bushmeat are high, and dried bushmeat is both a substantial component of the local economy and a significant source of protein to local communities (Loibooki 1997; Loibooki et al. 2002; Nyahongo 2004). Migratory herbivores supply the bulk of bushmeat illegally taken from the Serengeti ecosystem (Arcese et al. 1995; Hofer et al. 1996; Nyahongo et al. 2005). Of the approximately one million people living within 45 km of the western boundary of the PA (Campbell and Hofer 1995), an estimated 52,000 people are involved in illegal hunting activities inside the PA (Loibooki et al. 2002). It is likely that there is a link between poverty and illegal bushmeat hunting in the PA given that owners of livestock in communities west of the PA were significantly less likely to say that they practised bushmeat hunting than those that owned few or no livestock, and hunters arrested inside the SNP were predominantly young, poorly educated males that owned few or no livestock (Loibooki 1997; Loibooki et al. 2002). Arrested hunters may not be a representative sample of hunters operating in the SNP (possibly being younger, poorer and more likely to take risks that result in arrest than more experience hunters), but today's less experienced hunters may become tomorrow's experienced hunters that generate much of their annual income from illegal bushmeat (Loibooki 1997). A link between income and bushmeat hunting was also apparent in the Udzungwa Mountains in Tanzania, where people who said that they practised bushmeat hunting were from poorer households than those who said they were not bushmeat hunters (Nielsen 2006).

Does the Vast Size of the Serengeti Ecosystem Protect All Species?

The vast size of the PA clearly does provide resident herbivores that live at a distance from the PA boundary with protection against illegal bushmeat hunting. Equally clearly, the size of the PA is insufficient to provide large numbers of migratory herbivores and resident wildlife close to the western boundary of the PA with full protection against bushmeat hunters. This is particularly the case for areas of the SNP close to sections of the SNP boundary without a game reserve to act as a buffer-zone between the park and local communities (Nyahongo et al. 2005).

Wire snares are a non-selective hunting method that capture both target and non-target species. If wire snares are not cleared from an area they remain a potential threat to wildlife. Snares capture, injure or kill many non-target species as by-catch (Hofer et al. 1996) and are a particularly severe threat to spotted hyenas in the SNP (Fig. 8.2). This is because spotted hyenas regularly travel long-distances to feed in areas where bushmeat hunters are likely to set snares (Hofer et al. 1993). When large concentrations of migratory herbivores move through the western areas of the PA, bushmeat hunting in these areas significantly increases in intensity (Nyahongo et al. 2005) with the result that about 8% of spotted hyenas from social groups throughout the SNP are annually killed as a form of by-catch in wire snares (Hofer et al. 1993, 1996).

Conclusion

Currently, the vast size of the SNP and the fact that the park lies at the centre of an extensive network of protected areas is sufficient to provide populations of most resident species within the SNP with some protection against bushmeat hunters. Even so, migratory herbivores and wide-ranging species that venture close to the western boundary of the PA are particularly at risk from wire snares. By increasing the cost of travel by foot, buffer-zones around the SNP form a relatively effective distance barrier against extensive bushmeat hunting deep within the SNP. Two factors associated with bushmeat hunters are poverty and a low level of education. The development of environmentally appropriate income generating activities in collaboration with local communities where the number of livestock per inhabitant is low would help to reduce both poverty and bushmeat hunting. An improvement in the level of education within such disadvantaged communities would also likely achieve similar beneficial results. In comparison, and for the reasons that have already been outlined, fencing the SNP would result in a massive waste of money and be detrimental to this ecosystem. Given that the majority of the estimated 11,950 tons of wild herbivore meat illegally harvested each year from the PA (Hofer et al. 1996) comes from migratory species, any detrimental effect on migratory herbivore populations (which would be considerable should the SNP be fenced) would cause increased poverty and decreased nutrition in local communities. Fencing the PA boundary might also undermine tourist facilities established by communities close to the PA to generate income for the community, and generally heighten the perception of locals that wildlife in the PA was "reserved" for others.

Vaccine Barriers Against Disease Transmission

The transmission of diseases between people, domestic animals and wildlife can generate human–wildlife conflict, as is the case when farmers perceive that their income from livestock is reduced by diseases spread by wildlife. For example, some UK farmers believe that bovine tuberculosis (TB) caused by the bacterium *Mycobacterium bovis* in their cattle herds results from TB spread by infected badgers *Meles meles* (McDonald et al. 2008). People or their domestic animals can also transmit diseases to wildlife populations, thereby generating conflict with the conservation aims of protected areas, particularly when transmitted diseases threaten populations of endangered species. The spread of human diseases to great ape populations (Wallis and Lee 1999; Ferber 2000; Köndgen et al. 2008) illustrate this form of conflict.

Barriers of immunity against the spread of many infectious diseases can be created through the use of vaccines (Anderson and May 1991). Vaccines have been developed that immunise people and domestic animals against many diseases, but few vaccines have been developed for wildlife species. The application of vaccines developed for one species to another species can produce unwanted side effects, such as vaccine-induced disease (Carpenter et al. 1976; McCormack 1983).

To be effective, vaccine-induced immunity barriers require a sufficient proportion of susceptible individuals in a population to be immunised. For pathogens maintained in one host species, the establishment of an effective immunity barrier in that species will eliminate the pathogen from all other infected species that do not independently maintain infection. Vaccine-induced immunity barriers against rinderpest in domestic cattle and its subsequent elimination from wild ungulate species is an excellent example of this effect (see Section "Rinderpest"). In contrast, pathogens maintained in multiple host species require immunity barriers in these host species before the disease they cause can be controlled.

We initially outline a successful vaccine-induced barrier in cattle around the PA against rinderpest and why rinderpest control was essential for the health of the vast herds of migratory wild ungulates and thus the entire Serengeti ecosystem. Global control of rinderpest has been so successful that now rinderpest, like smallpox in humans, is thought to have been eradicated worldwide. We then discuss less successful attempts to establish vaccine-induced immunity barriers against rabies and CDV in the domestic dog population surrounding the PA aimed at protecting wild carnivores in the SNP, and against rabies in the African wild dog population in the PA.

Rinderpest

Rinderpest probably originated in Asia and was introduced to ungulates south of the Sahara in the 1880s, possibly by the movement of infected livestock. As both cattle and wild ungulates in sub-Saharan Africa lacked natural immunity to rinderpest, the virus rapidly spread through naïve cattle and wild ungulate populations (Sinclair 1979; Plowright 1982; McNaughton 1992). Rinderpest reached Tanzania by 1890 and informal observations suggest that by 1892 it had killed up to 95% of infected cattle. As a result, pastoralist communities lost their livelihoods and suffered famine and disease (Plowright and McCullough 1967; Plowright 1982; McNaughton 1992).

In the Serengeti ecosystem, rinderpest caused precipitous population declines in many wild ungulate species, including buffalo *Syncerus caffer* and giraffe *Giraffa camelopardalis*, and also in the large migratory wildebeest population. The substantial decline in the populations of wild herbivores caused by rinderpest resulted in major changes to ecosystem dynamics (McNaughton 1992). Wild herbivores that survived exposure to rinderpest gained lifelong immunity (as is thought to be true for other viruses in the genus *Morbillivirus* such as measles in humans and CDV infection in carnivores) and thus contributed to the prevailing level of herd immunity. By 1925, less virulent strains of rinderpest had appeared that chiefly caused mortality in yearlings which unlike calves, no longer received maternally derived antibodies (Plowright 1982; McNaughton 1992).

In the early 1950s, vaccination of cattle against rinderpest began in areas surrounding the SNP, but early vaccines increased calf mortality. The development of an attenuated vaccine solved this problem and lead to the start of extensive, regular, cattle vaccination programmes in the mid-1950s (Plowright 1982). By the mid-1960s, mass vaccination of cattle had controlled rinderpest in areas surrounding the PA, and resulted in a rapid decline in rinderpest antibodies in wild ungulates, with seroprevalence in adult wildebeest falling from 100 to 0% in 6 years, and in adult buffalo from 100 to 0% in 8 years. This demonstrated that rinderpest was maintained in cattle and not in wild ungulate populations, as was originally thought (Rossiter et al. 1983, 2006; McNaughton 1992). Once rinderpest was no longer spread to wild ungulate species from cattle, their populations grew rapidly in size. For example, the wildebeest population increased from approximately 250,000 in 1961 to about 1.4 million by the late 1970s and the buffalo population increased from approximately 30,000 in 1961 to about 63,000 in 1970 (Sinclair 1979; Campbell and Borner 1995). In response to this increase in prey, predator populations also significantly increased (Hanby and Bygott 1979; Hofer and East 1995). There was one fatal outbreak of rinderpest in buffalo in the SNP in 1982, possibly caused by the illegal movement of unvaccinated cattle through the park (Rossiter et al. 1983). If, as has been suggested, global control of rinderpest was achieved by 2010, mass vaccination of cattle surrounding the PA may no longer be required.

Rabies and Canine Distemper Virus (CDV)

Domestic dogs in areas surrounding the PA are infected with CDV and rabies (Alexander and Appel 1994; Cleaveland and Dye 1995; Cleaveland et al. 2000). Both these viruses also infect wild carnivore species in the PA (Harrison et al. 2004; Lembo et al. 2008; Goller et al. 2010) and SNP (Maas 1993; Haas et al. 1996; Roelke-Parker et al. 1996; East et al. 2001; Lembo et al. 2008). It has been suggested that both rabies and CDV are maintained in the domestic dog population but not in wild carnivore populations, and that the control of both these diseases in domestic dogs by mass vaccination would result in their elimination from wild carnivore populations (e.g. Alexander and Appel 1994; Cleaveland et al. 2000, 2006, 2007; Lembo et al. 2008). This idea has recently been queried by Alexander et al. (2010).

Although sporadic, limited domestic dog vaccination programmes occurred in various locations around the PA in the early 1990s (Alexander and Appel 1994; Cleaveland and Dye 1995; Cleaveland et al. 1999; Coleman and Saoli Ole Kina 2007). More recently, mass vaccination campaigns were conducted between 1996 and 2006 (Cleaveland et al. 2003; Lembo et al. 2008, their supplementary Appendix S2; Kaare et al. 2009) during which domestic dogs were simultaneously vaccinated with 1 ml of Nobivac rabies vaccine (Intervet) and 1 ml Puppy DP (Intervet), a vaccine against CDV and canine parvovirus (Cleaveland et al. 2003; Kaare et al. 2009), with Nobivac used as a solvent for Puppy DP (Cleaveland et al. 2003). More extensive domestic dog vaccination campaigns mounted by the Tanzanian government during the rabies epidemic (see Section "Mass Vaccination Campaigns from 2002 Onwards") administered only rabies vaccine.

To create a vaccine-induced immunity barrier against disease transmission, a critical proportion of susceptible animals should be immunised (Anderson and

May 1991), which for the control of rabies is 70% (Coleman and Dye 1996), and probably similar vaccine coverage is required to control CDV (Kommonen et al. 1997). The period during which vaccine induced rabies antibody titres persist at the level recommended by WHO (≥0.5 IU/ml) is influenced by many factors, including the period elapsed since the last vaccination, the age, nutritional status, parasite burden and breed of domestic dog, and the choice of vaccine (e.g. Kitala et al. 2001; Mansfield et al. 2004; Minke et al. 2009). Vaccination schedules against canine distemper typically recommend that domestic dog pups receive two separate doses of vaccine separated by several weeks, and that adult dogs receive vaccine every 2–3 years (e.g. Kommonen et al. 1997).

Mass Dog Vaccination Campaigns Between 1996 and 2001

Between October 1996 and February 2001, four mass domestic dog vaccination campaigns were conducted in Serengeti District (Cleaveland et al. 2003), an administrative area west of the SNP (Fig. 8.1). Each animal received vaccine against both rabies and CDV (see Section "Rabies and Canine Distemper Virus (CDV)"). The total rural dog population for these campaigns was estimated to be 11,664 dogs (Cleaveland et al. 2003), a substantially lower figure than the 21,635 dogs estimated for Serengeti District (19,838 dog in rural areas plus 1,797 dogs in the district's main town) for the period 1991 and 1993 (Cleaveland and Dye 1995). The reason for the extreme difference between these two dog population estimates is unclear. Comparison of the dog:human ratio of 1:5.9 for 1991–1993 (Cleaveland and Dye 1995; Hampson et al. 2009) update link for years and 1:6.3 for 1996–2001 (Cleaveland et al. 2003) does not indicate that a large decline in the dog population occurred between these two studies.

The estimated vaccine coverage for the campaigns, based on household surveys conducted after the completion of each campaign in June 1997, August 1998, August 1999 and February 2001 were 64.5, 61.6, 70.6, 73.7%, respectively (Cleaveland et al. 2003, their Table 2). These estimates differ from those reported by Lembo et al. (2008, their Appendix S2) who stated that relatively high vaccination coverage (>65%) was attained between 1997 and 1999 (quoting Cleaveland et al. 2003), and that coverage was lower (35–40%) and more patchy from 2000 to 2003. Hampson et al. (2009) also stated that vaccine coverage in Serengeti District in 2000 was low (35–40%) and patchy and provided no information on coverage in 2001 or 2002. We interpret information in both Lembo et al. (2008) and Hampson et al. (2009) to imply that vaccine coverage achieved by the campaign between May 2000 and February 2001 was well below the high coverage reported by Cleaveland et al. (2003). The reason for the significant reduction in the estimated coverage of the 2000–2001 campaign is not explained.

Cleaveland et al. (2003) assessed the efficacy of vaccination campaigns between 1997 and 2001 using the monthly incidence of suspected cases of rabid dogs recorded by wildlife officers between November 1996 and June 1999 in 15

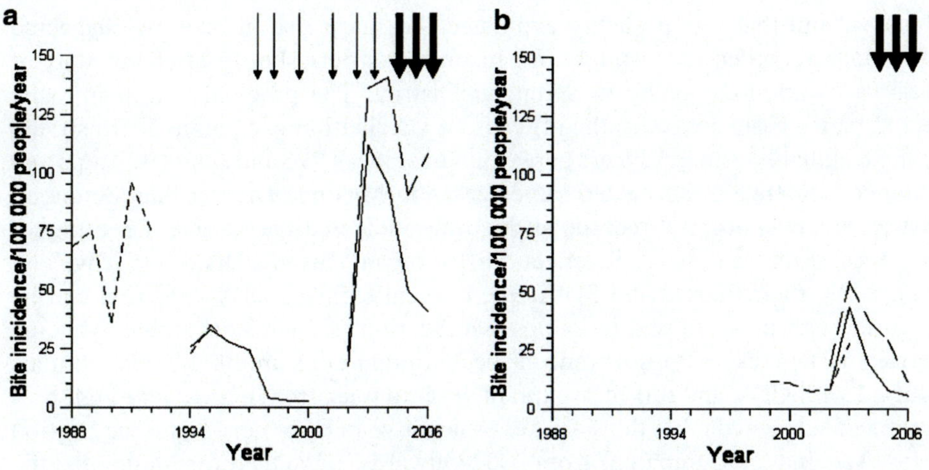

Fig. 8.3 Incidences of bites from suspected rabid domestic dogs (per 100,000 people per year) in two administrative districts, registered at local district hospitals. (**a**) Serengeti district: monthly incidence rates between 1988 and 1992 and between 1994 and 1996 (*short-dashed line*) were extracted from Lembo et al. (2008, their Fig. 4), and between 1994 and 2000 (*solid line*) from Cleaveland et al. (2003, their Fig. 3) and summed for each year. Because the data in Cleaveland et al. (2003, their Fig. 3) only start in February 1994, we added the value for January 1994 extracted from Lembo et al. (2008, their Fig. 4) to provide an annual estimate. Yearly incidence rates between 2002 and 2006 were obtained from Hampson et al. (2008, their Table 1). The upper estimate (*long-dashed line*) included all bite injuries whether or not the biting animal was a suspected rabies case. (**b**) Ngorongoro district: monthly incidence rates between 1999 and October 2003 (*dashed line*) were extracted from Lembo et al. (2008, their Fig. 4) and were tallied by year. Yearly incidence rates between 2002 and 2006 (*solid line*) were obtained from Hampson et al. (2008, their Table 1). The upper estimate (*long-dashed line*) included all bite injuries whether or not the biting animal was a suspected rabies case. *Arrows* indicate the end of vaccination campaigns when this date was known, otherwise point to the year in which they occurred. *Thick arrows* indicate extensive vaccination campaigns by the Tanzanian government

vaccination villages and ten control villages in an adjacent district (Musoma, Fig. 8.1) where no mass vaccination campaigns occurred, and bite cases by suspected rabid dogs recorded at local hospitals between September 1993 and December 2001 (both measures extrapolated to a monthly per capita incidence using human population census data; Cleaveland et al. 2003, their Figs. 2 and 3). Not all suspected cases of rabies were confirmed by laboratory tests (see Section "Does Rabies Spread from Domestic Dogs to Wild Carnivores in the SNP?"), instead reported incidences were judged as to whether or not they involved a rabid dog. It is plausible that recorders in Serengeti District would be far less likely to judge a bite-case as involving a rabid dog after mass vaccination campaigns had occurred than recorders in Musoma District, and hospitals in Musoma District would be more likely to administer post-exposure treatment for rabies to people bitten by dogs than hospitals in Serengeti District. Cleaveland et al. (2003) noted that the provision of human rabies vaccine by the project to district hospitals prompted increased administration of rabies vaccine to people bitten by dogs in

Musoma, and that this probably explained the sharp rise in bites by suspected rabid dogs recorded at hospitals in Musoma Districts shortly after the start of mass vaccination campaign in Serengeti District. The problem with using suspected rabies cases to assess the efficacy of vaccination campaigns is illustrated by the fact that livestock officers between November 1996 and June 1997 reported a higher incidence of suspected rabies cases in Musoma District than Serengeti District whereas hospital records at this time indicated the reverse, i.e. a higher incidence of rabies cases in Serengeti District than Musoma District (Cleaveland et al. 2003, their Figs. 2 and 3). As the clinical signs of CDV infection can be confused with those caused by rabies (see Section "Conclusion: Rabies Vaccine Barriers in Domestic Dogs") some suspected rabid dogs are likely to be animals infected with CDV and this may explain why samples from a relatively high proportion of suspected rabid dogs are rabies negative in laboratory tests (see Section "Mass Vaccination Campaigns from 2002 Onwards"). We therefore doubt that the prevalence of suspected rabies cases is a useful measure of the efficacy of rabies vaccination campaigns.

Mass Vaccination Campaigns from 2002 Onwards

Mass vaccination of domestic dogs in Serengeti District continued in 2002 and 2003, and reported vaccine coverage was low (35–40%) and geographically patchy (Lembo et al. 2008, their supplementary Appendix S2). In 2003, the focus of vaccination was shifted to villages within a 10 km zone adjacent to the entire western border of the SNP (Lembo et al. 2008, their supplementary Appendix S2; Hampson et al. 2009). Initial coverage between 2003 and 2005 (immediately after a campaign) ranging from 43 to 83% (Lembo et al. 2008, their supplementary Appendix S2) or between 40 and 80% (Hampson et al. 2009). Much of the western boundary of the SNP is flanked by game reserves that do not contain villages (and where domestic dogs are prohibited), thus many vaccination villages would have been more than 10 km from the SNP boundary particularly those outside the Maswa GR boundary (Fig. 8.1).

Evidence of an outbreak of epidemic rabies was evident in the domestic dog population in the Serengeti District in 2003. In an article discussing disease epidemics in wildlife (Livermore 2004), Kate Hampson is quoted as describing this epidemic in domestic dogs as "a rabies epidemic of unprecedented scale in recent Tanzanian history". Epidemic rabies persisted until 2006 along the western flank of the PA and was apparent between 2003 and 2005 on the eastern flank of the PA (Hampson et al. 2008, their Table1; Hampson et al. 2009, their Fig. 1b). Although Lembo et al. (2008) aimed to consider the long-term dynamics of rabies, their plot designed to illustrate long-term trends in suspected rabies incidence, plotted data from 1988 until the start of mass rabies vaccination campaigns in 1997, and excluded data that would have illustrated the failure of mass vaccinations between 1997 and 2002 to control rabies. To illustrate this point we have extracted data from published sources on people bitten by suspected rabid dogs (Cleaveland et al. 2003; Hampson et al. 2008; Lembo et al. 2008) and plotted these together with information on the timing of mass vaccination campaigns (Fig. 8.3a). The sharp increase in incidences in 2003 (Fig. 8.3a) no doubt reflects the increasing fear of rabies among people aware of an ongoing rabies epidemic and their increased effort to seek medical advice when bitten by a dog. Even so, more than 10% of suspected rabies exposure cases that attended medical facilities did not receive PEP because of shortages of human vaccine and some people died, despite receiving post-exposure treatment (Hampson et al. 2008).

Vaccination campaigns continued between 2003 and 2005 presumably in villages close to the western boundary of the PA and estimated coverage, immediately after vaccination campaigns, ranged between 43 and 83% (Lembo et al. 2008, their supplementary Appendix S2). Between 2004 and 2006 the Tanzanian government also conducted mass anti-rabies vaccination campaigns of domestic dogs in more extensive areas both east and west of the PA (Lembo et al. 2008, their supplementary Appendix S2) in an attempt to control the rabies epidemic (Fig. 8.3a, b).

Information on the extent of rabies vaccination campaigns along the eastern flanks of the SNP is limited. Before the start of the 2003 rabies epidemic, only small-scale, localised, campaigns occurred in one urban area east of the SNP (Lembo et al. 2008, their Appendix S2). Annual mass vaccination campaigns started in 2004 along the eastern flank of the SNP (Hampson et al. 2009), and vaccination coverage between 2004 and 2006 either exceeded 80% (Lembo et al. 2008, their Appendix S2), or varied between 19.2% (in villages with central point dog vaccinations) and 86.3% (in villages where central point vaccinations and community based strategies were combined, Kaare et al. 2009, their Table 2). Lembo et al. (2010, their Fig. 4) indicated a mass vaccination campaign east of the SNP that ended in the third quarter of 2003, which from previous published accounts appears to be an error. If so, the first extensive mass vaccination campaign east of the SNP ended in the third quarter of 2004, more than a year after the maximum peak in incidences of animal bite injuries by suspected rabid animals which occurred in the second quarter of 2003 (Lembo et al. 2010, their Fig. 4).

Before the start of mass rabies vaccination campaigns, the proportion of samples from domestic dogs showing neurological signs consistent with rabies that tested positive for rabies was 46.7% for 15 samples collected between 1991 and 1995 (Cleaveland and Dye 1995, their Table 2), and 48% of 29 samples collected between 1992 and 1996 (Kaare and Cleaveland 2007). Between 1997 and 2000, surprisingly 70.4% of 27 samples collected from suspected rabid domestic dogs were positive for rabies by laboratory diagnosis and no difference was found in the proportion of suspected dogs confirmed rabies-positive in the Serengeti District compared to non-vaccination districts (Cleaveland et al. 2003). These data do not suggest a significant decline in rabies in Serengeti District during the four vaccination campaigns (1996–2001), and taken together, data between 1991 and 2000 demonstrate that only 58% of suspected rabies cases were rabies positive by laboratory diagnosis. Between 2002 and 2006, a period including 4 years of epidemic rabies in domestic dogs (Hampson et al. 2008, their Table 1), 74.3% of 74 samples collected from suspected rabid animals were found positive.

Does Rabies Spread from Domestic Dogs to Wild Carnivores in the SNP?

Recently it was implied (Cleaveland et al. 2007) that mass vaccination of domestic dogs had controlled rabies in the domestic dog population outside the SNP and prevented the spread of rabies from domestic dogs to wildlife inside the SNP. It was also stated that the African wild dog population ("in the area") had increased. Does available information support these statements? The incidence of samples from wild carnivores inside the SNP that tested positive for rabies was certainly low before the start of mass domestic dog vaccination campaigns outside the SNP in 1997 (Lembo et al. 2008, their Fig. 3b). Most laboratory positive cases in wildlife in the SNP occurred in 1998 and 1999, and despite around 65 samples being tested between 2003 and 2006 only one laboratory rabies case was confirmed (Lembo et al. 2008, their Fig. 3b), during a period of epidemic rabies in the domestic dog population that Cleaveland et al. (2007) appear to have overlooked. Thus there is no evidence that epidemic rabies in the domestic dog population surrounding the PA spread into wild carnivores inside the SNP. There is, however, evidence that epidemic rabies spread from domestic dogs to wild carnivores in areas occupied by humans and their domestic dogs (Lembo et al. 2008, their Fig. 5).

We can find no evidence to support the claim by Cleaveland et al. (2007) that the size of the African wild dog population outside the SNP has increased as a result of mass rabies vaccination campaigns in domestic dogs. Since the extinction of African wild dogs inside the SNP in 1991 (Burrows 1992), this endangered canid has continued to persist in the area surrounding the SNP (Burrows et al. 1994), mostly in the LGCA and NCA to the east and south of the SNP (Fig. 8.1). As previously mentioned, mass rabies vaccinations of domestic dogs in the LGCA and NCA did not start until 2004, well after the start of a rabies epidemic in domestic dogs that persisted in these areas until 2005 (Fig. 8.3b). Thus, available information strongly suggests that African wild dogs in packs living along the eastern flanks of the SNP persisted *despite* epidemic rabies in sympatric domestic dogs, not because rabies in the domestic dog population was controlled.

Conclusion: Rabies Vaccine Barriers in Domestic Dogs

The establishment of an effective vaccine immunity barrier against the spread of rabies in domestic dogs surrounding the PA would decrease the threat of rabies to people, livestock and wildlife in rural and urban areas outside the PA. Available evidence does not lend strong support to the notion that rabies infection in domestic dogs spreads to wild carnivores inside the SNP. Why mass vaccinations of domestic dogs in the Serengeti Districts failed to prevent epidemic rabies is open to debate and the increased loss of human lives to rabies during the epidemic is highly regrettable. Achieving and maintaining sufficient vaccine coverage in domestic dog populations in rural Africa is an expensive and substantial challenge. Problems include catching dogs that are rarely if ever restrained, the decline in vaccination coverage caused by low survival and high recruitment of young unvaccinated pups, necessitating vaccination campaigns at 6 month intervals to maintain recommended coverage, ineffective vaccine induced immune response in poorly nourished dogs with high parasite burdens (a problem likely to be increased by the administration of multiple vaccines in one shot) and the existence of feral dogs (Kitala et al. 2001; Coleman and Saoli Ole Kina 2007; Kaare et al. 2009).

Barriers of Immunity Against Canine Distemper Virus

In 1993/1994, a CDV epidemic infected several wild carnivore species in the SNP and MMNR (Haas et al. 1996; Roelke-Parker et al. 1996; Carpenter et al. 1998; Kock et al. 1998; Harrison et al. 2004). In response to this outbreak, a project (Project LifeLion) was launched in Tanzania (Pain 1995) to vaccinate domestic dogs in Serengeti District against CDV, as it was proposed that CDV infection had spread from domestic dogs outside the PA to wild carnivores inside the SNP (Roelke-Parker et al. 1996). Vaccination of domestic dogs against CDV was thus included in the mass vaccination campaigns against rabies conducted between 1996 and 2001 (Cleaveland et al. 2003, see Section "Rabies and Canine Distemper Virus (CDV)"). Whether or not all the rabies vaccination campaigns between 2002 and 2006 also administered vaccines against CDV is not mentioned by Lembo et al. (2008) or Hampson et al. (2009). Kaare et al. (2009) report that both Nobivac and Puppy DP were administered to domestic dogs during their mass vaccinations in areas east and west of the SNP, but the years in which mass vaccinations occurred were not mentioned.

Neurological signs of CDV infection in domestic dogs can be confused with those of rabies infection. Thus, some samples from domestic dogs that exhibited neurological signs consistent with rabies that produced negative laboratory results for rabies infection (see Section "Mass Dog Vaccination Campaigns Between 1996 and 2001") may have come from CDV infected animals. Seroprevalence to CDV exposure in the domestic dog population in Serengeti District between 1992 and 1994 was high (Cleaveland et al. 2000) and this suggests that many people reporting to hospitals with bites from suspected rabid domestic dogs between 1988 and 1993 (Fig. 8.2a) may have mistaken the neurological signs of CDV for rabies.

Initially it was proposed that the 1993/1994 CDV epidemic in the SNP was caused by a new virulent biotype (Roelke-Parker et al. 1996). This idea has been questioned by evidence for several "silent" CDV epidemics that caused significant increases in seroprevalence without leading to clinical signs of disease in spotted hyenas in the MMNR between 2000 and 2001 (Harrison et al. 2004) and lions in the SNP in 1976, 1982, 1999 and 2007 (Munson et al. 2008). It is plausible that the pathological impact of CDV infections may be increased by factors such as concurrent infections with other pathogens. This has been proposed for lions co-infected with CDV and *Babesia* sp. (Munson et al. 2008), and for members of one pack of African wild dogs that were co-infected with CDV and several other pathogens, including parvovirus (Goller et al. 2010). This pack died of severe CDV infection in 2007 just outside the north-eastern boundary of the SNP. Interestingly, wild carnivore hosts inside the SNP that would be expected to be susceptible to CDV, such as young spotted hyenas (Haas et al. 1996) and lions (Roelke-Parker et al. 1996) did not display clinical signs of disease in that year (own unpublished observations), even though serological evidence indicated a "silent" wave of CDV infection in the SNP in 2007 (Munson et al. 2008).

Roelke-Parker et al. (1996) hypothesised that CDV-infected domestic dogs in Serengeti District transmitted the virus to spotted hyenas during their long-distance foraging trips and that the virus was then carried to the centre of the SNP. This is an unlikely scenario as commuting hyenas that forage in any given area will return to many different clan territories scattered throughout the SNP (Hofer and East 1993b) thereby transmitting CDV to many different locations in the SNP. Furthermore, CDV infected domestic dogs close to the western boundary of the PA would be expected to spread infection to resident wild carnivores in their area and CDV would then spread into the SNP across the park boundary. We know of no evidence to indicate that this happened. The scenario that CDV spreads from domestic dogs to wildlife inside protected areas is also contradicted by serum neutralising CDV antibody titres in spotted hyenas in the MMNR (Harrison et al. 2004), because animals in a territory that encompassed pastoralist settlements with domestic dogs on the edge of the reserve contained significantly fewer canine distemper antibody-positive hyenas than spotted hyenas from further inside the reserve and at a greater distance from contact with domestic dogs.

Conclusion: Canine Distemper Virus Vaccine Barriers in Domestic Dogs

The passage of "silent" canine distemper epidemics through the SNP lion population in 1999 and 2007 is not consistent with the suggestion (Cleaveland et al. 2006) that mass vaccination of domestic dogs against CDV conducted in Serengeti District since 1996 prevented epidemic waves of CDV infection in wild carnivores inside the SNP.

Vaccine Barriers Against Rabies in African Wild Dogs

The vaccination of members of a wildlife species thought to be at risk from a specific pathogen is similar in concept to the use of fences or cages around endangered plants (Hayward and Kerley 2009). One widely discussed, yet ultimately inconclusive example of a vaccination programme on an endangered species is provided by the sudden demise of all African wild dog packs studied in both the Kenyan and Tanzanian sectors of the Serengeti ecosystem between 1989 and 1991 (Burrows 1992, 1994; Burrows et al. 1994). Central to the debate was the possible negative effect that vaccination against rabies may have had on post-vaccination survival of vaccinated packs (Burrows 1992, 1994; Macdonald et al. 1992; Burrows et al. 1994, 1995; Ginsberg et al. 1995; East 1996; East and Burrows 2001; Woodroffe 2001).

African wild dogs vaccinated in the Tanzanian sector of the Serengeti ecosystem survived for significantly shorter periods after vaccination than those that were anaesthetised and radio-collared but not vaccinated (Burrows et al. 1994). The effect of vaccination on African wild dog packs studied in the Kenyan sector of the ecosystem has not been rigorously examined because Ginsberg et al. (1995) excluded rabies vaccination as a form of intervention in their analysis of survivorship (East 1996), even though it was evident that a minimum of 27 individuals were vaccinated between 1987 and 1991 (East and Burrows 2001; Woodroffe 2001). Information on the number and mode of vaccinations carried out on African wild dogs in Kenya is still unavailable (East and Burrows 2001; Woodroffe 2001). Rabies was diagnosed in all three packs from which tissue samples were submitted for laboratory diagnosis after pack deaths between 1989 and 1991 (Burrows et al. 1994; East and Burrows 2001; Woodroffe 2001). Thus, despite extensive vaccination against rabies, available evidence indicates that pack deaths between 1989 and 1991 were associated with rabies infection. Why did vaccination against rabies fail to create a barrier against rabies infection in African wild dogs inside the PA?

The rationale to vaccinate African wild dogs in the Serengeti ecosystem was based on several key assumptions. Most crucially, it was assumed that one standard dose of inactivated rabies vaccine, developed for domestic dogs, would induce a sufficient immune response to provide inoculated African wild dogs with protection against rabies infection. This assumption has been shown to be false and it is now recognised that African wild dogs require a course of multiple doses of inactivated domestic dog rabies vaccine to stimulate antibody titre levels likely to provide protection (East and Burrows 2001; Visee 2001; Woodroffe 2001). Furthermore, the time period during which vaccine-induced antibody protection was likely to persist in African wild dogs was also unknown when African wild dogs in the Serengeti ecosystem were vaccinated either by hand injection following anaesthesia, or by bounce-darting which entailed a plastic dart containing a dose of vaccine being fired at an animal from a dart-gun. Use of this last method assumed that there would be no dart malfunction caused by the failure of darts to inject an entire dose and that animals would not pull out darts before the delivery of the entire dose (Stearns and Stearns 1999). Thirdly, it was assumed that vaccinated African wild dogs lacked natural immunity to rabies, whereas serological data provided evidence for a degree of natural immunity, given that 5 of 12 animals serum-sampled in Tanzania before vaccination had rabies-specific antibody titres above the international standard (0.5 IU/ml) considered sufficient to provide protection against rabies infection (Gascoyne et al. 1993). Three of these five seropositive animals survived for 3 years, 2 years and 10 months, respectively, after serum sampling, suggesting that they survived rabies infection (Burrows 1992; Burrows et al. 1994).

Vaccination of African wild dogs in the Serengeti ecosystem drew criticism because there was a lack of adequate scientific design, such as the inclusion of unvaccinated packs to act as controls (Heinsohn 1992; Burrows et al. 1994). Post-treatment monitoring was insufficient, particularly of radio-collared individuals in Tanzania, and as a result no tissue samples for laboratory analysis were obtained to determine the cause of death of vaccinated Tanzanian packs (Heinsohn 1992; Burrows et al. 1994). High mortality among vaccinated animals in Tanzania was initially substantially under-reported (cf. comment by Burrows 1994 in response to Gascoyne et al. 1993), and vaccination by bounce-darting was an untested procedure that may have been physiologically stressful as darts were fired at each member of a pack in succession on the same occasion, which inevitably resulted in several hours of continual disturbance of larger packs (Burrows 1994; Stearns and Stearns 1999).

Conclusion: Vaccine Barriers Against Rabies in African Wild Dogs

Despite the loss of all vaccinated African wild dog packs, unvaccinated, dispersing animals and breeding packs have persisted since 1991 outside the SNP, particularly in the south and east of the ecosystem in areas inhabited by pastoralists and their domestic dogs (Burrows et al. 1994; Goller et al. 2010). The persistence of these packs, despite the occurrence of a substantial rabies epidemic between 2003 and 2006, suggests that unvaccinated African wild dog populations are not necessarily adversely affected by rabies epidemics in sympatric domestic dogs.

Conclusion

Non-physical barriers around the protected areas within the Serengeti ecosystem can be efficient in protecting natural resources, as demonstrated by the lack of illegal grazing inside protected areas because of park ranger patrols, and the absence of rinderpest infection in wild herbivores following an effective shield established by a highly efficacious livestock vaccination programme in northern Tanzania. The distance bushmeat hunters have to travel to hunting areas inside the protected areas limits their ability to utilise wild ungulate populations deep inside the SNP, because buffer zones around the edges of the park significantly increase travel costs for hunters. This situation might change should hunters use vehicles to access hunting areas. Mass vaccinations of domestic dogs against rabies did not prevent cycles of epidemic rabies outside the SNP and the PA. Furthermore, there is currently no evidence that epidemic rabies in rural domestic dog populations spreads to wild carnivores inside the SNP. Mass vaccination of domestic dogs against CDV in rural areas surrounding the SNP did not prevent the passage of "silent epidemics" of CDV infection in wild carnivores inside protected areas, suggesting either that CDV epidemics are not linked to infections in rural domestic dogs or that vaccination campaigns were ineffective in preventing CDV outbreak in domestic dogs which then spread to carnivores in the SNP. The death from CDV of one pack of African wild dogs in 2007 questions the efficacy of mass vaccination campaigns of domestic dogs to prevent fatal CDV in wild carnivores.

References

Alexander, K.A. & Appel, M.J.G. (1994) African wild dogs (*Lycaon pictus*) endangered by canine distemper epizootic among domestic dogs near the Masai Mara National Reserve, Kenya. *Journal of Wildlife Diseases*, **30**, 481–485.

Alexander, K.A., McNutt, J.W., Briggs, M.B., Standers, P.E., Fuston, P., Hemson, G., Keet, D. & van Vuuren, M. (2010) Multi-host pathogens and carnivore management in southern Africa. *Comparative Immunology, Microbiology and Infectious Diseases*, **33**, 249–265.

Anderson, R.M. & May, R.M. (1991) *Infectious diseases of humans: dynamics and control*. Oxford University Press, Oxford.

Arcese, P., Hando, J. & Campbell, K.L.I. (1995) Historical and present-day anti-poaching in Serengeti. *Serengeti II: Dynamics, management and conservation of an ecosystem* (eds. A.R.E. Sinclair & P. Arcese), pp. 506–533. Chicago University Press, Chicago.

Burrows, R. (1992) Rabies in wild dogs. *Nature*, **359**, 277.

Burrows, R. (1994) Rabies in African wild dogs of Tanzania. *Journal of Wildlife Diseases*, **30**, 297–299.

Burrows, R., Hofer, H. & East, M.L. (1994) Demography, extinction and intervention in a small population: the case of the Serengeti wild dog. *Proceedings of the Royal Society London, Series B*, **256**, 281–292.

Burrows, R., Hofer, H. & East, M.L. (1995) Population dynamics, intervention and survival in African wild dogs (*Lycaon pictus*). *Proceedings of the Royal Society London, Series B*, **262**, 235–245.

Campbell, K.L.I. & Borner, M. (1995) Population trends and distribution of Serengeti Herbivores: implications for management. *Serengeti II: Dynamics, management and conservation of an ecosystem* (eds. A.R.E. Sinclair & P. Arcese), pp 117–145. University of Chicago Press, Chicago.

Campbell, K.L.I. & Hofer, H. (1995) People and wildlife: spatial dynamics and zones of interaction. *Serengeti II: Dynamics, management and conservation of an ecosystem* (eds. A.R.E. Sinclair & P. Arcese), pp 534–570. University of Chicago Press, Chicago.

Carpenter, J.W., Appel, M.J.G., Erickson, R.C. & Novilla, K.N. (1976) Fatal vaccine induced canine distemper virus infection in black-footed ferrets. *Journal of the American Veterinary Medical Association*, **169**, 961–964.

Carpenter, M.A., Appel, M.J.G., Roelke-Parker, M.E., Munson, L., Hofer, H., East, M.L. & O'Brien, S.J. (1998) Genetic characterization of canine distemper virus in Serengeti carnivores. *Veterinary Immunology Immunopathology*, **65**, 259–266.

Chetkiewicz, C.L., St Clair, C.C. & Boyd, M.S. (2006) Corridors for conservation: integrating pattern and process. *Annual Review of Ecology, Evolution and Systematics*, **37**, 317–342.

Cleaveland, S. & Dye, C. (1995) Maintenance of a microparasite infecting several host species: rabies in the Serengeti. *Parasitology*, **111**, S33–S47.

Cleaveland, S., Barrat, J., Barrat, M.-J., Selve, M., Kaare, M. & Esterhuysen, J. (1999) A rabies serosurvey of domestic dogs in rural Tanzania: results of a rapid fluorescent focus inhibition test (RFFIT) and a liquid-phase blocking ELISA used in parallel. *Epidemiology and Infections*, **123**, 157–164.

Cleaveland, S., Appel, M.G.J., Chalmers, W.S.K., Chillingworth, C., Kaare, M. & Dye, C. (2000) Serological and demographic evidence for domestic dogs as a source of canine distemper virus infection for Serengeti wildlife. *Veterinary Microbiology*, **72**, 217–227.

Cleaveland, S., Kaare, M., Tiringa, P., Mlengeya, T. & Barrat, J. (2003) A dog rabies vaccination campaign in rural Africa: impact on the incidence of dog rabies and human dog-bite injuries. *Vaccine*, **21**, 1965–1973.

Cleaveland, S., Mlengeya, T., Kaare, M., Haydon, D., Lembo, T., Laurenson, M.K. & Packer, C. (2006) The conservation relevance of epidemiological research into carnivore viral diseases in the Serengeti. *Conservation Biology*, **21**, 612–622.

Cleaveland, S., Hampson, K. & Kaare, K. (2007) Living with rabies in Africa. *Veterinary Record*, **161**, 293–294.

Coleman, P.G. & Dye, C. (1996) Immunization coverage required to prevent outbreaks of dog rabies. *Vaccine*, **14**, 185–185.

Coleman, P.G. & Saoli Ole Kina, J. (2007) Domestic dog vaccination strategies around the Masai Mara National Reserve, Kenya. *Proceedings of the Southern and Eastern African rabies group meeting, Nairobi Kenya* (eds. P. Kitata, B. Perry, J. Barrat & A. King), pp 111–125. Éditions Fondation Méreux, France.

East, M.L. (1996) Survivorship in African wild dogs. *Conservation Biology*, **10**, 313–315.

East, M.L. & Burrows, R. (2001) The efficacy of inactivated rabies vaccine for the protection of African wild dogs (*Lycaon pictus*) against viral challenge: a review. *Verhandlungsberichte Erkrankungen der Zootiere*, **40**, 233–242.

East, M.L., Hofer, H., Cox, J H., Wulle, U., Wiik, H. & Pitra, C. (2001) Regular exposure to rabies virus and lack of symptomatic disease in Serengeti spotted hyenas. *Proceedings of the National Academy of Sciences USA*, **98**, 15026–15031.

Ferber, D. (2000) Human diseases threaten great apes. *Science*, **289**, 1277–1278.

Gascoyne, S.C., Laurenson, M.K., Lelo, S. & Borner, M. (1993) Rabies in African wild dogs *Lycaon pictus*, in the Serengeti regions, Tanzania. *Journal of Wildlife Diseases*, **29**, 396–402.

Ginsberg, J.S., Alexander, K.A., McNutt, W. & Mills, M.G.L. (1995) Handling and survivorship in the wild dog (*Lycaon pictus*), in five African ecosystems. *Conservation Biology*, **9**, 665–674.

Goller, K.V., Fyumagwa, R.D., Nikolin, V., East, M.L., Kilewo, M., Speck, S., Müller, T., Matzke, M. & Wibbelt, G. (2010) Fatal canine distemper infection in a pack of African wild dogs in the Serengeti ecosystem, Tanzania. *Veterinary Microbiology*, **146,** 245–252.

Haas, L., Hofer, H., East, M., Wohlsein, P., Leiss, P. & Barrett, T. (1996) Canine distemper virus infection in Serengeti spotted hyaenas. *Veterinary Microbiology*, **49**, 147–152.

Hampson, K., Dobson, A., Kaare, M., Dushoff, J., Magoto, M., Sindoya, E. & Cleaveland, S. (2008) Rabies exposures, post-exposure prophylaxis and deaths in a region of endemic canine rabies. *PLoS Neglected Tropical Diseases*, **2**, e339.

Hampson, K., Dushoff, J., Cleaveland, S., Haydon, D.T., Kaare, M., Packer, C. & Dobson, A. (2009) Transmission dynamics and prospects for the elimination of canine rabies. *PLoS Biology* **7**, e1000053.

Hanby, J.P. & Bygott, J.D. (1979) Population changes in lions and other predators. *Serengeti, Dynamics of an Ecosystem* (eds. A.R.E. Sinclair & M. Norton-Griffiths), pp 249–262. University of Chicago Press, Chicago.

Harrison, T.M., Mazet, J.K., Holekamp, K.E., Dubovi, E., Engh, A.L., Nelson, K., Van Horn, R.C. & Munson, L. (2004) Antibodies to canine and feline viruses in spotted hyenas (*Crocuta crocuta*) in the Masai Mara National Reserve. *Journal of Wildlife Diseases*, **40**, 1–10.

Hayward, M.W. & Kerley, G.L.H. (2009) Fencing for conservation: Restriction of evolutionary potential or a riposte to threatening processes. *Biological Conservation*, **142**, 1–13.
Heinsohn, R. (1992) When conservation goes to the dogs. *Trends in Ecology and Evolution*, **7**, 214–215.
Höner, O.P., Wachter, B., East, M.L., Runyoro, V.A. & Hofer, H. (2005) The effect of prey abundance and foraging tactics on the population dynamics of a social, territorial carnivore, the spotted hyena. *Oikos*, **108**, 544–554.
Hofer, H. & East, M.L. (1993a) The commuting system of Serengeti spotted hyaenas: how a predator copes with migratory prey. I. Social organization. *Animal Behaviour*, **46**, 547–557.
Hofer, H. & East, M.L. (1993b) The commuting system of Serengeti spotted hyaenas: how a predator copes with migratory prey. II. Intrusion pressure and commuters' space use. *Animal Behaviour*, **46**, 559–574.
Hofer, H. & East, M.L. (1993c) The commuting system of Serengeti spotted hyaenas: how a predator copes with migratory prey. III. Attendance and maternal care. *Animal Behaviour* **46**, 575–589.
Hofer, H. & East, M.L. (1995) Population dynamics, population size, and the commuting system of Serengeti spotted hyaenas. *Serengeti II: Dynamics, management and conservation of an ecosystem* (eds. A.R.E. Sinclair & P. Arcese), pp 332–363. University of Chicago Press, Chicago.
Hofer, H., East, M.L. & Campbell, K.L.I. (1993) Snares, commuting hyaenas and migratory herbivores: humans as predators in the Serengeti. *Symposia of the Zoological Society London*, **65**, 347–366.
Hofer, H., Campbell, K.L.I., East, M.L. & Huish, S.A. (1996) The impact of game meat hunting on target and non-target species in the Serengeti. *The exploitation of mammal populations* (eds. V.J. Taylor & N. Dunstone), pp 117–146. Chapman and Hall, London.
Hofer, H., Campbell, K L.I., East, M. L. & Huish, S.A. (2000) Modeling the spatial distribution of the economic costs and benefits of illegal game meat hunting in the Serengeti. *Natural Resource Modeling*, **13**, 151–177.
Hofer, H., Hildebrandt, T.B., Göritz, F., East, M.L., Mpanduji, D.G., Hahn, R., Siege, L. & Baldus, R.D. (2004) *Distribution and movements of elephants and other wildlife in the Selous-Niassa wildlife corridor, Tanzania*. Deutsche Gesellschaft für Technische Zusammenarbeit (GTZ) GmbH, Eschborn.
Holmern, T., Nyahongo, J. & Røskaft, E. (2007) Livestock loss caused by predators outside the Serengeti National Park, Tanzania. *Biological Conservation*, **135**, 534–542.
Kaare, M. & Cleaveland, S. (2007) Targeting rabies control in the Serengeti: The rational and design of a domestic dog vaccination campaign *Proceedings of the Southern and Eastern African rabies group meeting, Nairobi Kenya* (eds. P. Kitata, B. Perry, J. Barrat & A. King), pp 126–132. Éditions Fondation Méreux, France.
Kaare, M., Lembo, T., Ernest, E., Estes, A., Mentzel, C. & Cleaveland, S. (2009) Rabies control in rural Africa: Evaluating strategies for effective domestic dog vaccination. *Vaccine*, **27**, 152–160.
Kissui, B.M. (2008) Livestock predation by lions, leopards, spotted hyenas and their vulnerability to retaliatory killing in the Maasai steppe, Tanzania. *Animal Conservation*, **11**, 422–432.
Kitala, P., McDermott, J., Kyule, M., Gathuma, J., Perry, B. & Wandeler, A. (2001) Dog ecology and demography information to support the planning of rabies control in Machakos District, Kenya. *Acta Tropica*, **78**, 217–230.
Kock, R., Chalmers, W.S., Mwanzia, J., Chillingworth, C., Wambua, J., Coleman, P.G. & Baxendal, W. (1998). Canine distemper antibodies in lions of the Masai Mara. *Veterinary Record*, **142**, 662–665.
Kommonen, C.Ek., Sihvonen, K., Pekkanen, U., Rikula, U. & Nuotio, L. (1997) Outbreak of canine distemper in vaccinated dogs in Finland. *Veterinary Record*, **141**, 380–383.
Köndgen, S., Kühl, H., N'Goran, P.K., Walsh, P.D., Schenk, S., Ernst, N., Biek, R., Formenty, P., Mätz-Rensing, K., Schweiger, B., Junglen, S., Ellerbrok, H., Nitsche, A., Briese, T., Lipkin, I.H., Pauli, G., Boesch, C. & Leendertz, F.H. (2008) Pandemic human viruses cause decline of endangered Great Apes. *Current Biology*, **18**, 1–5.
Lembo, T., Hampson K., Haydon, D.T., Craft M., Dobson, A., Dushoff, J., Ernest, E., Hoare, R., Kaare, M., Mlengeya, T., Mentzel, C. & Cleaveland, S. (2008) Exploring reservoir dynamics:

a case study of rabies in the Serengeti ecosystem. *Journal of Applied Ecology*, **45**, 1246–1257.

Lembo, T., Hampson, K., Kaare, M.T., Eblate, E., Knobel, D., Kazwala, R.R., Haydon, D.T. & Cleaveland, S. (2010) The feasibility of canine rabies elimination in Africa: Dispelling doubts with data. *PLoS Neglected Tropical Diseases*, **4**, e626.

Ling, S. & Milner-Gulland, E.J. (2008) When does spatial structure matter in models of wildlife harvesting? *Journal of Applied Ecology* **45**, 63–71.

Livermore, B (2004) Crunching numbers, outsmarting disease. Scientists study wildlife epidemics, and safeguard human health. *Princeton Alumni*, http://www.princeton.edu/paw/archive_new/PAW03-04/15-0609/features1.html.

Loibooki, M.T. (1997) People and poaching: the interactions between people and wildlife in and around the Serengeti National Park, Tanzania. M.Sc. Thesis, Reading University, UK.

Loibooki, M.T., Hofer, H., Campbell, K.L.I. & East, M.L. (2002) Bushmeat hunting by communities adjacent to the Serengeti National Park, Tanzania: the importance of livestock ownership and alternative sources of protein and income. *Environmental Conservation*, **29**, 391–398.

Maas, B. (1993) Bat-eared fox behavioural ecology and the incidence of rabies in the Serengeti National Park. *Onderstepoort Journal of Veterinary Research*, **60**, 389–393.

Macdonald, D.W., Artois, M., Aubert, M., Bishop, D.L., Ginsberg, J. R., King, A., Kock, N. & Perry, B.D. (1992) Causes of wild dog deaths. *Nature*, **360**, 633–634.

Mansfield, K.L., Burr, P.D., Snodgrass, D.R., Sayer, R., & Fooks, A.R. (2004) Factors affecting the serological response of dogs and cats to rabies vaccination. *Veterinary Microbiology*, **154,** 423–426.

McCormack, A.E. (1983) Canine distemper in African hunting dogs (Lycaon pictus) – possibly vaccine induced. *Journal of Zoo Animal Medicine*, **14**, 66–71.

McDonald, R.A., Delahay, R.J., Carter, S.P., Smith, G.C. & Cheeseman, C.L. (2008). Perturbing implications of wildlife ecology for disease control. *Trends in Ecology and Evolution*, **23**, 53–56.

McNaughton, S.J. (1990) Mineral nutrition and seasonal movements of African migratory ungulates. *Nature*, **345**, 613–615.

McNaughton, S.J. (1992) The propagation of disturbance in savannas through food webs. *Journal of Vegetation Science*, **3**, 301–314.

McNaughton, S.J. & Banyikwa, F.F. (1995) Plant communities and herbivory. *Serengeti II: Dynamics, management and conservation of an ecosystem* (eds. A.R.E. Sinclair & P. Arcese), pp 49–70. University of Chicago Press, Chicago.

Mduma, S.A.R., Hilborn, R. & Sinclair, A.R.E. (1998) Limits to exploitation of Serengeti wildebeest and implications for its management. *Dynamics of tropical communities* (eds. D.M. Newbury, H.H.T. Prins & N. Brown), pp 243–265. Blackwell Science, Oxford.

Mills M.G.L. & Hofer, H. (1998) *Hyaenas, status survey and action plan*. World Conservation Union, Gland.

Minke, J.M., Bouvet, J., Cliquet, F., Wasniewski, M., Guiot, A.L., Lemaitre, L., Cariou, C., Cozette, V., Vergne, L. & Guigal, P.M. (2009) Comparison of antibody responses after vaccination with two inactivated rabies vaccines. *Veterinary Microbiology*, **133**, 283–286.

Munson, L., Terio, K.A., Kock, R., Mlengeya, T., Roelke, M.E., Dubovi, E., Summers, B., Sinclair, A.R.E. & Packer, C. (2008) Climate extremes promote fatal co-infections in African lions. *PLoS One*, **6**, e2545.

Newmark, W.D. (1996) Insularization of Tanzanian parks and the local extinction of large mammals. *Conservation Biology*, **10**, 1549–1556.

Nielsen, M.R. (2006) Importance, cause and effect of bushmeat hunting in the Udzungwa Mountains, Tanzania: Implications for community based wildlife management. *Biological Conservation*, **128**, 509–516.

Nowell, K. & Jackson, P. (1996) *Wild cats, status survey and conservation action plan*. World Conservation Union, Gland, Switzerland.

Nyahongo, J.W. (2004) Impact of human activities on the carnivore populations in the western corridor of the Serengeti ecosystem. MSc. Thesis, University of Dar es Salaam, Tanzania.

Nyahongo, J.W., East, M.L., Mturi, F.A. & Hofer, H. (2005) Benefits and costs of illegal grazing and hunting in the Serengeti ecosystem. *Environmental Conservation*, **32**, 326–332.

Ogada, M.O., Woodroffe, R. Oguge, N.O. & Frank, L.G. (2003) Limited depredation by African carnivores: the role of livestock husbandry. *Conservation Biology*, **17**, 1521–1530.

Pain, S. (1995) Jabs for dogs may save Serengeti lions. *New Scientist*, **1993**, 10.

Plowright, W. (1982) The effect of rinderpest and rinderpest control on wildlife in Africa. *Symposia of the Zoological Society London*, **50**, 1–28.

Plowright, W. & McCullough, B. (1967) Investigations on the incidence of rinderpest virus infection in game animals of N. Tanganika and S. Kenya 1960/63. *Journal of Hygiene*, **65**, 343–358.

Puth, L.M. & Wilson, K.A. (2001) Boundaries and corridors as a continuum of ecological flow control: Lessons from rivers and streams. *Conservation Biology*, **15**, 21–30.

Rackham, O. (2003) *The Illustrated History of the Countryside*. Weidenfeld & Nicolson: London.

Roelke-Parker, M.E., Munson, L., Packer, C., Kock, R., Cleaveland, S., Carpenter, M., O'Brien SJ, Pospischil, A., Hofmann-Lehmann, R., Lutz, H., Mwamengele, G.L.M., Mgasa, M.N., Machange, G.A., Summers, B.A. & Appel, M.J. (1996) A canine distemper virus epidemic in Serengeti lions (*Panthera leo*). *Nature*, **379**, 441–445.

Rossiter, P.B., Karstad, L., Jessett, E.M., Yamamoto, T., Dardiri, A.H. & Mushi, E.Z. (1983) Neutralizing antibodies to rinderpest virus in wild animal sera collected in Kenya between 1970 and 1981. *Preventative Veterinary Medicine*, **1**, 257–264.

Rossiter, P., Wamwayi, H. & Ndungu, E. (2006) Rinderpest seroprevalence in wildlife in Kenya and Tanzania 1982–1993. *Preventative Veterinary Medicine*, **75**, 1–7.

Schaller, G.B. (1972) *The Serengeti lion: A study of predator-prey relations*. Chicago University Press, Chicago.

Sinclair, A.R.E. (1979) The Serengeti environment. *Serengeti, Dynamics of an Ecosystem* (eds. A.R.E. Sinclair & M. Norton-Griffiths), pp 31–45. University of Chicago Press, Chicago.

Sinclair, A.R.E. (1995) Serengeti past and present. *Serengeti II: Dynamics, management, and conservation of an ecosystem* (eds. A.R.E. Sinclair & Arcese, P.), pp. 3–30. University of Chicago Press, Chicago.

Smith, J.L.D., Ahern, S.C. & McDougal, C. (1998) Landscape analysis of tiger distribution and habitat quality in Nepal. *Conservation Biology*, **12**, 1338–1346.

Spinage, C.A. (1992) The decline of the Kalahari wildebeest. *Oryx*, **26**, 147–150.

Stearns, B.T. & Stearns, S.G. (1999) Handle with care: A wild dog story. *Watching from the edge of extinction* pp. 147–187. Yale University Press, Yale.

Treves, A. & Karanth, K.U. (2003) Human-carnivore conflict and perspectives on carnivore management worldwide. *Conservation Biology*, **17**, 1491–1499.

Turner, M. (1987) *My Serengeti years*. Elm Tree Books, London.

Visee, A.M. (2001) Distemper, rabies and parvovirus vaccination in a captive breeding programme for the African wild dog (*Lycaon pictus*) in Northern Tanzania. *Verhandlungsberichte Erkrankungen der Zootiere*, **40**, 243–250.

Wallis, J. & Lee, D.R. (1999) Primate conservation: the prevention of disease transmission. *International Journal of Primatology*, **20**, 803–826.

Williamson, D. & Williamson, J. (1984) Botswana's fences and the depletion of Kalahari wildlife. *Oryx*, **18**, 218–222.

Woodroffe, R. (2001) Assessing the risk of intervention: immobilization, radio-collaring and vaccination of African wild dogs. *Oryx*, **35**, 234–244.

Chapter 9
Barriers, the Beef Industry and Unnatural Selection: A Review of the Impact of Veterinary Fencing on Mammals in Southern Africa

Michelle E. Gadd

Introduction

History and Purpose of Veterinary Cordon Fences

Veterinary cordon fences (VCFs) zigzag across the southern African savannah. The fences are intended to separate disease-free livestock from infected livestock and their closest wild relatives, buffalo *Syncerus caffer* and to restrict the movement of antelope that could carry diseases of concern. Livestock production is an important aspect of many African nations, economically and culturally, locally and nationally, for both subsistence and commercial producers. However, the fences do not discriminate between targets and non-targets, and create obstacles for many large mammals.

Over the past 130 years, fences have been constructed at various times to control Foot-and-Mouth Disease (FMD), Contagious Bovine Pleuropneumonia (CBPP), trypanosomiasis, rinderpest and other diseases that can affect livestock (see Taylor and Martin 1987; Bengis et al. 2002; Hargreaves et al. 2004; Kock 2005; Mapitse 2008; Osofsky et al. 2008).

Some diseases pose a serious threat to livestock and, in turn, to food security and human livelihoods. Others, particularly FMD, do not significantly affect livestock production nor suitability for human consumption, but are controlled in order to meet conditions set by the World Organization for Animal Health (OIE, formerly the Office International des Epizooties) for trade on the more lucrative international market. The actual losses caused by FMD to subsistence pastoralists are low (Kock 2005), but financial losses to cattle exporters are high because of stringent

M.E. Gadd (✉)
Division of International Conservation, U.S. Fish and Wildlife Service,
4401 N. Fairfax Drive, Arlington, VA 22203, USA
e-mail: megadd@alumni.rice.edu

M.J. Somers and M.W. Hayward (eds.), *Fencing for Conservation: Restriction of Evolutionary Potential or a Riposte to Threatening Processes?*,
DOI 10.1007/978-1-4614-0902-1_9,

processing and handling requirements (Taylor and Martin 1987; Mapitse 2008). European policies and tariffs have buoyed cattle export from some African countries, and have subsidised further fence building; the 1976 Lomé Convention and its successor, the 2000 Cotonou Agreement, gave Botswana preferred trading status for beef, guaranteeing prices 25% higher than the global average (Environmental Investigation Agency 2004; Nair 2007; Mapitse 2008). Donors have funded fences under the guise of poverty alleviation and economic development, but evidence indicates that the majority of revenue from beef export is captured by elites (Perkins 1996; Nair 2007; Mapitse 2008; Scoones and Wolmer 2008).

FMD is particularly challenging to control in situ. The virus that causes FMD is hardy and may be spread through the air (when infected or carrier animals cough or sneeze), through fomites (inanimate objects or substances) and possibly by feeding in the same area (when grazing mammals feed on grass that has been fed on by an infected animal). The virus enters hosts by inhalation or ingestion. Weather is also believed to be a factor, spreading more readily during cool, damp spells (du Toit 2005). However, aerial transmission is unlikely in southern Africa under the prevailing dry, hot weather (Sutmoller 2002). The virus can be found in a diverse range of hosts including hedgehogs, artiodactyls, primates, armadillos and rodents. Cattle *Bos taurus* and *Bos indicus*, pigs *Sus scrofa*, sheep *Ovis aries* and goats *Capra hircus,* are the domesticated species most seriously affected (Bengis et al. 2002). In Africa, impala *Aepyceros melampus,* kudu *Tragelaphus strepsiceros,* wildebeest *Connochaetes taurinus* and sable *Hippotragus niger* have low to negligible mortality from the FMD virus, but are known to carry it (Kock 2005). Buffalo have historically been regarded as the most important wild host for FMD virus and the most likely to interact with livestock and transmit infection, which led to the intentional extermination of buffalo in many cattle-producing areas in Zimbabwe (Taylor and Martin 1987), Namibia (Martin 2005) and Botswana (Albertson 1997). Cattle and buffalo become long-term carriers of FMD viruses, whereas antelope do not (Hargreaves et al. 2004). In order to satisfy OIE trade conditions, many southern African nations employ a combination of fences and vaccination regimes. In countries that have secured disease-free zones, outbreaks are controlled with slaughter of livestock to confine and eradicate the disease before it becomes economically devastating.

Fence Purposes and Designs

Veterinary cordon fences vary in strength and penetrability depending upon the disease targeted. Simple wire-strand cattle fences, 1–2 m high, suffice for CBPP control, and are built to restrict the movement of cattle and buffalo only. Most veterinary cordon fences are comprised of horizontal wires only, without the vertical or subterranean mesh that would be required to stop crawling or digging animals.

Stronger double-cordon fences are constructed to reduce the transmission of FMD. FMD fences are intended to exclude potentially infected or reservoir species, and to create a mammal-free gap of 10 m or more between the infected zone and animals in vaccination or quarantine zones (Taylor and Martin 1987). Vegetation between the fences is cleared manually or mechanically, and roads to facilitate maintenance or patrolling are cut along one or both sides of the fence. To reinforce the role of the fences in preventing disease transmission, trespassing livestock are destroyed. When animals that could be disease transmitters get on the "wrong" side of fences, well-financed wildlife departments may actively chase the animals back into wildlife areas (e.g. Kruger National Park in South Africa [F. Jori pers. comm.] and along the Northern Buffalo fence in Botswana [Albertson 1997]). More often, countries cull potential carriers that manage to get through (instituted in Zimbabwe [Taylor and Martin 1987] and Botswana [Albertson 1997]). Some countries have proposed "shoot to kill" policies for all wild animals that cross cordons, regardless of whether they are disease risks or not, as proposed for a 300-km fence along Namibia's border with Angola (see Gadd 2007). Because of the wide swath of bush that has to be cleared, the height, and the double rows of fencing, FMD-fences create a greater hurdle to wildlife than simple cattle or game fences.

The third and strongest type of fence prevalent in southern Africa along international borders serves multiple purposes, including preventing illegal movement of people and restricting animal (livestock and wildlife) movement. Long stretches of the borders shared by Botswana, South Africa, Mozambique, Zimbabwe and Namibia are fortified with razor wire or electrified, carrying 7–12 kV of electricity (enough to deter elephants and to jolt humans), and may be actively patrolled.

Southern African countries have experienced an upsurge of disease outbreaks in recent years (FAO 2005), including within fenced zones. Although the efficacy of fences has long been questioned (Owen and Owen 1980; Ross 2003), fences are still regarded as an essential component of disease control because they provide partial protection: "the rate of spread of disease is proportional to the amount of animal traffic, which fences facilitate holding to a low level" (Taylor and Martin 1987). This paper does not aim to examine the efficacy of veterinary cordon fences in controlling diseases nor the economic costs and benefits of fencing. Nor does it address the social impact of fencing on human residents whose movements and livelihoods may be adversely affected (Albertson 1997; Gupta 2004; Mapitse 2008; Pierson and Gadd 2008). Instead, the following pages focus on the ecological costs of veterinary cordon fences, with an eye to anticipating future ramifications. Expected impacts on various ecological levels (individual, population, species, community and ecosystem) are outlined. A review of observed effects is presented, limitations of existing data are evaluated and general trends are summarised. Approaches that could improve our understanding of the impact of fences and actions that could lessen detrimental effects are highlighted.

Trend Towards More and Stronger Fencing

Fences are already a prominent feature of southern African rangeland and fencing is increasing exponentially across Africa for myriad reasons: increasing human population, shifting from pastoralism to agriculture, changing land ownership policies (including privatisation and redistribution), sub-division of existing large blocks of land into smaller privately owned fragments, and escalating human-wildlife conflict. Globally, livestock numbers are expected to increase dramatically to satisfy the increasing demand for meat worldwide (due to growing human populations and increasing wealth per capita, enabling more people to afford meat). Africa is projected to be a net supplier to meet this increasing demand, and more regions will undoubtedly seek to control diseases that diminish production or jeopardise their ability to export to international markets (Kock et al. 2002).

Disease outbreaks, increasing instability in neighbouring countries, and increases in fence breakage (by people and by wildlife) have triggered the fortification of existing VCFs. Fences are being heightened, electrified and elephant-proofed, which substantially increases the obstruction they pose to wildlife. In 1995, an outbreak of CBPP among cattle around Botswana's northern border with Namibia precipitated the hasty construction of three parallel east–west fences (Samochimo, Ikoga and Setata) to try to limit the spread (Ross 2003). In spite of these measures, the disease quickly jumped the fences (by means of illegal cattle movement through the fences) and the government culled all 320,000 cattle in Ngamiland district to prevent an even more costly spread of the disease to export zones further south. To prevent incidents of this magnitude, Botswana upgraded and electrified its Caprivi fence to better barricade against livestock covertly entering from Namibia and Angola (Weaver 1997; Albertson 1998; Martin 2005). More recently, as Zimbabwe's government disintegrated and its ability to maintain disease controls came into question, Botswana took steps to defend its eastern border, adding a second row of fencing and electrifying it to prevent breakage by elephants (Gadd 2001).

Why We Might Expect Adverse Effects on Wildlife

Extent of Fencing

The sheer extent of fencing in southern Africa makes fencing a substantial modifier of the landscape (Fig. 9.1). For example, Botswana's perimeter is less than 3,700 km but within the country (including border fences), more than 5,000 km of fenceline protect the cattle industry (calculated from Williamson 2002; Environmental Investigation Agency 2004, 2005, 2007). Fences span hundreds of kilometres without any openings or gaps to allow passage of wildlife (e.g. Botswana's 300 km Kuke fence [Ross 2003]).

Alignment of Fencing

Fences have been aligned according to political decisions, not ecological ones. Many of the fences run east–west (e.g. Botswana's Kuke fence, Namibia's existing

"Red Line" fence and its proposed 250 km fence along the border with Angola), cutting directly across habitat types, without any regard for the distribution of natural resources or wildlife. Fences cut through wilderness areas, hemming mammals into whichever side of the fence they happen to be on at the time of construction (Albertson 1997; Gadd 2001). They do not accommodate predictable seasonal movements of migratory species, nor wet season range expansion, or dispersal of adolescent animals leaving their natal territories. Fences often join at acute angles, unintentionally funnelling wildlife into blind corners with no outlet. Other fences jut out across miles of pristine wilderness before coming to an abrupt, seemingly arbitrary, end. When deciding on fence alignment around water points, wildlife usually loses out, with access to water being given to cattle owners. Countries have fenced extensive parts of their perimeters. Where these borders coincide with major river systems, e.g. the Limpopo, Shashe, and Kavango rivers, fencing must be wholly contained within one country, therefore, the water source (and, sometimes, its riparian buffer) is fenced entirely in or entirely out, separating wildlife from vital water supplies.

Fences are not amenable to changing land uses or disease patterns. Fences are sometimes built as an emergency response to an active outbreak. When the threat has passed or the fence has failed and no longer serves any disease control purpose, the fences are abandoned. Without maintenance, fences may even become more of a death trap for wildlife: unchecked conversion of wire to snares, broken dangling wires ensnaring animals, and as they decay over time, becoming less visible but equally impenetrable. Some fences go through areas of dubious disease transmission importance, with little or no cattle (Albertson 1997). In others, the value of wildlife-based industries already exceeds the value of the livestock industry. In parts of sub-Saharan Africa where wildlife is a profitable use of marginal land, some landowners are shifting away from pure livestock towards multi-species systems (du Toit 2005; Mapitse 2008; Osofsky et al. 2008). The recently established Kavango Zambezi Transfrontier Conservation Area (KAZA), spans more than 250,000 km^2 in five southern African countries and holds great promise to become a premiere tourism destination; however, it is littered with fences, particularly in its southern reaches. Tangles of active, redundant and defunct fences compartmentalise areas and prevent animals from expanding into others. For communities anxious to partake in the new wildlife-based development plans, fences thwart their hopes of mammal re-establishment and recovery.

Expected Effects

Based on theoretical ecology and on evidence observed from analogous structures elsewhere, we can anticipate certain consequences of barriers on wildlife (Table 9.1). Wherever humans occur, manmade objects interrupt and alter the landscape. Even porous objects like settlements and roads have significant effects on local ecology. A growing body of research shows the undesirable effects of roads on ecology (see

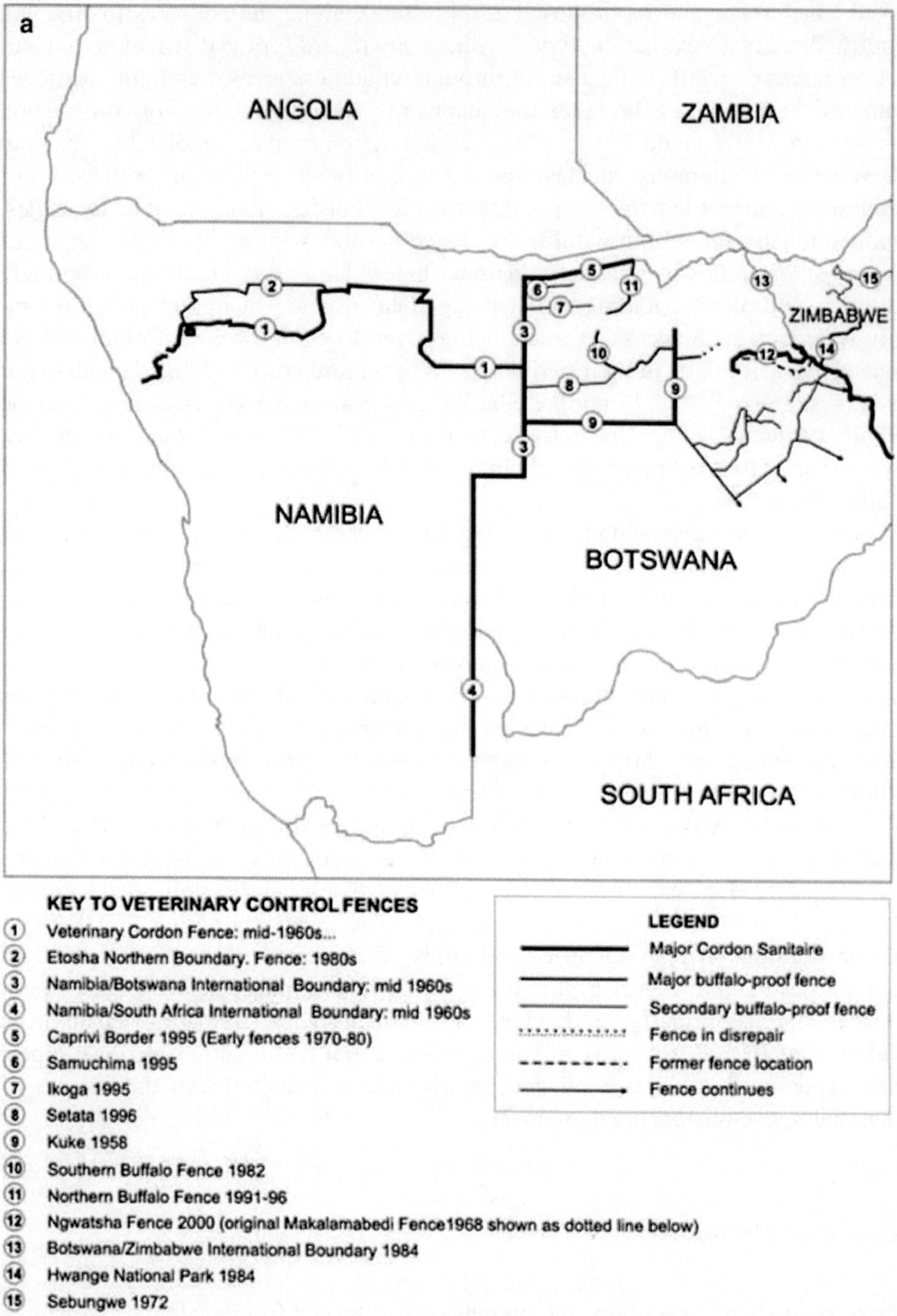

Fig. 9.1 (**a**) Major fences in southwestern Africa before 2000 (Martin 2005). (**b**) Fences in Botswana before 1997 (Williamson 2002)

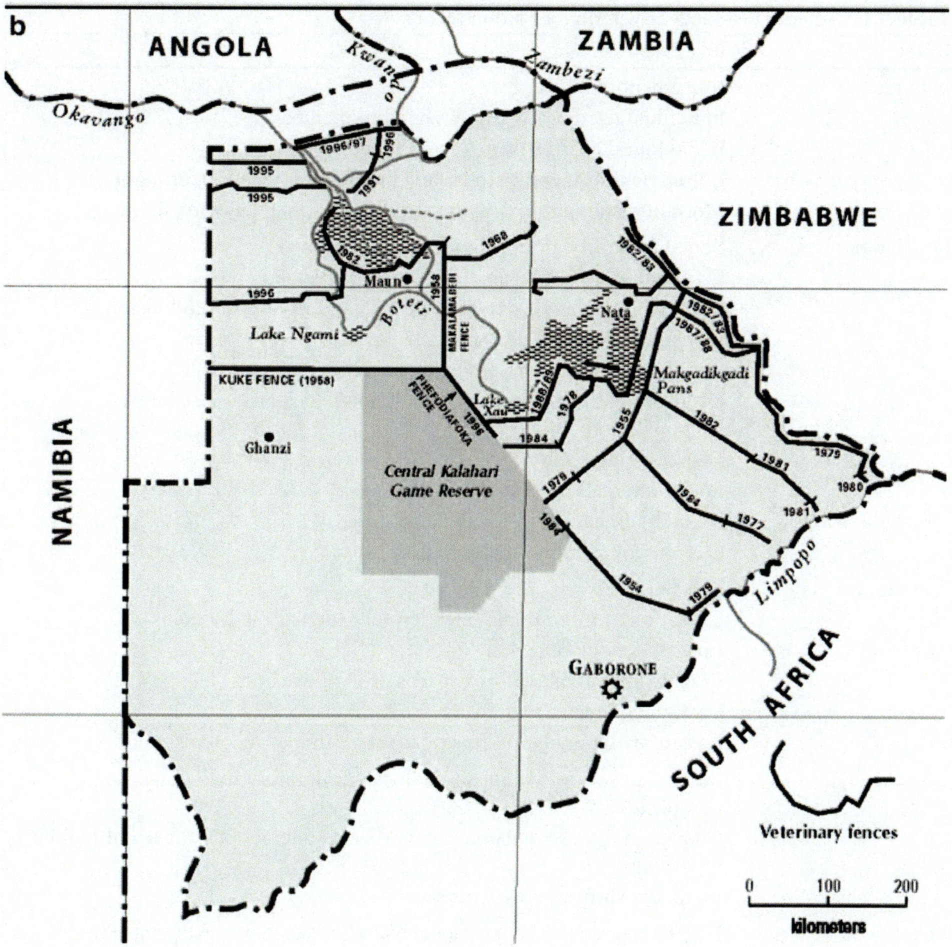

Fig. 9.1 (continued)

Forman et al. 2003). Even though roads would appear to be a minimal obstacle to large-bodied, wide-ranging mammals like elephants, recent research on collared elephants in the Congo Basin indicates that forest elephants avoid crossing roads outside protected areas (Blake et al. 2008). Roads, and the human activities that accompany them, may artificially restrict elephant movement and sub-divide populations. We would expect fences, which are specifically designed to stop animal movement to have an even greater impact than roads, but empirical research is thin (discussed below in Methods). Like other barriers, fences may cause ecological effects directly and indirectly, immediately and over the long-term. If fences are non-porous, they may function as hard boundaries, fragmenting the landscape into small, disconnected patches. Habitat loss and fragmentation are major drivers behind the current wave of species extinctions (McGarigal and Cushman 2002). Patches isolated by fences could function as land-locked islands, subject to the

Table 9.1 Expected impacts of veterinary cordon fences

Level	Effect
Individual	Movement impeded
	Individual territory or home range fragmented
	Behavioural change to cope with fence
	Entrapment or inability to escape fire, flood, predation, drought
	Mortality: starvation, dehydration, entanglement, electrocution
Population	Social or family groups divided or fragmented
	Essential daily or seasonal movement prevented
	Effective population size reduced by mortality or by subdivision
	Increased predation pressure
	Shift in prey selection, predation success
	Disappearance of migratory population/persistence of sedentary population
	Dispersal inhibited
	Social interactions restricted
	Breeding behaviour altered
	Overcrowding, abnormally high density
Species	Mass mortality
	Cessation of migrations, selection for sedentary individuals
	Loss of genetic potential
	Disease spread in confined spaces or high densities
	Lack of connectivity between groups/conspecifics
	Loss of metapopulation function: prevention of recolonisation
Community	Change in species composition: loss of migratory individuals and species
	Change in disease dynamics: new species interactions at sites of limited resources
	Predators shifting prey species
Ecosystem	Habitat degradation due to local overabundance when animals are constrained
	Depressed local primary productivity
	Nutrient depletion
	Avenues for invasive species
	Provides access for humans to remote areas
	System resilience lowered: potential for recovery diminished
	Expansion of incompatible human land uses

predictions of island biogeography theory (MacArthur and Wilson 1967). Isolating pieces of land may result in the loss of species and eventual species relaxation, as predicted for mammals in protected areas in Tanzania (Newmark 1996) and witnessed in Ghana (Brashares 2003).

Fences physically divide contiguous populations of mammals into separate, smaller, isolated populations. Small populations are inherently more at risk of extinction via stochastic events than large populations (reviewed in Caughley 1994). Small populations are more vulnerable to demographic failures (e.g. inbreeding or inability to find mates), and are less able to recover from disasters such as drought, flood, and fire. In metapopulations (where local extirpations periodically occur and local colonisation or recolonisation events re-establish sub-populations) (reviewed in Hanski 1998), fences prevent the recolonisation and recovery of satellite sub-populations. Individual stressors may combine synergistically, pushing populations or species beyond their ability to recover, to the point of extinction. Perturbations can trigger an extinction vortex: a mutually reinforcing cycle of biotic and abiotic processes that drive population size further downward toward extinction (Brook et al. 2008).

Habitat fragmentation and physical barriers have been called "the greatest obstruction to maintaining species diversity and ecological integrity" (Clevenger and Waltho 2000). Fragmenting a landscape reduces the heterogeneity within each fragment. Confining wild or domestic herbivores in finite areas within a larger landscape reduces the variation in vegetation type, quality and quantity available to them. Empirical evidence from livestock production areas in southern Kenya indicates that sub-division of land resulted in numerous small plots of relatively uniform quality, with lower overall carrying capacity and mammalian biomass production than when it was a contiguous heterogeneous unit (Boone and Hobbs 2004).

Methods

Meta-Analysis

I categorised the effects of veterinary cordon fences on wildlife from 34 published and unpublished reports (Table 9.2). Of these, 25 contained primary data or included first-hand eyewitness accounts. Those articles that did not contain primary data were popular articles or position papers, summarising fieldwork conducted by others. In cases where the same species, events, sites and years were referred to by multiple authors, I made every effort to identify and cite only the first or original source in order to avoid double counting. Only one report was an environmental impact assessment conducted prior to fence construction, weighing various proposed fence alignment options (Scott Wilson Resource Consultants 2000). The vast majority of documents assessed impact during or after fence construction (Albertson 1997, 1998, 2005, 2008; Scott Wilson Resource Consultants 2000; Gadd 2001, 2003; Gupta 2004). One article reviewed fence strengths and purposes (Hoare 1992), and one proposed alternative fence designs and mitigation measures (Kalikawe 1997).

Attempts to correlate impact with fence type, length and age proved impossible due to the limitations of the data (see below) and incomplete details on fence attributes. Quantifying the scope and magnitude of fencing proved impossible: fence types and lengths are not available from a central, updated source. Previously published maps provided the best record of major fence lines in specific regions, but these are now more than 10 years out of date (Williamson 2002: Botswana fences constructed prior to 1997; Ross 2003; Martin 2005: Namibia, Botswana and north-western Zimbabwe fences constructed prior to 1996), cover limited locations and did not specify fence type.

Table 9.2 Known impacts of veterinary cordon fences on wildlife in southern Africa. Type of event, species affected, detail (quantification where possible), country and specific location, method, date of event, and reference

Event	Species	Detail	Country	Location	Method	Date	Source
Fence encounter/ passage rate	Buffalo	1/6 broke out of GS, 5/696 broke out of Sengwa, 1 walked around end	Zimbabwe	Gokwe South/Sengwa boundary	Spoor count, cumulative over 1 year	1992–1993	Booth et al. (1998)
	Elephants	416/790 broke out of GS, 209/988 broke out of Sengwa, 603 walked around end	Zimbabwe	Gokwe South/Sengwa boundary	Spoor count, cumulative over 1 year	1992–1993	Booth et al. (1998)
	Antelope	60–600 individuals hop over fence each year	Zimbabwe	Save valley conservancy	Interview of patrol personnel	2002	Sutmoller (2002)
Trapped between fences but alive	Elephants	Herd	Botswana	Caprivi	Direct observation	1997	Albertson (1998)
	Wildebeest	Subadult alive inside Kedia fence	Botswana	Central Kalahari Game Reserve, northern boundary	Direct observation	1997	Albertson (1998)
Death due to starvation, dehydration or entanglement	Wildebeest		Botswana	Kuke fence	Direct observation	1960s	Silberbauer (1965) in Williamson and Mbano (1988)
	Wildebeest		Botswana	Kalahari fences	Not described	1979–1980	Owen and Owen (1980)
	Giraffe, impala, sable, kudu, ostrich		Zimbabwe	Hwange and Gonarezhou	Not described		Taylor and Martin (1987)
	Flamingos	Adults and chicks entangled as water receded below fenceline	Botswana	Nata sanctuary	Direct observation	1994	Williamson (2002)
	Hartebeest, kudu, gemsbok, wildebeest, giraffe	7 giraffe, 8 gemsbok, 2 wildebeest, 2 hartebeest, 3 kudu	Botswana	Setata fence	Community report	1995	Ludbrook pers.comm. in Albertson (1998, 2005)
	Zebra, wildebeest		Botswana	Setata fence	Not described	1996	Mughogho pers. comm. in Albertson (2005)
	Giraffe, ostrich	Death by entanglement	Botswana	Setata fence	Community report	1996	Albertson (1998)
	Hartebeest, gemsbok, kudu	Death after running into fence or tripping over it	Botswana	Setata fence	Community report	1996	Albertson (1998)

	Gemsbok, eland, kudu, ostrich, wildebeest	Vet report confirmed 2 kudu, 1 eland, 1 ostrich	Botswana	Ikoga	Community report	1996	Albertson (1998)
	Kudu, eland, sable, roan, giraffe, elephant, ostrich, duiker, steenbok	During 1997, mostly June, July: 5 kudu, 2 eland, 1 sable, 1 roan, 5 giraffe, 1 elephant, UNK ostrich, UNK duiker, UNK steenbok	Botswana	Caprivi fence	Community report	1997	Albertson (1998)
	Ostrich		Botswana	Central Kalahari game reserve, northern boundary	Direct observation	1997	Albertson (1998)
	Giraffe, buffalo, elephant, roan	During 1997: 7 giraffe, 2 buffalo, 2 elephant, UNK roan	Botswana	Northern Buffalo fence, Okavango	Community report	1998	Albertson (1998)
	Giraffe, hartebeest, wildebeest, gemsbok, ostrich	7 giraffe, 5 hartebeest, 7 wildebeest, 3 gemsbok, 2 ostrich	Botswana	Setata fence, Okavango	Community report	1998	Kavadimba (1998)
	Giraffe, gemsbok, hartebeest	4 giraffe, 1 gemsbok, 2 hartebeest	Botswana	Setata fence, Okavango	Direct observation by Ludbrook	1998	Kavadimba (1998)
	Buffalo		Namibia	Caprivi	Personal communication NAPHA chair	1988	Martin (2005)
Fragmentation of individual territory or home range	Eland, roan, sable, tsessebe, giraffe		Botswana	Northern Buffalo fence	Inference	1997	Albertson (1998)
Division of family/social groups	Wildebeest		Botswana	Kalahari fences	Not described	1979–1980	Owen and Owen (1980)
	Gemsbok, hartebeest	Exhausted calves on one side, herds on other	Botswana	Setata fence	Direct observation	1997	Albertson (1998)
	Elephants		Botswana	Caprivi fence	Direct observation	1997	Albertson (1998)
	Elephant, roan, eland, tsessebe		Botswana	Northern Buffalo fence, Okavango	Direct observation	1997	Albertson (1998)

(continued)

Table 9.2 (continued)

Event	Species	Detail	Country	Location	Method	Date	Source
	Tsessebe, giraffe	Herds seen on both sides of fence, with smaller ones on one side	Botswana	Northern Buffalo fence, Okavango	Direct observation	1997	Albertson (1998)
	Elephants	Subadults stuck on one side, all stressed and in poor condition	Botswana	Nxai Pan buffalo fence	Direct observation	1997	Albertson (1998)
Prevention of essential daily or seasonal movement	Wildebeest		Botswana	Kalahari fences	Not described	1979–1980	Owen and Owen (1980)
	Hartebeest, wildebeest	Unable to reach Limpopo River for water	Botswana	Limpopo River, between Lephepe to Dibete	Not described		Owen and Owen (1980)
	Gemsbok, eland, kudu, ostrich	Older spoor	Botswana	Ikoga	Spoor	1996	Albertson (1998)
	Elephant, sable, roan, eland, giraffe, kudu	Significant fragmentation and movement obstruction within 60 km stretch	Botswana	Caprivi fence	Direct observation	1997	Albertson (1998)
	Zebra, buffalo, wild dog	Some fragmentation and movement obstruction within 60 km stretch	Botswana	Caprivi fence	Direct observation	1997	Albertson (1998)
	Sable, zebra, buffalo, elephants	Unable to reach dry season range	Botswana	Northern Buffalo fence	Not described	1997	Albertson (1997)
	Elephants, giraffe		Botswana	Nxai Pan buffalo fence	Not described	1997	Albertson (1998)
	Wildebeest, hartebeest		Botswana	Phefodiafoka fence, northeast CKGR	Direct observation	1997	Albertson (1998)
	Elephants		Zimbabwe	Sebungwe fence at Matusadona NP			Taylor and Martin (1987)
	All mammals	Animals jump fence to get water, are shot for disease control zone	Zimbabwe	Chirisa Safari area	Interview of patrol personnel		Taylor and Martin (1987)
	Elephant		Namibia	West Caprivi Game Park (Bwabwata)	Visit to fenceline	1997	Weaver (1997)
	Elephants		Botswana	Shashe fence	Repeated spoor counts		Gadd (2001)

Forced change in migration route	Wildebeest	Migrating animals unable to reach Okavango and Boteti River, turn east to Lake Xau (not previously a migratory destination)	Botswana	Kuke, Ngamiland fences	Comparison to historical record	Post-1955	Williamson and Mbano (1988)
	Wildebeest	Tried to go north, stopped by fence and walked east	Botswana	Setata	Community report	1996	Albertson (1998)
	Zebra, wildebeest, elephants	Seasonal migration pattern changed	Botswana	Northern Buffalo fence	Comparison with recorded natural history	1997	Albertson (1998)
Fragmentation and division of populations	Wildebeest, hartebeest, eland		Botswana	Kalahari	Aerial survey, model	1980s	Spinage and Matlhare (1992)
	Gemsbok, eland		Botswana	Setata fence	Not described	1995	Albertson (2005)
	Wildebeest, hartebeest, eland, ostrich	Restriction into smaller ranges (and artificial waterpoints) increased mortality rates during droughts	South Africa	Central Kalahari	Aerial survey	1995	Knight (1995)
	Roan, sable, tsessebe		Namibia	West Caprivi Game Park (Bwabwata)	Visit to fenceline	1997	Weaver (1997)
	Roan, oribi, sable, wild dog		Botswana	Northern Buffalo fence	Not described	1997	Albertson (1997)
	Rhino		Zimbabwe	Gonarezhou, Save Valley			du Toit (2005)

(continued)

Table 9.2 (continued)

Event	Species	Detail	Country	Location	Method	Date	Source
Mass mortality event	Wildebeest	300,000 died	Botswana	Lake Xau, Mopipi Dam	Aerial survey, carcass count	1963	Child (1972)
	Wildebeest		Botswana	Kuke fence, Lake Xau	Not described	1961, 1964, 1970, 1979	Owen and Owen (1980)
	Hartebeest, wildebeest	"Thousands"	Botswana	Limpopo River, between Lephepe to Dibete	Not described		Owen and Owen (1980)
	Wildebeest	52,000 carcasses estimated at end of dry season	Botswana	Lake Xau, Mopipi Dam: 1982–1983	Carcass count on ground, sampled areas, stratified by estimated carcass density	1982–1983	Williamson and Mbano (1988)
	Wildebeest	80,000 carcasses estimated	Botswana	Lake Xau, Mopipi Dam: 1982–1983	Aerial survey, carcass count	1982–1983	Parry (1987)
	Hartebeest	10,000 hartebeest died	Botswana	Ghanzi fences	Carcass count	1981–1987	Spinage (1992)
	Zebra	60,000 died	Botswana	Kalahari fences	Not described	1980s	Ross (2003)
	Roan, elephant, and other wildlife		Botswana	Northern Buffalo fence	Not described	1997	Albertson (1998)
Population crash	Wildebeest	90% decline, from 262,000 to 16,000	Botswana	Kalahari to Okavango	Comparison with past aerial counts	1979-late1980s	Spinage (1992)
	Hartebeest	70% decline	Botswana	Kalahari to Okavango	Comparison with past aerial counts	1979-late1980s	Spinage (1992)
	Wildebeest	30,000 animals lost	Namibia	Etosha moving north			Berry and Siegfried (1979)
	Wildebeest and hartebeest	Wildebeest declined by 97%, hartebeest by 86%: drought, lack of migration and die-offs	Botswana	Central Kalahari game reserve	Aerial survey and model	1979–1986	Spinage and Matlhare (1992)
	Springbok		South Africa	Karoo: 1896	Review of historical documents	1896	Roche (2008)

Cessation of a mass migration	Springbok	Megatrek: est'd migrations ranged from 100,000–1,000,000	South Africa	Karoo: 1896	Review of historical documents	1896	Roche (2008)
	Wildebeest, hartebeest, eland		Botswana	Kalahari		1980s	Spinage and Matlhare (1992)
	Wildebeest	Estimate: 1964–1983 100,000 wildebeest lost	Botswana	Kgalagadi to Okavango, confined at Lake Xau by Kuke and Ngamiland fences	Aerial survey, carcass count		Williamson and Mbano (1988)
Entrapped and killed by fire while confined by fences	All species	Fires killed thousands of wild animals trapped against a fence	Botswana	Central Kalahari game reserve, northern boundary	Not described	1996	Albertson (1997, 1998)
Abnormally high density of animals due to confinement by fences	Wildebeest	80,000 wildebeest crowded into 125 km^2 area	Botswana	Lake Xau	Not described	1979	Owen and Owen (1980)
	Wildebeest	50,000 wildebeest concentrated in an area	Botswana	Lake Xau	Aerial survey	1983	Williamson and Mbano (1988)
Habitat degradation due to animals constrained by fences	Springbok		South Africa	Karoo fences	Review of historical documents	1896	Roche (2008)
	Elephant		Zimbabwe	Chirisa Safari Area	Direct observation	1987	Taylor and Martin (1987)
	Elephants		Botswana	Northern Buffalo fence	Direct observation	1997	Albertson (1998)
	Elephants		Botswana	Caprivi fence	Direct observation	1997	Albertson (1998)
	Elephants, eland, sable, roan, buffalo		Namibia	West Caprivi Game Park (Bwabwata)	Not described	1997	Albertson (1998)
	Kudu, wildebeest, giraffe, elephant		Botswana	Northern Buffalo fence	Spoor	1997	Albertson (1998)
	Buffalo	Overcrowding on south side of fence	Botswana	Northern Buffalo fence	Direct observation	1997	Albertson (1998)

(continued)

Table 9.2 (continued)

Event	Species	Detail	Country	Location	Method	Date	Source
Prevention of recolonization	Eland	Population decline since 1980s. Without immigration, local extinction likely	Botswana	Central Kalahari game reserve	Aerial survey and model	1995	Spinage and Matlhare (1992)
	Roan, sable, tsessebe		Namibia	Caprivi	Inference		Martin (2005)
Isolation of national parks from one another	All, including elephants		Zimbabwe	Throughout	Comparison to historical record		Taylor and Martin (1987)
	All		Botswana	Moremi, Chobe and West Caprivi separated from wildlife areas to west and north	Inference	1997	Albertson (1997)
	All	Transboundary seasonal movement prevented	Botswana-Namibia	Caprivi fence	Inference	1997	Albertson (1998)
	All		Namibia	Caprivi parks	Inference	1997	Weaver (1997)
	All		Namibia	Western Caprivi, Mahango and Khaudum	Inference		Martin (2005)
Excess concentration of one herbivore leading to decline of another herbivore	Cattle deplete grazing, wildebeest starve		Botswana	Kuke fence, Lake Xau	Not described	1979	Owen and Owen (1980)
	Cattle deplete grazing, wildebeest starve	Grazing had been severely depleted by an enormous concentration of domestic livestock	Botswana	Lake Xau, Mopipi Dam	Direct observation	1982–1983	Williamson and Mbano (1988)
	Elephant degrading habitat for all, especially rare antelope		Botswana, Namibia	Okavango, Caprivi	Inference from declining game counts	1995, 1998, 2002 data	Martin (2005)

Shift in mammalian community composition	Wildebeest and eland will decline while gemsbok and springbok will persist	Migratory species will decline because of fencing and drought while animals that are not dependent on water will persist	Botswana	Kalahari	Aerial survey data and model		Spinage and Matlhare (1992)
	Wildebeest and eland will decline while gemsbok and springbok will persist	Species have declined because of droughts and migration routes impeded by settlement and fences. Legal and illegal offtake may alter or accelerate the decline	Botswana	Kalahari	Aerial survey data and model		Knight (1995)
Loss of nutrient input due to herbivore crash	All, especially wildebeest, hartebeest		Botswana	Khutse Game Reserve, Central Kalahari	Comparative soil samples	Unknown	de Queiroz (1993)
Decreased carrying capacity	All ungulates	Observed wild biomass 25% of expected. 439 kg/km^{-2}, vs. 1,833 kg/km^{-2}	Botswana	Kalahari	Aerial survey compared to biomass model	1980	Williamson and Williamson (1981)
Predators hunting along fence	Lion, wild dog	Lion and wild dog activity noted	Botswana	Ikoga	Community report	1996	Albertson (1998)
	Lions	Godikwa residents report increased predators attracted by congregation of game against western side of NBF	Botswana	Northern Buffalo fence	Community report	1997	Albertson (1998)

(continued)

Table 9.2 (continued)

Event	Species	Detail	Country	Location	Method	Date	Source
Hunting by humans while confined by fences	Springbok	With firearms	South Africa	Karoo: 1896	Review of historical documents	1896	Roche (2008)
	Wildebeest	Men in trucks, on horseback and on foot using packs of domestic dogs. The dogs run the herds until the exhausted wildebeest can only stand while being disembowelled	Botswana	Kuke fence, Lake Xau	Not described	1979	Owen and Owen (1980)
	Wildebeest	"Remorselessly harassed by hunters"	Botswana	Lake Xau, Mopipi Dam	Direct observation	1982–1983	Williamson and Mbano (1988)
	All species	Poaching from vehicles along the fence cutlines	Botswana	Setata fence	Spoor, animal flight distance	1997	Albertson (1998)
	All species	By poachers from other communities	Botswana	Setata fence	Community report	1998	Kavadimba (1998)
	Giraffe	Shot by government employees against fence and eaten	Botswana	Caprivi fence	Community report	1997	Albertson (1998)
	Elephants	Poaching increased because of accessibility afforded by VCF cutline	Namibia	West Caprivi Game Park (Bwabwata)	Report from professional hunter	1997	Albertson (1998)
Conversion of fences into snares	All species	2,000 snares collected between February and July 1979, all made from tsetse control fence	Zimbabwe	Chirisa Safari Area		1979	Conway (1984) in Taylor and Martin (1987)
	All species		Zimbabwe				Booth et al. (1998)
	All species		Zimbabwe	Gonarezhou			Mail and Guardian newspaper
	All species		Zimbabwe	Zambezi valley	Direct observation		L Osborne, pers.comm.
	Rhinos		Zimbabwe	Save valley, Bubye, Bubiana Conservancies	Direct observation		du Toit, pers. comm.

Escalation of human-wildlife conflict	Elephants	Crop raiding	Zimbabwe	Chirisa Safari area	Not described		Taylor and Martin (1987)
	Elephants	Crop raiding	Namibia	Caprivi fence	Community report	1997	Weaver (1997)
	Elephants	Crop raiding	Namibia	Outside western Caprivi game park, due to Caprivi fence	Community report	1997	Albertson (1998)
	Elephants	9 elephants shot by Wildlife Department for breaking a decommissioned VCF	Botswana	Nxai Pan buffalo fence	Community report	1996	Albertson (1998)
	Elephants	Crop raiding	Zimbabwe	Gokwe North, Nyaminyami, Chawarura			Booth et al. (1998)
	Elephants	Water	Botswana	Makgadigadi fence	Interview of residents	2005	Gupta (2004)
	Elephants	Water, crop raiding	Botswana	Shashe fence	Direct observation, community report		Gadd (2001)
	Lions	Godikwa residents report increased predation on dogs, horses and donkeys due to increased predators attracted by congregation of game against western side of NBF	Botswana	Northern Buffalo fence	Community report	1997	Albertson (1998)

Type of event, species affected, detail (quantification where possible), country and specific location, method, date of event, and reference

Limitations

Ideally, ecological experiments should have replicated, controlled, paired samples (Hurlbert 1984; McGarigal and Cushman 2002). Wildlife research rarely takes place under ideal conditions, and monitoring fence effects on wildlife in Africa is no exception. Unfortunately, most assessments were based on single visits. Driving or walking a portion of fence line and counting spoor or carcasses was the most common method employed. In five cases, aerial surveys were done (Child 1972; Williamson and Williamson 1981; Williamson and Mbano 1988; Spinage and Matlhare 1992; Knight 1995). Regrettably, game counts did not have optimally paired or repeated pre- and post-construction datasets. Many reports did not attempt to address and eliminate other potential causative factors, such as change in precipitation or increased livestock densities, etc.

Fences were not visited systematically, thoroughly, repeatedly, or frequently; therefore, rates of encounter and entanglement, frequency of individual and mass mortalities and total cumulative effects remain unknown. Carcass counts delineated the range of species affected, but because they were conducted only over finite areas and limited time periods, provide only the absolute minimum estimate of mortalities. Community reports and interviews with fence maintenance staff may undercount (disappearance of carcasses due to decay, scavenging or human interference) or overcount (by multiple residents reporting the same carcass or authors retelling the same event) actual impact.

The quality of veterinary fence impact studies and long-term monitoring has been severely hampered by political pressures. Biologists and conservation advocates were not informed or consulted before fence construction. Lack of communication within governments was also at play: fence construction and maintenance lie within the mandate of veterinary departments and wildlife departments are not always consulted. Fence construction was often rushed and impact assessments waived. Where assessments were conducted beforehand (Scott Wilson Resource Consultants 2000), recommendations on optimal alignment were ignored (Albertson 2008; Environmental Investigation Agency 2004, 2005, 2007). Governments discouraged or denied proposals to monitor fences. Researchers have been threatened, denied entry or expelled for their conduct related to fence effects. In some cases, government employees actively interfered with data collection: hiding, removing, burning or burying carcasses to prevent researchers from documenting wildlife mortality events (Albertson 1998). Concerns about wildlife may be downplayed for several reasons: a tendency to undervalue the contributions of wildlife to the national economy, powerful lobbying by wealthy cattle owners, pro-cattle cultural values in local and national political arenas, local support for the jobs created by fencing, a lack of interest or awareness from consumers, the desire for income generation and foreign exchange by donors and politicians, and donor preference for short-term, tangible deliverables.

Results

Individual

Animal Encounter/Passage Rates

How often African mammals attempt to cross fences and fail remains unknown. Heavily travelled game trails along veterinary fences indicate that animals approach the fence and are forced to turn left or right. Fence maintenance personnel in Zimbabwe visited a veterinary fence separating a designated cattle area from a wildlife area and kept a running tally of animal tracks approaching and successfully passing through the fence in either direction over 1 year (Booth et al. 1998). Buffalo rarely escaped from the wildlife area to the cattle area (691 out of 696 approaches were rebuffed). Elephants broke in and broke out of the wildlife area with some success: 20% of attempts to get out of the wildlife area succeeded (209 of 988), while 52% (416 of 790) approaches to get in succeeded. A more successful strategy was found by 603 elephants, which followed the fence to its terminus and walked around it. By contrast, only one buffalo found its way around the end. In northern Botswana, well-worn paths indicated that elephants frequently walked the length of the Nxai Pan fence to reach the open end (Albertson 2008). Upon encountering a newly electrified fence, elephants in eastern Botswana walked alongside it for several kilometres, until they reached the last white insulator and then broke through (Gadd 2001). In another study, 60–600 antelope were estimated to jump over veterinary fencing per year in Save Valley, Zimbabwe (Sutmoller 2002).

Individual Behaviour

None of the reports assessed the effects of fencing on individual mammals through focal animal observation or behavioural studies before- and after- fence construction. Anecdotal evidence suggests that hippos, *Hippopotamus amphibious,* are quick to accept the boundaries demarcated by fences (Booth et al. 1998; Hoare 1992). Others, notably elephants and buffalos, challenge fences (Hoare 1992). Elephants may "retaliate" (Hoare 1992), removing large sections of fenceposts and wire after being shocked, or after youngsters stray inside fenced areas (pers. obs.). Giraffe, *Giraffa camelopardalis,* are among the slowest to learn the risks of fencing (Goodwin 1985 cited in Hoare 1992), and also the most reluctant to cross over fence lines when wires have been removed (Albertson 2005).

The effects of fences on small, sedentary, territorial species were rarely mentioned. Small ungulates (e.g. dik dik *Madoqua* spp., steenbok *Raphicerus campestris,* duikers *Cephalophus* spp. and *Sylvicapra grimmia)* are probably adept at slipping under fences or between wires, if spaced adequately, and not electrified at low level. At an experimental fence intended to exclude wild mammals in Kenya, steenbok occurred in higher densities inside fenced areas than outside, indicating that the exclusion of predators resulted in increased survival or that individuals outside actively sought refuge or lower herbivore competition and immigrated inside (Young et al. 2005). Suids and digging animals are difficult to deter (Hoare 1992; Booth et al. 1998; Schumann et al. 2006) and may be less affected by VCFs.

How veterinary fences alter carnivore behaviour and distribution is not well documented, but many species squeeze through or dig under. Wild dogs *Lycaon pictus* in northern Botswana readily crossed veterinary fences, and lions *Panthera leo* forced their way through, although certain lions more often than others (McNutt, pers. comm.). Evidence from experimental fenced plots indicates that predators can penetrate ten-strand electrified game-proof fencing: cheetahs, *Acinonyx jubatus,* frequently went inside and lions were encountered inside once (pers. obs.). Predation incidents by lion and spotted hyena *Crocuta crocuta* remained unchanged before and after an electric fence was installed around Hwange National Park, indicating that their passage was unimpeded (Booth et al. 1998).

Impeding Movement and Dividing Groups

Wildebeest, gemsbok *Oryx gazella,* roan *Hippotragus equinus,* tsessebe *Damaliscus lunatus,* giraffe, and elephants were seen stranded on opposite sides of the fence from their conspecifics (Owen and Owen 1980; Albertson 1998). The smallest youngsters probably wandered under or through (elephants, sable, roan, eland) and adults were unable to follow. In other cases, the largest individuals may leap over, walk over, or push through and the smallest individuals are left behind.

Mortalities

Reptiles, birds and mammals were among the fence casualties. Flamingos *Phoenicopterus* spp., were entangled as they attempted to follow receding waterlines in drying pans traversed by fences (Williamson 2002). Ostriches, *Struthio camellu,* have been found with necks, wings or legs entangled in fences (Taylor and Martin 1987; Albertson 1998; Kavadimba 1998). Tortoises can become stuck under the lowest wires of fences (pers. obs. in Northern Kenya, EIA unpublished photo at Boteti, Botswana).

Mammal carcasses found along, entangled in, or trapped between fences included virtually all medium and large ungulates and sub-ungulates in the vicinity: duiker, steenbok, springbok *Antidorcas marsupialis,* impala, hartebeest *Alcelaphus buselaphus,* wildebeest, sable, kudu, zebra *Equus quagga,* gemsbok, eland *Taurotragus oryx,* buffalo, giraffe and elephants. Although not definitively known, entanglement or confinement and subsequent dehydration were probably the cause of death.

Surprisingly, veterinary cordon fences pose a serious obstacle to elephants, which are notoriously difficult to restrain. Elephant carcasses were found along fence lines on paths to water, families were seen on separate sides of fences, individual elephants paced fence lines or walked kilometres to find a gap (Albertson 1998; Gadd 2001), and high concentrations of elephants were found against fence lines.

In other circumstances, particularly where rewards on the other side of fences are high, elephants doggedly find ways through fences: short-circuiting fences with their tusks (Hoare 1992), bringing felled trees to drop across electric wires (pers. obs.), or pushing smaller elephants through. Cutlines and roads alongside VCFs may add to their barrier effect: elephants can be reluctant to cross abrupt or unnatural changes in vegetation cover (pers. obs., F.L. Osborn pers. comm.). Individual behaviour and experience may play a role: elephants that have been electrocuted on fencing elsewhere may avoid all fencing (even non-electrified) (Gadd 2001), while others become adept fence-breakers, regularly breaking out of protected areas into cultivated crops (Craig 2007).

Where fence wires have been dismantled, some animals still shy away. Giraffe and eland retreated from a section where wire had been removed, possibly due to the visual barrier of the cutline or the fence posts, or the smell of creosote-treated posts (Albertson 2005, 2008). Elephants trapped in an enclosure refused to cross the fence line even after it had been dismantled to release them (pers. obs.). Elephants are reluctant to cross manmade roads (Blake et al. 2008) and clearings (F.L. Osborn, pers. comm.) in other circumstances and may cross unfamiliar gaps only when highly motivated to do so.

Predation

Cleared cutlines alongside fences provide easy access for humans to remote areas and create opportunities for predators to track or chase and corner wildlife. Some animals charge directly into fences when startled by humans (on foot or in vehicles) (pers. obs.), but data are lacking on which species are most likely or most affected. Carnivores follow fencelines and roads, and may use fences to their advantage. In South Africa, wild dogs killed larger prey than usual (adult male kudus and waterbuck) by using Pilanesberg National Park's perimeter fences (van Dyk and Slotow 2003). Domestic dogs in Kenya used game fences to corner a zebra (pers. obs.). Numerous cases were documented where humans capitalised on "easy pickings" offered by veterinary fences: slaughtering hundreds of trapped springbok against fences in South Africa in the 1890s (Roche 2008); and wildebeest in Botswana (Owen and Owen 1980; Williamson and Mbano 1988), hunting on foot with dogs (Owen and Owen 1980) or shooting from vehicles (Albertson 1998). Veterinary fences provide inroads for poachers from other areas in community owned hunting areas (Albertson 1998; Kavadimba 1998) and in commercial safari areas (Weaver 1997). Fence maintenance personnel have also been implicated in cornering and killing wildlife (Albertson 1998). Wherever wire fences are built, snares soon follow. People deftly convert fence wire into snares for bushmeat (pers. obs., Booth et al. 1998). Over a 6-month period, 2000 snares were picked up in the Chirisa Safari Area, all made from the nearby tsetse control fence (Conway 1984 cited in Taylor and Martin 1987). How many new snares are made available annually by veterinary fences is unknown, but the practice is ubiquitous.

Population Effects

Isolation of Populations

Populations that were once contiguous and interbreeding have been severed by fences. Fencing is believed to be a causal factor in the long-term decline of wildebeest, hartebeest and eland in the Kalahari (Spinage and Matlhare 1992; Knight 1995); roan, sable and tsessebe in the Caprivi, Namibia (Weaver 1997; Martin 2005); and roan, oribi, *Ourebia ourebi,* and sable in northern Botswana (Albertson 1997). Fencing enforced hard edges on rhinos, *Diceros bicornis minor,* in Zimbabwe, preventing natural dispersal and intermingling and necessitating active metapopulation management (du Toit 2005).

Failure to Recolonise or Recover

Namibia's Caprivi is a thin strip of land bounded by the major rivers of the Kavango and the Zambezi. Wildlife populations in the Caprivi and in Angola were heavily depleted by war and anthropogenic pressures prior to 1990. Botswana has provided a source of recolonizing ungulates making their way north again, particularly buffalo, elephants, roan, sable, and tsessebe. However, since 1995, when Botswana's northern border fences were fortified, roan, sable and tsessebe have markedly declined. Examining all other factors (including rainfall, law enforcement, patrol effort, and human population trends), Martin (2005) concluded that the declines can be attributed to the fortified veterinary fences which prevent immigration of ungulates, and possibly, to elephant-induced habitat change.

Mass Mortalities

Mass mortality events were most common in migratory ungulates, with die-offs numbering in the tens of thousands. In South Africa, enormous herds of springbok once migrated across the Karoo region. Although the deaths of hundreds of thousands of springbok in the 1890s was previously blamed on rinderpest, an examination of newspaper reports reveals that not long after fences were constructed to protect grazing resources for domestic livestock, springbok became confined in high densities, died of starvation, and were slaughtered by settlers (Roche 2008). Shortly after Botswana's Kuke and Ngamiland veterinary fences were constructed in the late 1950s, wildebeest and hartebeest began dying in great numbers at the ephemeral Lake Xau (Child 1972; Owen and Owen 1980; Williamson and Mbano 1988; Spinage 1992). It is believed that when wildebeest and hartebeest migrating from the Kalahari Desert to the inundated Okavango Delta encountered fences blocking their traditional northerly migration, they turned east and followed the fence line for hundreds of kilometres, possibly drawn by the scent of water. In subsequent years, the wildebeest repeated their ill-fated migration to the new destination: with successive wildebeest die-offs witnessed in 1961, 1963–1964, 1970, and 1982–1983. In the dry season of 1963, an estimated 300,000 wildebeest died (Child 1972). Thirty-five years after fence construction, mass mortalities continued with a further 52,000–80,000 wildebeest dying at Lake Xau in 1982–1983 (Parry 1987; Williamson and Mbano 1988). By 1986, there was no migration, and a 1987 aerial count found only 260 wildebeest in the Central Kalahari (Ross 2003), down from 262,000 in 1979 (DHV survey data in Spinage 1992).

Selective Pressures

Eventually, the toll taken on migratory populations and species has led to a shift in animal behaviour and in community composition. As mentioned above, South Africa's largest mass migration disappeared after springbok were confined by fence-lines and died *en masse* in consecutive years in the 1890s. Within a decade, the springbok migration vanished and only small herds and scattered individuals persisted (Roche 2008). After the construction of fences across the northern boundaries of Etosha National Park in Namibia, the migration of 30,000 wildebeest disappeared. A smaller, sedentary population of wildebeest survives within Etosha, but is susceptible to episodic declines, probably due to disease outbreaks (Berry 1983). In the Kalahari, migratory wildebeest, hartebeest and zebra perished when they failed to reach their annual destination (Spinage 1992; Spinage and Matlhare 1992; Ross 2003), but some non-migratory individuals survived (Knight 1995). Within a matter of years, the repeated selection against migratory individuals has led to the predominance of sedentary individuals. Community structure may also shift, away from species reliant upon water (e.g. wildebeest and eland) to species that are not migratory or water dependent (e.g. gemsbok and springbok) (Spinage and Matlhare 1992; Knight 1995).

Ecosystem Effects

Confinement of herbivores by VCFs can lead to habitat degradation, depressed primary production and, eventually, decreased carrying capacity. Restriction of springbok, kudu, wildebeest, and giraffe by VCFs to areas of very high grazing and browsing are blamed for population declines in the Caprivi (Albertson 1998). Models using aerial survey data indicate that ungulate biomass in the Kalahari was only 25% of its predicted level after fence construction (1,833 vs. 439 kg/km^2) (Williamson and Williamson 1981). Soil samples in and around Khutse Game Reserve, Botswana linked low wildebeest and hartebeest biomass to a downward spiral in carrying capacity: free-ranging wild herbivores are necessary for nutrient input and in their absence, soil nutrition may decline (de Queiroz 1993).

Artificially high concentrations of elephants can cause rapid habitat change, so their confinement by fences can have marked effects. Elephants are keystone species, capable of causing shifts in vegetation structure and in species composition (plant and animal) (see Caughley 1976; Cumming et al. 1997; Ogada et al. 2008 for reviews). Trampling by elephants along fence lines caused local habitat degradation in Zimbabwe (Taylor and Martin 1987). Exceptionally high densities of elephants in northern Botswana may be due in part to range restriction imposed by veterinary fences (Albertson 1998). In combination with lack of recolonisation opportunities, the resulting elephant-induced changes in vegetation structure may contribute to the marked decline of buffalo, roan, sable and tsessebe in the Caprivi (Martin 2005).

Human-Wildlife Conflict

Although fences can be built specifically to curtail human-wildlife conflict, VCFs can intensify conflict. Where fences separate wildlife, especially elephants, from water supplies, human-wildlife conflict escalates (Taylor and Martin 1987; Gadd 2001). Animals denied access to natural watercourses are forced to seek water elsewhere, often resorting to waterholes in close proximity to humans and their livestock. Fences can facilitate land use practices that are incompatible with wildlife conservation. Where fences are built along the boundary of a wildlife zone, agriculture tends to expand right up to the boundary, even if a buffer zone has been designated (Taylor and Martin 1987). Fences that dead-end near agricultural settlements unintentionally funnel elephants in, increasing crop raiding (Booth et al. 1998). Fences remove incentives for people to actively herd their cattle, and cattle wander unattended, rendering them more susceptible to predation, stock theft and accidentally wandering through broken fences into disease zones (Gadd 2001). When elephants break VCFs and unaccompanied cattle escape, hostility toward elephants increases and cattle owners may demand compensation or extermination of elephants (Albertson 1998; Gadd 2001, 2003).

Benefits of VCFs to Wildlife

In addition to preventing disease spread, some veterinary cordon fences do confer advantages to wildlife. Fences made it possible to keep commercially valuable, disease-free buffalo in Zimbabwe prior to the collapse of the central government in the early 2000s (Taylor and Martin 1987; du Toit 2005) and in Namibia's Nyae Nyae Conservancy (pers. obs.). Fences can also exclude domestic stock from wilderness areas: e.g. the Southern Buffalo Fence skirts the southern edge of Botswana's Okavango Delta, keeping wildlife to the north and domestic stock to the south (Ross 2003).

Synergies

The most devastating and long-lasting impacts occurred when fences combined with other factors. Fences, disease, drought, confinement and resulting high local densities, and subsequent predation by people are blamed for the disappearance of hundreds of thousands of springbok from South Africa's Karoo (Roche 2008). Fences, loss of migration routes, and disease outbreaks depressed the wildebeest population in Etosha, Namibia to a fraction of its original size (Berry 1983). Fences, human hunting, drought and competition with cattle precipitated the demise of the Kalahari's wildebeest (Owen and Owen 1980; Williamson and Mbano 1988; Spinage 1992). Fences and drought contributed to the declines of hartebeest and zebras in the Kalahari in the 1980s (Spinage 1992; Ross 2003). Fences, provisioning of artificial waterholes and drought contributed to the decline of wildebeest, hartebeest, eland, and ostrich in Kalahari Gemsbok National Park (Knight 1995). Fences, excessively high concentration of herbivores and subsequent die-offs may cause depressed productivity in the Khutse area of the Kalahari (de Queiroz 1993). Fences, high elephant densities which depleted browsing and grazing, and the failure of immigrants to replenish northern satellite populations explain the decline of local ungulates in Caprivi, Namibia (Martin 2005). Fencing of the Shashe River, human-elephant conflict around agriculture and water points, and targeted hunting are likely to eliminate the last few dozen elephants around Mmadinare, Botswana (Gadd 2001, 2003).

Mitigation

Few attempts have been made to lessen the impact of veterinary cordon fences, either by leaving openings or by designing fences that are more permeable to wildlife. Give-and-go fences that could be triggered by elephants have been suggested (Kalikawe 1997) but not field-tested. Carnivores showed some proclivity for learning where passage points were in a ranch fence in Namibia (Schumann et al. 2006). Attempts to curtail crop raiding by elephants could yield insights that could be applied to veterinary fences. Managers at Ol Pejeta Conservancy in Kenya identified points in their perimeter fence that elephants often broke through. At these breakage points, managers fortified the fences and where the property joined wildlife-friendly areas, managers removed sections of fence to encourage passage. Preliminary evidence from collared individuals and spoor counts indicates that elephants abandoned strengthened sections and quickly learned to use gaps (Craig 2007).

Progress and Restoration

Although pre-construction wildlife surveys were rarely used to decide upon the alignment of fences, governments recently agreed to remove specific sections of fence line after construction, after local and international outcry. To the acclaim of conservationists and local residents dependent on wild products and non-export livestock, Botswana's Department of Agriculture dismantled 210 km of the Setata fence (west of the Okavango) and 66 km of the Nxai Pan Buffalo Fence (east of the Okavango) in 2003 and 2004. Within weeks, elephants, zebras and wildebeest traversed the old line and moved into their former range (Albertson 2005). After being absent for decades, a hippo and a rhino made their first forays south of the old Setata line (Albertson 2008). In spite of these successes, Botswana's Department of Animal Health reversed its standpoint in 2007 and made the unilateral decision to rebuild the Setata fence (Environmental Investigation Agency 2007), selecting the alignment least recommended by the environmental impact assessment. However, progress was made again in 2008, when the Botswana government announced that it would leave a 100-km section of fence to allow wildlife to move unencumbered and would follow a less objectionable alignment option (Botswana Office of the President 2008).

Discussion

Trends Across Studies

Data on the impact of veterinary fences have been collected over finite sampling periods and covered only a fraction of the existing fences; therefore, they vastly underestimate the true toll. However, they provide incontrovertible evidence that veterinary cordon fences have played a significant role in the deaths of thousands of mammals, the disappearance of mass migrations and the collapse of local populations.

Effects have been recorded at all levels of organisation (individual to ecosystem) and across time scales (short- and long-term). With the exception of small mammals and carnivores, most mammals in the vicinity experienced individual entanglement and mortality. However, large, migratory ungulates have been the most severely affected, experiencing repeated mass mortalities and population crashes.

In the short term, animals in the immediate vicinity are separated from vital resources, family groups and conspecifics. Populations are restricted into habitat fragments and isolated into smaller populations. During drought periods, animals are unable to reach their dry season destination. Diverted by fences, animals may crowd into blind corners or confined areas, reaching unnaturally high concentrations, where they suffer exhaustion and stress and may succumb to dehydration, starvation, and predation. Human hunters take advantage of these aggregations, quickly and easily slaughtering confined, exhausted animals.

Deaths continue decades after fence construction, as new generations try to disperse, degraded habitat or climatic conditions force animals to take new paths or attempt to restore old ones. Over a few years, migratory individuals may perish, and over generations, migratory behaviours may disappear. The relict population may be more sedentary, isolated and susceptible to extinction. Community structure may shift, favouring sedentary species. In the long-term, the instinct and the ability to migrate could be lost.

Fences create feast and famine situations; too many herbivores confined on one side can lead to overgrazing, trampling, disease outbreaks, habitat degradation, starvation, and eventual decrease in carrying capacity. Too few on one side can lead to local extinctions. Fences bisect habitat and divide populations, blocking corridors and severing connectivity of metapopulations and of wildlife areas. Over time, populations dependent on inflow of new individuals from source populations may crash.

When combined with other forces, including natural environmental stressors (drought, fire, disease), competition with other herbivores (particularly cattle and elephants) and hunting by humans, fence effects can be catastrophic.

Learning from Our Mistakes

We have every reason to expect the prevalence of fences and the negative impacts of fencing on wildlife to increase in the coming years. In addition to new fences, existing fences are being strengthened to withstand disease threats and breakage by elephants in some regions and unwanted wildlife in other regions, and to resist the uncontrolled movement of people.

We should anticipate that climate change may exacerbate fence impacts. Animals will need to move in response to changing rainfall patterns and resource distribution, and will be forced to track shifting prey species or habitat distribution. Animals may find their habitat or prey species moving north or south of fences. Fence sections which have not had significant impact may suddenly be in the way of vital range shifts. Disease dynamics and animal interactions are bound to change. Existing fences may not be properly placed to accommodate changing land uses and disease control needs. There may be demand for a new generation of fences on the new frontlines of the wildlife/livestock interface; therefore, it is imperative that we have a realistic understanding of the effects of existing fences and a strategy to minimise undesirable consequences.

Improved Planning and Monitoring

For too long, disease control policies have been planned and executed in isolation. A new approach that considers economic productivity, ecosystem function, biodiversity and human health is essential (Osofsky et al. 2008). Diseases are a very real threat, but control plans should be developed at the regional level, in consultation with key stakeholders including cattle producers, veterinary departments, human health advocates, wildlife agencies and local residents.

Policies and subsidies need to be examined closely for hidden costs. It has been suggested that the standards for international trade are antiquated and that different protocols need to be explored (Nair 2007; Scoones and Wolmer 2008; Thomson 2008), which could alleviate the need for FMD fencing. For example, advocates of commodity-based trade argue that hygienically processing healthy cattle in Africa (by removing bones and lymph nodes) before export would serve the dual purpose of eliminating the risk of disease transmission and adding value to the commodity, thus removing the need for physical barriers and allowing African nations to retain more revenue from their exported beef (Scoones and Wolmer 2008; Thomson 2008). International trade agreements with consumer countries, aid from developed nations and policies within African countries that encourage sub-division and fencing, need to be examined for their unintended effects.

Conscientious Consumers and Educated Donors

Consumers are becoming more conscientious about the ecological footprint of the food they buy, yet few European consumers are aware of where their beef comes from, and even fewer are cognizant of the links to African wildlife. Campaigns to label the country of origin of animal products and to publicise the ecological cost could result in improved consumer awareness and more discerning buyers.

New fences are often paid for under the aegis of economic development. Cost/benefit analyses need to appropriately value the wildlife resources affected. Donor nations have the right, and the obligation to conduct environmental, social and economic impact assessments before fencing.

Mitigation

Several simple steps could improve our understanding of existing fences. Firstly, we need to take stock of the existing fences. An accurate, centralised, spatially explicit database of all fences (including information on key fence attributes, like type of fence, condition, etc.) is critical to understanding the current fence network and to prioritising future construction and removals.

Impediments to research need to be removed so that impact assessments can be carried out in a transparent, objective manner. Researchers have been unable to collect paired before- and after- or long-term data, and advice on minimizing impact to wildlife has been ignored. When new fences are deemed absolutely necessary, objective, independent environmental impact assessments must be conducted prior to fence construction and recommendations adhered to. At a minimum, for any proposed new fencing, mammal distribution, migration pathways, resource distribution (including vegetation types, rainfall gradient and water sources), historic knowledge of movements and migrations, altitudinal gradients, present and proposed land uses should be taken into account. Repeated, systematic paired assessments should be conducted before and after fence construction to see how wildlife distribution and abundance changes. Longitudinal studies to determine how individuals and species are affected at a given location need to be initiated. Focal animals should be studied before and after construction in order to answer questions about how individuals respond to altered resource availability, home ranges or territories, and divided social groups.

Existing fences must be managed adaptively. Fence encounter and passage rates can be calculated through simple yet systematic, frequent monitoring of fences. With regular monitoring, we can better understand mortality patterns and, most importantly, events that trigger them. By anticipating when and where essential wildlife movements are likely to happen, and where the disease control trade-off is reasonable, proactive measures like fence removal can be undertaken. Identifying locations that are likely to be used by wildlife that could be left open without compromising restriction of cattle movement would be a vast improvement. Where mammal traffic or mortality is high along VCFs or where threatened species occur, changes need to be made. In high impact areas, important corridors, or at times of year when migration is essential, alternatives to fencing must be explored.

Fence design has been relatively unchanged for the last 50 years. Simple innovations that allow the passage of some animals while preventing others could lessen the effects of fences without compromising their disease control functions. For example, where cattle and buffalo are the species of concern, cattle grids could be installed and effectively maintained. These will prevent the passage of cattle and buffalo while elephants and other wildlife could move across, unimpeded. The removal of fences within emerging transfrontier conservation areas, such as Kruger National Park's border with Mozambique, provide a unique opportunity to study how animals respond to fence removal and to test new wildlife friendly fencing sections, in a controlled, replicated manner.

For decades, the impact of veterinary fences on wildlife has been swept under the rug. It is time to take an honest accounting of the impact thus far and to take steps to prevent such unnecessary damage in future.

Acknowledgements Many thanks to the people who have shared their knowledge, photos and data over the years,especially Arthur Albertson, Karen Ross, Mary Rice, Steve Osofsky and the AHEAD Greater Limpopo Working Group, and to Rowan Martin for permission to use Figure 9.1.

References

Albertson, A. (1997) *Botswana: death by cattle*. The Ntsu Foundation, Gabarone.

Albertson, A. (1998) *Northern Botswana Veterinary Fences*: Critical Ecological Impacts. The Wild Foundation/Kalahari Conservation Society, Gabarone.

Albertson, A. (2005) *Removal of decommissioned veterinary fences in Ngamiland District, Botswana*. Kalahari Conservation Society, The Sierra Club and The Wild Foundation, Gabarone.

Albertson, A. (2008) Botswana's Ministry of Agriculture's Proposal to re-erect the Setata fence: points for urgent consideration. *Cleft Stick* 2.

Bengis, R.G., Kock, R.A. & Fischer, J. (2002) Infectious animal diseases: the wildlife/livestock interface. *Revue Scientifique et Technique de l Office International des Epizooties*, **21**, 53–65.

Berry, H. (1983) The blue wildebeest problem at Etosha National Park. *African Wildlife*, **39**, 192–197.

Berry, H.H. & Siegfried, W.R. (1979) Stop-watch Wildebeest. *African Wildlife*. **33**(1), 17–21.

Blake, S., Deem, S.L., Strindberg, S., Maisels, F., Momont, L., Isia, I.-B., Douglas-Hamilton, I., Karesh, W.B. & Kock, M.D. (2008) Roadless wilderness area determines forest elephant movements in the Congo Basin. *PLoS ONE*, **3**, 1–9.

Boone, R.B. & Hobbs, N.T. (2004) Lines around fragments: effects of fencing on large herbivores. *African Journal of Range & Forage Science*, **21**, 147–158.

Booth, V., Hoare, R.E., & Mackie, C. (1998) Use of electric fences in communal areas of Zimbabwe. *Southern Africa regional workshop on fences* . In: K. Ross (ed.) Conservation International, Maun, Botswana. pp. 14–32.

Brashares, J.S. (2003) Ecological, behavioral, and life-history correlates of mammal extinctions in West Africa. *Conservation Biology*, **17**, 733–743.

Brook, B.W., Sodhi, N.S. & Bradshaw, C.J.A. (2008) Synergies among extinction drivers under global change. *Trends in Ecology and Evolution*, **23**, 453–460.

Caughley, G. (1976) The elephant problem - an alternative hypothesis. *East African Wildlife Journal*, **14**, 265–283.

Caughley, G. (1994) Directions in Conservation Biology. *The Journal of Animal Ecology*, **63**, 215–244.

Child, G. (1972) Observations on a wildebeest die-off in Botswana. *Arnoldia*, **5**, 1–13.

Clevenger, A.P. & Waltho, N. (2000) Factors Influencing the Effectiveness of Wildlife Underpasses in Banff National Park, Alberta, Canada. *Conservation Biology*, **14**, 47–56.

Conway, A.J. 1984. Anti-poaching measures in Chirisa Safari Area, Zimbabwe. in D.H.M Cumming and P. Jackson (eds.). *The Status and Conservation of Africa's Elephants and Rhinos*. International Union for the Conservation of Nature and Natural Resources, Gland, Switzerland. pp. 164–181.

Craig, B. (2007) *Wildlife Surveillance and Elephant Monitoring in Ol Pejeta Conservancy.* Final report to U.S. Fish & Wildlife Service.

Cumming, D.H.M., Fenton, M.B., Rautenback, I.L., Taylor, R.D., Cumming, G.S., Cumming, M.S., Dunlop, J.M., Ford, A.G., Hovorka, M.D., Johnson, D.S., Kalcounis, M., Mahlangu, Z. & Portfors, C.V.R. (1997) Elephants, woodlands and biodiversity in southern Africa. *South African Journal of Science*, **93**, 231–236.

de Queiroz, J.S. (1993) Range degradation in Botswana: myth or reality? *Pastoral Development Network Paper*, **35b**, 1–17. Overseas Development Institute, London.

du Toit, R. (2005) Foot and Mouth Disease management and land-use implications in the Zimbabwean Lowveld: the rationale for creating a Biosphere Reserve. *Conservation and development interventions at the wildlife/livestock interface: implications for wildlife, livestock and human health* (ed S. Osofsky), pp. 109–111. IUCN, Gland.

Environmental Investigation Agency (EIA) (2004) *A line in the sand: the unsustainable expansion of Botswana's beef industry at the expense of local communities and the Okavango Delta.* EIA, London.

Environmental Investigation Agency (EIA) (2005) *Dead end: a briefing document on the construction of a game-proof fence around the Makgadikgadi National Park, Botswana.* EIA, London.

Environmental Investigation Agency (EIA) (2007) *Background and implications of the decision to re-erect the Setata fence in Ngamiland.* EIA, London.

FAO (2005) Foot-and-mouth disease status in southern Africa. *EMPRES Transboundary animal disease bulletin*, **28**, 2–6.

Forman, R.T.T., Sperling, D., Bissonette, J.A., Clevenger, A.P., Cutshall, C., Dale, V.H., Fahrig, L., France, R., Goldman, C.R., Heanue, K., Jones, J.A., Swanson, F.J., Turrentine, T. & Winter, T.C. (2003) *Road Ecology* Island Press, Washington DC.

Gadd, M.E. (2001) *Final report to Botswana Department of Wildlife and National Parks: 2000–2001 Eastern Botswana Elephant Project.* Unpublished report.

Gadd, M.E. (2003) *Elephant ecology and conservation in African rangelands.* Ph.D. Thesis. University of California, Davis.

Gadd, M.E. (2007) *Preliminary comments on the potential effects of a new veterinary cordon fence across Northern Namibia/Southern Angola.* US Department of the Interior, Interior Technical Assistance Program (ITAP) for the Millennium Challenge Corporation (MCC).

Goodwin, J. (1985) Control of African Wildlife. Paper presented at 2nd Gallagher Wildlife Seminar. Dubbo, Australia. Gallagher Electronics Ltd. P/Bag, Hamilton, New Zealand. 24 pp.

Gupta C. (2004). *A study of the decision to construct the Makgadikgadi Wildlife Fence and its subsequent impacts on the Makgadikgadi Pans region.* (Research Report). Dartmouth College, Hanover.

Hanski, I. (1998) Metapopulation dynamics. *Nature*, **396**, 41–49.

Hargreaves, S.C.F., Anderson, E.C., Bastos, A.D.S., Thomson, G.R., Ferris, N.P. & Knowles, N.J. (2004) An investigation into the source and spread of foot and mouth disease virus from a wildlife conservancy in Zimbabwe. *Scientific and Technical Review of the Office International des Epizooties.*, **23**, 783–790.

Hoare, R.E. (1992) Present and future use of fencing in the management of larger African mammals. *Environmental conservation*, **19**, 160–164.

Hurlbert, S.H. (1984) Pseudoreplication and the design of ecological field experiments. *Ecological Monographs*, **54**, 187–211.

Kalikawe, M.C. (1997) *Wildlife friendly fencing for Botswana: a presentation of recommended procedures, specifications and important considerations.* Department of Wildlife and National Parks, Gabarone.

Kavadimba, S. (1998) Community case studies: impact of fences on wildlife resource use. *Southern Africa regional workshop on fences* (ed K. Ross), pp. 38–39. Conservation International, Maun.

Knight, M.H. (1995) Drought-related mortality of wildlife in the southern Kalahari and the role of man. *African Journal of Ecology*, **33**, 377–394.

Kock, R. (2005) What is this infamous "wildlife/livestock disease interface?" A review of current knowledge for the African continent. *Conservation and development interventions at the wildlife/livestock interface: implications for wildlife, livestock and human health* (ed S.A. Osofsky), pp. 1–13. IUCN, Gland.

Kock, R., Kebkiba, B., Heinonen, R. & Bedane, B. (2002) Wildlife and Pastoral Society; Shifting Paradigms in Disease Control. *Annals of the New York Academy of Sciences*, **969**, 24–33.

MacArthur, R.H. & Wilson, E.O. (1967) *The Theory of Island Biogeography.* Princeton University Press, Princeton.

Mapitse, N. (2008) *Botswana's foot-and-mouth disease and beef trade policy.* Institute of Development Studies, Brighton.

Martin, R.B. (2005) The influence of veterinary control fences on certain wild large mammal species in the Caprivi, Namibia. *Conservation and development interventions at the wildlife/livestock interface: implications for wildlife, livestock and human health* (ed S. Osofsky), pp. 27–39. IUCN, Gland.

McGarigal, K. & Cushman, S.A. (2002) Comparative evaluation of experimental approaches to the study of habitat fragmentation effects. *Ecological Applications*, **12**, 335–345.

Nair, M.K.S. (2007) Governance and upgrading opportunities in beef industry in Botswana: a value chain approach. *The Second International Conference on Management of Globally Distributed Work, Session 7: Capacity Management and Economics*, pp. 431–440.

Newmark, W.D. (1996) Insularization of Tanzanian parks and the local extinction of large mammals. *Conservation Biology*, **10**, 1549–1556.

Ogada, D., Gadd, M.E., Ostfeld, R., Young, T.P. & Keesing, F. (2008) Impacts of large herbivorous mammals on bird diversity and abundance in an African savanna. *Oecologia*, **156**, 387–397.

Osofsky, S.A., Cumming, D.H.M. & Kock, M.D. (2008) Transboundary management of natural resources and the importance of a 'One Health' approach: perspectives on Southern Africa. *State of the Wild 2008-2009: A Global Portrait of Wildlife, Wildlands, and Oceans* (ed E. Fearn), pp. 89–98. Island Press, Washington, D. C.

Owen, M. & Owen, D. (1980) The fences of death. *African Wildlife*, **34**, 25–27.

Parry, D. (1987) Wildebeest (*Connochaetes taurinus*) mortalities at Lake Xau, Botswana. *Botswana Notes and Records*, **19**, 95–101.

Perkins, J.S. (1996) Botswana: fencing out the equity issue. Cattleposts and cattle ranching in the Kalahari Desert. *Journal of Arid Environments*, **33**, 503–517.

Pierson, O. & Gadd, M.E. (2008) To fence or not to fence: use of wildlife data to influence U.S. Government foreign assistance decision-making in Northern Namibia. XXI Meeting of the Society for Conservation Biology (SCB), Chattanooga.

Roche, C. (2008) 'The fertile brain and inventive power of man': anthropogenic factors in the cessation of springbok treks and the disruption of the Karoo ecosystem, 1865–1908. *Africa*, **78**, 157–188.

Ross, K. (2003) *Okavango: Jewel of the Kalahari.* Struik Publishers, Cape Town.

Schumann, M., Schumann, B., Dickman, A., Watson, L.H. & Marker, L. (2006) Assessing the use of swing gates in game fences as a potential non-lethal predator exclusion technique. *South African Journal of Wildlife Research*, **36**, 173–181.

Scoones, I. & Wolmer, W. (2008) *Foot-and-mouth disease and market access: challenges for the beef industry in southern Africa.* Institute of Development Studies, Brighton.

Spinage, C.A. (1992) The decline of the Kalahari wildebeest. *Oryx*, **26**, 147–150.

Spinage, C.A. & Matlhare, J.M. (1992) Is the Kalahari cornucopia fact or fiction? A predictive model. *Journal of applied Ecology*, **29**, 605–610.

Sutmoller, P. (2002) The fencing issue relative to the control of Foot-and-Mouth Disease. *Annals of the New York Academy of Sciences*, **969**, 191–200.

Scott Wilson Resource Consultants (2000) *Environmental assessment of veterinary fences in Ngamiland – summary report.* Consultants' report to the Department of Animal Health and Production of Botswana. Scott Wilson Resource Consultants in association with The Environment and Development Group, Basingstoke.

Taylor, R.D. & Martin, R.B. (1987) Effects of veterinary fences on wildlife conservation in Zimbabwe. *Environmental Management*, **11**, 327–334.

Thomson, G. (2008) *A short overview of regional positions on foot-and-mouth disease control in southern Africa.* Institute of Development Studies, Brighton.

van Dyk, G. & Slotow, R. (2003) The effects of fences and lions on the ecology of African wild dogs reintroduced to Pilanesberg National Park, South Africa. *African Zoology*, **38**, 79–94.

Weaver, L.C. (1997) Background and potential impacts resulting from construction of a game and livestock proof fence by the Government of Botswana south of the West Caprivi Game Reserve. WWF/LIFE Programme.

Williamson, D. (2002) Dimensions and dilemmas of conservation in the Kalahari. *Open Country*, **4**, 15–24.

Williamson, D. & Mbano (1988) Wildebeest mortality during 1983 at Lake Xau, Botswana. *African Journal of Ecology*, **26**, 341–344.

Williamson, D. & Williamson, J. (1981) An assessment of the impact of fences on large herbivore biomass in the Kalahari. *Botswana Notes and Records*, **13**, 107–110.

Young, T.P., Palmer, T.M. & Gadd, M.E. (2005) Competition and compensation among cattle, zebras, and elephants in a semi-arid savannah in Laikipia, Kenya. *Biological Conservation*, **122**, 351–359.

Chapter 10
Modelling the Effect of Fences on the Viability of Spatially Structured Populations of African Wild Dogs

Michael J. Somers, Markus Gusset, and Fredrik Dalerum

Introduction

Fences have become a common management tool to counteract the increasing pressure on protected areas and their animals from encroachment of an expanding human population, increased prevalence of diseases and increased spread of alien invasive species. However, although fences often have positive effects by protecting land from urban sprawl, and from poaching or encroaching livestock, they may also have negative effects on animal movement and demographics (reviewed in Hayward and Kerley 2009). Of these negative effects, one of the most important is that fences often restrict the movement of the enclosed animals (Newmark 2008), thereby creating a network of isolated populations. If fences are impenetrable, there will be no interpopulation movement, which has demographic consequences (e.g. Somers et al. 2008).

Movement between subpopulations is an important process in the dynamics of spatially structured populations, because it provides the linkage between subpopulations and enables the colonization of unoccupied areas by dispersing individuals (Brachet et al. 1999; Thomas 2000; Revilla and Wiegand 2008). Movement also ensures the long-term persistence of metapopulations (Hanski 1998). Because conservation areas cover only a small fraction of the total land area, the management of matrix habitat in between isolated populations becomes increasingly important

M.J. Somers (✉)
Centre for Wildlife Management, Centre for Invasion Biology, Mammal Research Institute, University of Pretoria, Pretoria, 0002, South Africa
e-mail: mjs@up.ac.za

M. Gusset
Wildlife Conservation Research Unit, Department of Zoology, University of Oxford, Oxford, UK

F. Dalerum
Centre for Wildlife Management, Mammal Research Institute, Department of Zoology and Entomology, University of Pretoria, Pretoria, South Africa

M.J. Somers and M.W. Hayward (eds.), *Fencing for Conservation: Restriction of Evolutionary Potential or a Riposte to Threatening Processes?*,
DOI 10.1007/978-1-4614-0902-1_10, © Springer Science+Business Media, LLC 2012

(Prugh et al. 2008; Franklin and Lindenmayer 2009; Prevedello and Vieira 2010; Watling et al. 2011). For instance, Akçakaya et al. (2007) emphasized that the viability of a population may depend on surrounding populations, in which case metapopulation processes influence or determine reserve design and management options.

Large carnivores are often persecuted in rural areas because of the damage they do to livestock, the fact that they compete with humans over game or the direct threat they pose to humans (Kruuk 2002). This has led to heavy persecution, and subsequently for many species to local, regional or global extinction (Dalerum et al. 2009). Because of their body size and potential conflict with humans, many populations of large carnivores are enclosed by fences, which often have led to detrimental levels of population fragmentation (Hayward and Kerley 2009). This fence-driven fragmentation process has been accentuated by an increase in large carnivore reintroductions (Hayward and Somers 2009). Although substantial efforts are devoted to keep fences secure, many large carnivores are able to penetrate them, possibly decreasing the negative demographic effects of fences. Such breakouts often occur following floods when fences placed across rivers are washed away, when poachers break fences to gain access to conservation areas or simply through holes made by other species such as warthogs *Phacochoerus africanus* (Somers, unpublished data).

The African wild dog *Lycaon pictus* is a large (20–30 kg) canid that lives in complex social groups of up to 24 individuals (Creel and Creel 2002; Somers et al. 2008). It occurs throughout central, eastern and southern Africa. Following heavy persecution, the species became endangered during the twentieth century and it remains at a fraction of its previous population size. In South Africa, wild dog conservation has focused on reintroducing packs within fenced conservation areas and on facilitating movement between these isolated populations through translocations (Davies-Mostert et al. 2009; Gusset 2010). Wild dogs usually disperse in single-sex groups at the age of 1–2 years. Individuals rarely breed if they remain in their natal packs beyond this age because wild dogs usually do not breed with close relatives (Somers et al. 2008; Spiering et al. 2011). Therefore, unless animals of dispersal age are actively translocated (Gusset et al. 2009), they are effectively demographically lost to the South African managed metapopulation.

Although persecution of wild dogs outside of conservation areas in South Africa may be substantial (Gusset et al. 2008), it is potentially beneficial if dispersing wild dogs manage to penetrate fences since they then may meet unrelated dispersers of the opposite sex and form new packs. This may be inside conservation areas or elsewhere. If such spontaneous dispersal is sufficiently common, a natural metapopulation will form (Hanski 1998). This has large demographic and genetic benefits compared to isolated populations where the only method of dispersal between subpopulations would be by human translocation (Frankham 2009). Therefore, evaluating the influence of fences on wild dog demography and population persistence is central to the conservation of the species particularly in southern Africa, where fences are prevalent and populations are fragmented. Here, we use stochastic population models to investigate the demographic effects of varying levels of fence penetrability on the viability of spatially structured wild dog populations.

Methods

We modified a stochastic population model previously described in Dalerum et al. (2008). The model is temporally discrete with a simple sex and age structure, and can be formalized as:

$$P_i = AM_i + AF_i + SM_i + SF_i + J_i, \tag{10.1}$$

where P_i, AM_i, AF_i, SM_i, SF_i and J_i are population size, number of adult males and females, number of subadult males and females and number of juveniles at year i, respectively. For our purposes, we regarded adults as animals 2 years or older, subadults as animals from 1 to 2 years and juveniles as young of the year (Somers et al. 2008). We calculated the number of animals in each age and sex category as:

$$J_i = AF_i \cdot f_i, \tag{10.2}$$

$$\begin{aligned} SM_i &= 0.5 \cdot J_{i-1} \cdot js_i + IMS_i - EMS_i, \\ SF_i &= 0.5 \cdot J_{i-1} \cdot js_i + IFS_i - EFS_i, \end{aligned} \tag{10.3}$$

$$\begin{aligned} AM_i &= AM_{i-1} \cdot mas_i + SM_{i-1} \cdot mss_{i-1} - EMA_i + IMA_i, \\ AF_i &= AF_{i-1} \cdot fas_i + SF_{i-1} \cdot fss_{i-1} - EFA_i + IFA_i, \end{aligned} \tag{10.4}$$

where f_i is annual birth rate (number of pups per female per year after weaning; weaning sex ratio is estimated to be 1:1), js_i is juvenile survival, mss_i and fss_i are subadult survival for males and females, mas_i and fas_i are adult survival for males and females, respectively, EMS_i, EFS_i, EMA_i and EFA_i are the net number of subadult and adult emigrating males and females, and IMS_i, IFS_i, IMA_i and IFA_i are the net number of subadult and adult immigrating males and females.

We ran the model with discrete 1-year time steps, and for each year we drew female fecundity from a normal distribution, and survival from a binomial distribution. We calculated the number of emigrants by multiplying the number of animals in each age and sex category with binomial probabilities of dispersal. We calculated the number of immigrants from emigrating individuals from connected populations.

To evaluate the effect of fences, we multiplied the number of immigrants from a particular population with a scaling factor ranging from 0 to 1, so that a scaling factor of 0 would result in no immigrants and a scaling factor of 1 would result in all potential immigrants entering the population. We regarded this scaling factor as a theoretical proxy for fences of varying penetrability, so that a scaling factor (i.e. fence penetrability) of 0 would represent a completely wild dog proof fence and a fence penetrability of 1 would represent an unfenced population boundary. This approach controlled only the net number of immigrants, but not the number of emigrants. We used this "one way" fence approach (i.e. restricting immigration but not emigration) since we modelled relatively small populations and any prospective

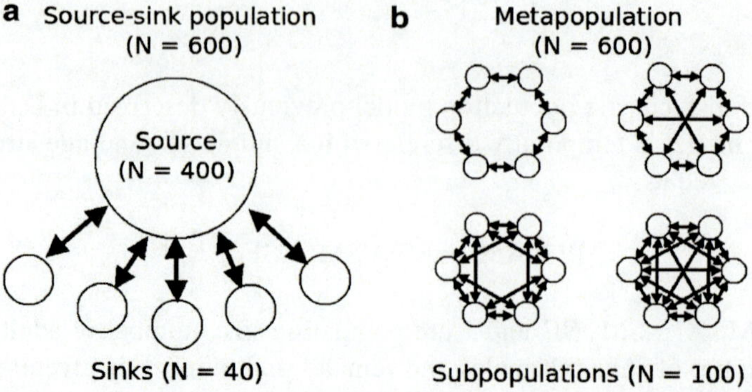

Fig. 10.1 Conceptual description of model scenarios. We explored the effect of fences on (**a**) source–sink populations consisting of one source population with 400 individuals and five sink populations with 40 individuals each, and (**b**) four different metapopulation scenarios, each with six subpopulations consisting of 100 animals each but with varying connectivity (ranging from each subpopulation being connected to its two nearest neighbours to each subpopulation being connected to all subpopulations in the metapopulation)

emigrants that would have been locked in by a fence would most likely not be demographically active (Somers et al. 2008). We repeated each subset of simulations over the full range of fence penetrability values 1,000 times, and for each simulation we coded population size at 25 years as either below the initial population size or not, or extinct or not-extinct. Based on these binary codes we used logistic regression models to calculate probabilities of population decline and extinction over varying levels of fence penetrability (e.g. McCarthy et al. 1995, 1996; Cross and Beissinger 2001; Dalerum et al. 2008).

We considered two different scenarios of wild dog populations. Both are relevant from a conservation perspective (see Davies-Mostert et al. 2009). First, we considered a source–sink scenario with one large source and five small sink populations, with each sink population being 10% of the size of the source population (Fig. 10.1a). Secondly, we considered a metapopulation scenario with six evenly sized subpopulations (Fig. 10.1b). For the metapopulation scenario, we altered the connectivity of the subpopulations to range from one to five (i.e. the number of connected populations for each subpopulation), and also the number of connected populations that were affected by the fence variable (ranging from one to the maximum number of connected populations for each scenario).

Fecundity, survival, demographic structure and dispersal parameters underlying the simulations are given in Table 10.1. To avoid unlimited exponential growth, we capped source and sink populations at a carrying capacity of 500 individuals, and subpopulations within the metapopulaton at 200 individuals (see below for descriptions of the two demographic scenarios). However, for simplicity we did not include density-dependent effects on demographic parameters until the carrying capacity was reached.

Table 10.1 Parameters underlying our modelling approach to evaluate the effect of fences on the viability of spatially structured wild dog populations

Parameter	Value	Source
Adult males (%)	29.86	Creel and Creel (2002)
Adult females (%)	27.09	Creel and Creel (2002)
Subadult males (%)	8.03	Creel and Creel (2002)
Subadult females (%)	9.84	Creel and Creel (2002)
Juveniles (%)	25.18	Creel and Creel (2002)
Fecundity mean (offspring/female/year)	2.4	Estimated from Creel and Creel (2002)
Fecundity SD (offspring/female/year)	1.2	Estimated from Creel and Creel (2002)
Juvenile female survival (annual)	0.75	Creel and Creel (2002)
Juvenile male survival (annual)	0.66	Creel and Creel (2002)
Subadult female survival (annual)	0.84	Creel and Creel (2002)
Subadult female survival (annual)	0.99	Creel and Creel (2002)
Adult female survival (annual)	0.69	Creel and Creel (2002)
Adult male survival (annual)	0.73	Creel and Creel (2002)
Subadult male migration (proportion animals)	0.30	Creel and Creel (2002)
Subadult female migration (proportion animals)	0.49	Creel and Creel (2002)
Adult male migration (proportion animals)	0.10	Creel and Creel (2002)
Adult female migration (proportion animals)	0.11	Creel and Creel (2002)
Sink population size mean	40	Estimated from Davies-Mostert et al. (2009)

Results

In the source–sink scenario, completely wild dog proof fences (i.e. with zero penetrability) generated substantial probabilities of population decline in all five sink populations (>75%), and over 20% probability of having up to three sink populations going extinct within 25 years. Fences with less than 25% penetrability generated almost 100% probabilities of population decline in at least one of the sink populations over 25 years (Fig. 10.2a), and generated associated risks above zero of having at least one sink population going extinct (Fig. 10.2b). There appears to be a threshold of fence penetrability at about 50%; populations enclosed by fences with higher penetrability had substantially lower probabilities of decline and almost zero probability of extinction (Fig. 10.2).

In the metapopulation scenario, both connectivity within metapopulations and number of fenced population connections affected how important fences were for wild dog population viability. At least half of the population connections had to be fenced for fences to have a substantial effect on the probabilities of population decline (Fig. 10.3a), and all population connections had to be fenced for fences to have any effect on subpopulation extinction probabilities (Fig. 10.3b). Even with all connections affected by fences, as is typically the case in South Africa, there appears to be a threshold at approximately 50% penetrability for fences to have substantial effects on the

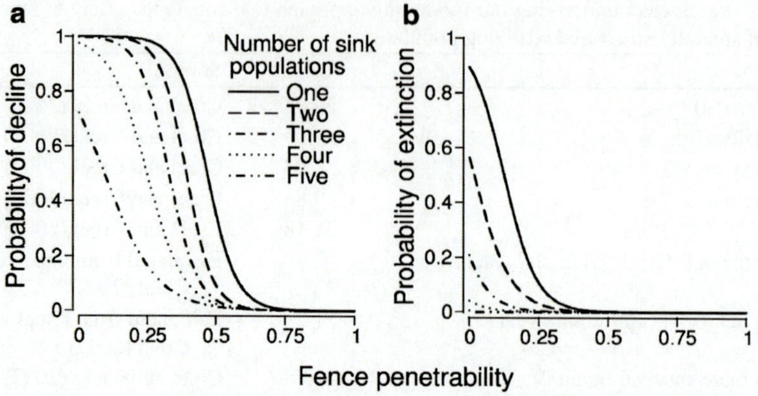

Fig. 10.2 Probability of (**a**) decline and (**b**) extinction after 25 years of simulations of theoretical wild dog populations, each consisting of a source population with 400 animals as well as five sink populations with 40 animals each. Each *line* represents the probabilities of decline and extinction in one to five sink populations. The model assumes that animals not being able to disperse were not demographically active while remaining in their population of origin

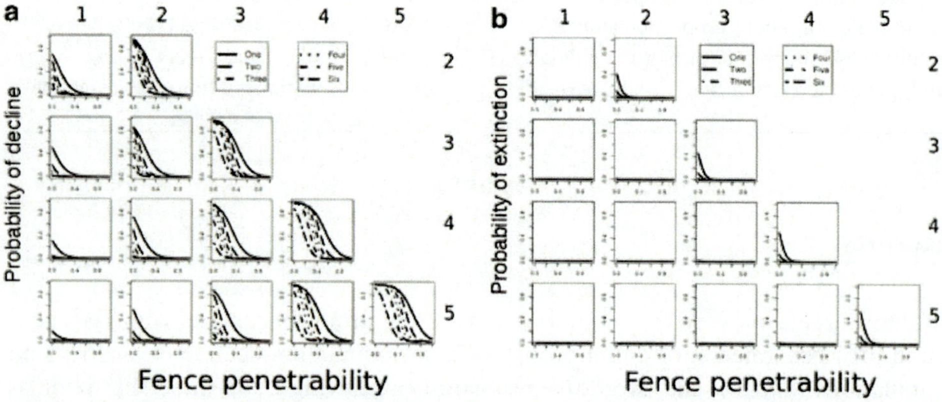

Fig. 10.3 Probability of (**a**) decline and (**b**) extinction after 25 years of simulations of theoretical wild dog metapopulations, each consisting of six subpopulations with 100 animals each. Each *row* represents the number of connecting subpopulations (ranging from two to five) and each *column* represents the number of fenced subpopulation connections (ranging from one to the maximum number of connections for each model scenario)

probabilities of subpopulation decline (Fig. 10.3a), and a threshold at approximately 20% penetrability for fences to substantially affect extinction probabilities (Fig. 10.3b).

The effects of fences on the probabilities of both decline and extinction in at least one subpopulation were higher in the source–sink than in the metapopulation scenario, unless all subpopulation connections in the metapopulation were fenced (Table 10.2). Moreover, metapopulations were more sensitive to the effects of fences if they had a large number of connections, so that fences in metapopulations with a large number of connections could have higher penetrability but still generate substantial probabilities of decline (Table 10.2).

Table 10.2 Fence penetrability (ranging from 0 being complete wild dog proof fence to 1 being no fence), generating 25, 50 and 75% probabilities of decline and extinction after 25 years of simulations in at least one subpopulation of a source–sink population and in metapopulations with varying levels of connectivity and with different numbers of fenced subpopulation connections

Population scenario	Number of connections	Number of fenced connections	Decline 25%	Decline 50%	Decline 75%	Extinction 25%	Extinction 50%	Extinction 75%
Source–sink			0.53	0.47	0.41	0.19	0.12	0.06
Metapopulation	2	1	0.23	0.10	0	0	0	0
		2	0.44	0.31	0.18	0.04	0	0
	3	1	0.13	0	0	0	0	0
		2	0.34	0.21	0.09	0	0	0
		3	0.56	0.43	0.30	0.05	0	0
	4	1	0.03	0	0	0	0	0
		2	0.24	0.12	0	0	0	0
		3	0.46	0.33	0.20	0	0	0
		4	0.67	0.54	0.41	0.06	0	0
	5	1	0	0	0	0	0	0
		2	0.15	0.02	0	0	0	0
		3	0.36	0.23	0.10	0	0	0
		4	0.57	0.44	0.32	0	0	0
		5	0.79	0.66	0.53	0.07	0.01	0

Discussion

Our results highlight the fact that fences can substantially affect the viability of source–sink populations of wild dogs through constraints in dispersal, and that fences efficient in limiting wild dog movements can generate substantial extinction probabilities in such populations. Our results further indicate that the connectivity within metapopulations influences the effects of fences on population viability. In both scenarios, there appears to be two crude thresholds of fence penetrability; fences with about 50% penetrability or less seem to generate substantial probabilities of decline, whereas fences with 20% penetrability or less seem to generate probabilities of extinction that are above zero. Although our analyses should be interpreted in a largely qualitative context, these results still suggest that entirely predator-proof fences are detrimental for the conservation of large carnivore species such as wild dogs, unless the animals are periodically translocated between fenced conservation areas (Gusset et al. 2009). However, the penetrability of fences typically varies owing to varying levels of maintenance, flooding or destruction by animals such as elephants *Loxodonta africana*, so that completely predator-proof fences are, in reality, probably absent or very rare.

The demographic impacts of fences will depend both on the level of density-dependent regulation in fenced populations and the mortality rates in unfenced areas. Therefore, fences may not have detrimental effects on all wild dog populations since they might prevent animals from suffering persecution and other human-related mortality outside of fenced areas. For instance, fence length, a surrogate for the level of fence maintenance, was negatively related to the survival of reintroduced wild dogs in South Africa (Gusset et al. 2008). Furthermore, the only reintroduced pack within the South African managed metapopulation that had a mortality risk higher than the population average was released into the only area that was not entirely fenced (Gusset et al. 2010). The major cause for the increased mortality among these wild dogs was snaring immediately outside of the reserve. However, contrasting these results are findings that wild dogs succeed in dispersing over long distances, and occasionally between fenced conservation areas (e.g. 100 km between Hluhluwe-iMfolozi Park to Ithala Game Reserve; Somers, unpublished data). These inconsistent results regarding the effect of fences on wild dog demographics highlight the complexities in evaluating the consequences of fences for populations of large carnivores. Therefore, we stress that our suggestion that a certain level of fence penetrability may be beneficial for large carnivore conservation does not apply to all species and for all management scenarios. Instead, the influence of dispersal barriers such as fences should optimally be evaluated for each species and situation separately. It is only in such context-dependent evaluations that species-specific dispersal behaviour can be considered and weighted against estimated mortality risks in a matrix habitat.

To conclude, our results suggest that fences can generate substantial probabilities of decline and extinction probabilities above zero in both source–sink populations and metapopulations of wild dogs. However, these suggestions of purely negative effects of fences are contradicted by empirical data from South Africa where wild dogs suffered higher mortality in areas with higher fence penetrability due to human-related mortality outside of conservation areas (Gusset et al. 2008, 2010). We argue that these inconsistencies are caused by the relative effects of the levels of density-dependent constraints on population growth inside fenced reserves (Somers et al. 2008) and mortality rates in matrix habitat. However, we still acknowledge that large protected areas probably are the best way to protect biodiversity, especially wide-ranging species such as large carnivores (Mills 2005; Hayward and Kerley 2009).

Acknowledgements We are grateful to Mark Boyce, Matt Hayward and Kyran Kunkel for helpful comments on this chapter.

References

Akçakaya, H.R., Mills, M.G.L. & Doncaster, C.P. (2007) The role of metapopulations in conservation. In: *Key topics in conservation biology* (eds. Macdonald, D.W. & Service, K.), pp. 64–84. Oxford: Blackwell Publishing.

Brachet, S., Olivieri, I., Godelle, B., Klein, E., Frascaria-Lacoste, N. & Gouyon, P.-H. (1999) Dispersal and metapopulation viability in a heterogeneous landscape. *Journal of Theoretical Biology*, **198**, 479–495.

Creel, S. & Creel, N.M. (2002) *The African wild dog: behavior, ecology, and conservation*. Princeton, NJ: Princeton University Press.

Cross, P.C. & Beissinger, S.R. (2001) Using logistic regression to analyze the sensitivity of PVA models: a comparison of methods based on African wild dog models. *Conservation Biology*, **15**, 1335–1346.

Dalerum, F., Shults, B. & Kunkel, K. (2008) Estimating sustainable harvest in wolverine populations using logistic regression. *Journal of Wildlife Management*, **72**, 1125–1132.

Dalerum, F., Cameron, E.Z., Kunkel, K. & Somers, M.J. (2009) Diversity and depletions in continental carnivore guilds: implications for prioritizing global carnivore conservation. *Biology Letters*, **5**, 35–38.

Davies-Mostert, H.T., Mills, M.G.L. & Macdonald, D.W. (2009) A critical assessment of South Africa's managed metapopulation recovery strategy for African wild dogs. In: *Reintroduction of top-order predators* (eds. Hayward, M.W. & Somers, M.J.), pp. 10–42. Oxford: Wiley-Blackwell.

Frankham, R. (2009) Genetic considerations in reintroduction programmes for top-order, terrestrial predators. In: *Reintroduction of top-order predators* (eds. Hayward, M. W. & Somers, M. J.), pp. 371–387. Oxford: Wiley-Blackwell.

Franklin, J.F. & Lindenmayer, D.B. (2009) Importance of matrix habitats in maintaining biological diversity. *Proceedings of the National Academy of Sciences of the USA*, **106**, 349–350.

Gusset, M. (2010) The re-introduction of African wild dogs in South Africa. In: *Global re-introduction perspectives: additional case-studies from around the globe* (ed. Soorae, P. S.), pp. 220–224. Abu Dhabi: IUCN/SSC Re-introduction Specialist Group.

Gusset, M., Ryan, S.J., Hofmeyr, M., van Dyk, G., Davies-Mostert, H.T., Graf, J.A., Owen, C., Szykman, M., Macdonald, D.W., Monfort, S.L., Wildt, D.E., Maddock, A.H., Mills, M.G.L., Slotow, R. & Somers, M.J. (2008) Efforts going to the dogs? Evaluating attempts to re-introduce endangered wild dogs in South Africa. *Journal of Applied Ecology*, **45**, 100–108.

Gusset, M., Jakoby, O., Müller, M.S., Somers, M.J., Slotow, R. & Grimm, V. (2009) Dogs on the catwalk: modelling re-introduction and translocation of endangered wild dogs in South Africa. *Biological Conservation*, **142**, 2774–2781.

Gusset, M., Stewart, G.B., Bowler, D.E. & Pullin, A.S. (2010) Wild dog reintroductions in South Africa: a systematic review and cross-validation of an endangered species recovery programme. *Journal for Nature Conservation*, **18**, 230–234.

Hanski, I. (1998) Metapopulation dynamics. *Nature*, **396**, 41–49.

Hayward, M.W. & Kerley, G.I.H. (2009) Fencing for conservation: restriction of evolutionary potential or a riposte to threatening processes? *Biological Conservation*, **142**, 1–13.

Hayward, M.W. & Somers, M.J. (eds) (2009) *Reintroduction of top-order predators*. Oxford: Wiley-Blackwell.

Kruuk, H. (2002) *Hunter and hunted: relationships between carnivores and people*. Cambridge: Cambridge University Press.

McCarthy, M.A., Burgman, M.A. & Ferson, S. (1995) Sensitivity analysis for models of population viability. *Biological Conservation*, **73**, 93–100.

McCarthy, M.A., Burgman, M.A. & Ferson, S. (1996) Logistic sensitivity and bounds for extinction risks. *Ecological Modelling*, **86**, 297–303.

Mills, M.G.L. (2005) Large carnivores and biodiversity in African savanna ecosystems. In: *Large carnivores and the conservation of biodiversity* (Ed. by Ray, J. C., Redford, K. H., Steneck, R. S. & Berger, J.), pp. 208–229. Washington, DC: Island Press.

Newmark, W.D. (2008) Isolation of African protected areas. *Frontiers in Ecology and the Environment*, **6**, 321–328.

Prevedello, J.A. & Vieira, M.V. (2010) Does the type of matrix matter? A quantitative review of the evidence. *Biodiversity and Conservation*, **19**, 1205–1223.

Prugh, L.R., Hodges, K.E., Sinclair, A.R.E. & Brashares, J.S. (2008) Effect of habitat area and isolation on fragmented animal populations. *Proceedings of the National Academy of Sciences of the USA*, **105**, 20770–20775.

Revilla, E. & Wiegand, T. (2008) Individual movement behavior, matrix heterogeneity, and the dynamics of spatially structured populations. *Proceedings of the National Academy of Sciences of the USA*, **105**, 19120–19125.

Somers, M.J., Graf, J.A., Szykman, M., Slotow, R. & Gusset, M. (2008) Dynamics of a small re-introduced population of wild dogs over 25 years: Allee effects and the implications of sociality for endangered species' recovery. *Oecologia*, **158**, 239–247.

Spiering, P.A., Szykman Gunther, M., Somers, M.J., Wildt, D.E., Walters, M., Wilson, A.S. & Maldonado, J.E. (2011) Inbreeding, heterozygosity and fitness in a reintroduced population of endangered African wild dogs (*Lycaon pictus*). *Conservation Genetics*, **12**, 401–412.

Thomas, C.D. (2000) Dispersal and extinction in fragmented landscapes. *Proceedings of the Royal Society B*, **267**, 139–145.

Watling, J.I., Nowakowski, A.J., Donnelly, M.A. & Orrock, J.L. (2011) Meta-analysis reveals the importance of matrix composition for animals in fragmented habitat. *Global Ecology and Biogeography*, **20**, 209–217.

Chapter 11
Towards a True Ecology: Exploring the Implications for Conservation of the Human and Social Dimensions of Fencing in the Subtropical Thicket Biome, South Africa

Andrew T. Knight and Richard M. Cowling

Introduction

Regional-scale conservation planning initiatives are undertaken to mobilise collective action by communities to achieve strategic nature conservation goals at sites of conservation importance (e.g., Cowling and Pressey 2003; Rouget et al. 2006; Smith et al. 2008). Increasingly, spatial prioritisation techniques are applied by conservation planning initiatives (Pressey 2002) to explicitly locate and design areas where optimal suites of conservation instruments (Young et al. 1996) can be most effectively and efficiently implemented to ensure the conservation and sustainable management of nature. Despite controversy surrounding the design and efficacy of local-scale conservation corridors, due, for example, to uncertainty surrounding actual use by target species and their potential to facilitate spread of invasive alien species (Hobbs 1992; Simberloff et al. 1992; Dobson et al. 1999), regional-scale corridors are increasingly designed to achieve conservation goals. This is a response to the rising awareness of the importance of protecting environmental processes which ensure the persistence of biodiversity and their links to the complex and systemic functioning of social–ecological systems. Environmental processes such as migration, speciation and climate change have been included in both target-driven (e.g., Cowling et al. 1999, 2003; Carroll et al. 2001; Rouget et al. 2006) and non-target-driven (Conservation International 2000; Dinerstein et al. 2000;

A.T. Knight (✉)
Department of Botany, Nelson Mandela Metropolitan University, P.O. Box 77000, Port Elizabeth, South Africa

Department of Conservation Ecology and Entomology, Stellenbosch University, Private Bag X1, Matieland, 7602, South Africa
e-mail: tawnyfrogmouth@gmail.com

R.M. Cowling
Department of Botany, Nelson Mandela Metropolitan University, P.O. Box 77000, Port Elizabeth, South Africa

M.J. Somers and M.W. Hayward (eds.), *Fencing for Conservation: Restriction of Evolutionary Potential or a Riposte to Threatening Processes?*,
DOI 10.1007/978-1-4614-0902-1_11,

Sanderson et al. 2002; Muruthi 2004; Beier et al. 2008) spatially explicit regional conservation corridor designs.

Practical conservation planning initiatives aim to deploy a Landscape Management Model (Knight et al. 2006a) that comprises a diverse but optimal suite of conservation instruments: establishing formally protected areas, entering into voluntary and/or binding agreements with private land managers, offering incentives, restoring habitat and supporting institutions (Young et al. 1996). Conservation corridors are often designed to link existing protected areas across privately owned land (e.g., Rouget et al. 2006; Beier et al. 2008). Private lands are typically managed as small land parcels (cadastres) which may be managed in groups as properties by numerous individual land managers. Land managers typically have diverse management goals and activities and, though regulated by government, often manage autonomously. Properties and cadastres are managed at a scale far finer than the scale at which these landscapes function (Briggs 2001), and are often separately and independently fenced. Depending upon the management approach, individual cadastres may be further internally subdivided with fencing. This fine-scale partitioning of the landscape by fences disrupts the functioning of environmental processes, countering the goal of conservation corridor initiatives. A vexing question for those designing and implementing conservation corridors is: "How do we promote collective action by land managers to maintain environmental processes across fenced landscapes dominated by private tenure"? There are multiple significant challenges to providing a long-term solution to this question.

Designing Corridors Using Conservation Opportunity

When aiming to effectively implement a regional-scale conservation corridor, it is essential that it be planned using a spatial prioritisation technique (Knight et al. 2006b; Rouget et al. 2006). Historically, these techniques have applied ecological data alone, sometimes coupled with vulnerability data, to determine conservation priorities for areas (Pressey 1997; Pressey and Taffs 2001). However, mapping opportunities for, and constraints upon, implementing effective conservation action has been promoted as a more effective approach to improve the likelihood of these initiatives achieving their conservation goals (Cowling et al. 2004; Knight and Cowling 2007; Knight et al. 2010).

Mapping conservation opportunity provides an explicit approach for planning the integration of different types of capital. Much as environmental processes are essential for ensuring the persistence of nature managed within conservation corridors, human and social capital must also be explicitly understood and managed, because these determine the processes and locations of land-use pressures, and the feasibility of effective implementation. We hypothesise that human and social capital interact together to facilitate connectivity throughout a corridor. Human capital (e.g., the knowledge, willingness and capacity of individual people) forms the basic building-blocks essential for collective action. If individuals are unable or unwilling to implement conservation action, then the collective action essential for mobilising the hundreds or thousands of people required to effectively implement regional-scale conservation corridors is impossible. Social capital provides the "glue" that forges the partnerships essential to mobilise individuals collectively towards a common conservation goal. Factors defining social capital include social networks, confidence in governance, a local sense of belonging (Grootaert and van Bastelaer 2001) and willingness to collaborate. In turn, social capital is underpinned by, but also influences, a range of human capital factors that enable the processes which maintain social capital to function effectively. These are hypothesised to include multiple factors (Knight et al. 2010) such as whether an individual person (often a land manager): suffers burnout (Maslach et al. 1996); is recognised as a local or regional champion (someone who provides leadership); is entrepreneurial; has the financial capacity to participate (as conservation goals are typically decided by society, but costs often realised by individual land managers); fosters strong interpersonal relationships; is willing to collaborate with other stakeholders; and/or is willing to participate in implementing specific conservation instruments (e.g., a covenant). Other factors may also be important, depending on the context. Actively managing human and social capital is as important as managing other capital (e.g., natural, financial) if effective conservation action and, ultimately, adaptive management are to be implemented. This can be achieved through processes that enhance connectivity between people, notably institutions which facilitate collaboration, negotiation and conflict resolution, social learning and consilience (the fusion of knowledge traditions; Wilson 1998).

Our conceptual framework recognises five dimensions of conservation opportunity: (1) conservation value; (2) vulnerability; (3) economic costs of implementation; (4) human capital; and (5) social capital (Knight et al. 2010). These factors are formulated in the context of an optimal suite of conservation instruments (Young et al. 1996) useful for securing important areas, and are collectively constituted as a Landscape Management Model (Knight et al. 2006a). Multiple factors define each of these dimensions: for example, vulnerability may comprise exposure, intensity and impact (Wilson et al. 2005). Factors defining individual dimensions vary from region to region through differences in regional characteristics and available data, meaning specific factors, are best defined on a case-by-case basis.

Fencing and Human and Social Capital Within Corridors

Research into the ecological impact of fencing confirms that fencing inhibits the functioning of environmental processes (e.g., Hayward and Kerley 2009; de Tores and Marlow 2012). In contrast, studies examining the influence of human and social capital on ensuring connectivity across landscapes are rare. This is unfortunate, and perhaps surprising, given the interplay which exists between ecological connectivity (as impacted by fencing) and social connectivity. The ecological functioning of a corridor is directly impacted by land managers' goals and activities (Cowling and

Pressey 2003). In a landscape partitioned into many small cadastres through private land tenure, and its associated diversity of management activities, ecological connectivity is enhanced when social connectivity is improved. Fences demarcate land ownership, a partitioning of resources, separating "mine" from "yours". They also simplify management of individual cadastres, as individual owners manage their cadastres autonomously. This removes the need to consult or collaborate with neighbours, which cannot occur effectively without strong social capital (i.e., trust, a shared sense of place, and social networks) (Brunckhorst 2002; Brunckhorst and Coop 2003). In practice, social capital (through strong relationships) "transcends" fencing, and so is a prerequisite for ensuring ecological connectivity on private fenced land. Having multiple land managers remove fences between, and within, their properties is a highly complex process as it requires that they: (1) have a common vision of future landscape management; (2) practice land-uses compatible with an absence of fencing; (3) be willing to embrace collective action; (4) have the capacity to make the required changes to their land management activities; and (5) be willing to establish an institution through which to orchestrate their activities.

Unfortunately, conservation corridor initiatives rarely evaluate the human and social capital required to identify a conservation opportunity that can promote effective connectivity. The mapping of a suite of human and social capital indices (for example, Table 11.1) specifically to inform the design and establishment of a conservation corridor provides insights into factors defining the likelihood of proposed conservation actions being effective (Knight et al. 2010). Human factors, such as burnout and willingness-to-participate, are hypothesized to influence such social processes as collaboration and social learning which determine whether conservation goals will be achieved. Mapping conservation opportunity provides a technique to promote landscape-scale understanding of land managers' values and behaviour and to ensure that the design and implementation of a conservation corridor is feasible, cost-effective and reduces conflict.

A Case Study from the Subtropical Thicket Biome

The Subtropical Thicket biome is located primarily within South Africa, and forms the south-western portion of the Maputaland–Pondoland–Albany "hotspot", one of three identified for southern Africa (Mittermeier et al. 2004). It houses over 1,550 plant species, 20% of which are endemic (Vlok et al. 2003). The majority of farming ventures comprise commercial pastoralism, primarily goats and sheep, though farming of indigenous game for both eco-tourism and meat and trophy hunting has expanded rapidly in recent times (Langholz and Kerley 2006). Stock farms occupy extensive areas, commonly 3,000–5,000 ha, whilst private game reserves are substantially larger, up to 22,000 ha. Irrigated cropping of citrus, pineapples, vegetables and chicory and plantation forestry occur along small riverine areas and along the coast where rainfall is comparatively high. In the early to mid 1900s, extensive areas were degraded through over-grazing, with 42% of the total solid thicket vegetation,

Table 11.1 Criteria applied for defining local-scale "conservation opportunity" in a portion of the Fish-Kowie conservation corridor, Eastern Cape, South Africa

Criteria		Measure	Rationale/assumptions
1. Human capital			
1a	Conservation knowledge	Knowledge of nature conservation and ecologically sustainable land management issues and processes	• Knowledge comprises the cognitive component of the tripartite model for describing attitudes (Bohner and Wanke 2002) • Land managers with better knowledge of conservation theory and practice may be more likely to adopt conservation practices (Sanz and Grajal 1998; Rhodes et al. 2002; Holmes 2003; Steinmetz et al. 2006)
1b	Conservation behaviour	Participation in conservation-friendly activities such as alien invasive plant removal	• Behaviour comprises a component of the tripartite model for describing attitudes (Bohner and Wanke 2002) • Behaviour is a better reflection of values than attitudes, e.g., a strong stewardship ethic is not linked to increased adoption of best practice land management (Curtis and de Lacy 1998) • Land managers already practicing conservation-friendly activities may require fewer incentives to maintain these practices, though certainty of long-term funding is typically required
1c	Entrepreneurial orientation	Characteristics of entrepreneurship exhibited by land managers	• Skills and traits required to initiate a small business are probably useful for private land conservation initiatives, for example an ability to recognise and seize opportunities, be self-motivated and innovative, and/or are prepared to take calculated business risks (Lumpkin and Dess 1996)
1d	Local champion – personal	Characteristics of leadership and drive exhibited by a land manager	• Champions are fundamental to leading private land conservation initiatives (Knight et al. 2003; ten Kate et al. 2004; Shanley 2006) and for mainstreaming conservation into other sectors (Cowling and Pressey 2003; Knight et al. 2006b)
1e	Local champion – peers	Land manager well-regarded by his/her peers	• A champion must be capable of building social capital and promoting collective action amongst his peers if landscape-level conservation is to be effective
1f	Willingness-to-participate	Identifies the conservation instruments and incentives a land manager will and will not engage with, and the level of reduced production they will accept	• Private land conservation initiatives are often voluntary and so rely on incentives and encouragement rather than on coercion or enforced involvement (Young et al. 1996; Byron and Curtis 2002), which requires we have a better understanding of the social and economic factors that underpin land managers' willingness to engage in land management initiatives (Curtis et al. 2001)

(continued)

Table 11.1 (continued)

Criteria		Measure	Rationale/assumptions
2. Social capital			
2a	Local sense of belonging	Land managers' level of trust and the strength of norms of reciprocity and sharing	• Land managers who trust and have confidence in each other will probably work more effectively together, and will likely require less input to foster collective action
2b	Confidence in governance	Land managers' level of trust in governance systems	• Civil and political liberties, political stability, the absence of political violence and measures of contract enforcement, expropriation risk, corruption and the quality of government bureaucracy impact upon economies (Grootaert and van Bastelear 2001). Land managers' willingness to invest in a conservation initiative may reflect confidence in these components
2c	Local networks	Land managers' level of involvement in community institutions and organisations, and his/her social networks	• Effective private land conservation initiatives will probably require common property resource management, where multiple land managers manage their properties collectively (Ostrom 1990; Briggs 2001; Grootaert and van Bastelear 2001; Brunckhorst 2002)
2d	Broader networks	Land managers' "connectedness" with regional, provincial or national institutions and networks	• Local-scale conservation initiatives may derive significant benefits from people and resources found beyond the local area, for example, scientists or like-minded land managers trialling innovative ways of ensuring sustainability, whose access may be enhanced through a local contact
2e	Willingness-to-collaborate	Identifies the agencies or organisations a land manager will and will not engage and their preparedness to work with them	• Collaboration is fundamental to effective conservation (Yaffee and Wondolleck 2000). Some land managers are disillusioned with Government, and/or may have had negative experiences with other land management organisations and non-government organisations. To be effective, conservation initiatives must be sure with whom land managers are prepared to collaborate
2f	Willingness-to-sell	Identifies land manager to engage and agency or organisations	• Conservation agencies have funds available for land acquisition; however, land manager willingness-to-sell is known to be variable (Tans 1974; Knight et al. 2010)

Criteria were identified from an extensive literature review, are context specific and will probably differ between planning regions and 77% of the mosaic thicket vegetation types transformed (i.e., degraded through human activities resulting in a decline in extant habitat, species diversity and ecological function) (Lloyd et al. 2002). As a result, conservation, ecotourism and animal production values have been severely compromised (Hoffman and Cowling 1990; Lechmere-Oertel et al. 2005). Abatement of rural poverty and unemployment are the primary government goals throughout the Subtropical Thicket biome.

In July 2000, the Subtropical Thicket Ecosystem Planning (STEP) Project was funded by the Global Environment Facility to raise awareness of the destruction of Subtropical Thicket and present a strategy for its conservation. The collective vision of the STEP Project was, as agreed by stakeholders, that "The people of the Thicket Biome take custodianship of their unique living landscapes and work together to conserve, enhance and use their natural resources to ensure sustainable ecological processes and livelihoods, now and in the future" (Knight et al. 2003). Phase One concluded in December 2003 having delivered (1) an operational model for regional conservation planning (Knight et al. 2006a); (2) a public participation programme targeting key implementing organisations; (3) a spatial prioritisation, identifying seven priority conservation corridors (Rouget et al. 2006); (4) conservation planning products, including training of land-use decision-makers (Pierce et al. 2005); and (5) an implementation strategy (Knight et al. 2003). Phase Two – the implementation stage – began in January 2004. It focused upon mainstreaming the STEP planning products into local government decision-making and the strategic implementation of proactive conservation initiatives, particularly within the proposed Fish-Kowie conservation corridor, and the building of social learning institutions to promote effective conservation management (Knight et al. 2003).

The proposed Fish-Kowie corridor occupies some 394,000 ha and comprises 1,027 individual cadastres, many of which are individually fenced, primarily because they stock sheep and goats which are actively rotated between individual cadastres or between smaller internally fenced camps within cadastres over periods of days or months to manage grazing pressure. These fences include both low stock fences and high game fences, some of which are electrified. Groups of properties managed as conservancies – a voluntary agreement between multiple private land managers to manage their natural resources in an environmentally friendly manner (NACSA 2003) – are also game-fenced, so as to secure a Certificate of Adequate Enclosure, a prerequisite for legal hunting. However, nature conservation is not the primary goal. Internal fencing is typically retained, as is individual management by land managers. Several conservancies are situated within the study area, and all include stock and game. Several have re-introduced a range of endemic and exotic antelope species. Accordingly, some parts of the landscape contain many fences. In contrast, formally protected areas and private game reserves are fenced to enclose game and megaherbivores such as elephant *Loxodonta africana* and black rhinoceros *Diceros bicornis*. These reserves often comprise scores of cadastres whose internal fences have been removed to enable the natural movement of wildlife. The operation of environmental processes, such as species movement and migration and the associated impacts of altered herbivory, seed dispersal and predation, throughout the landscapes of the

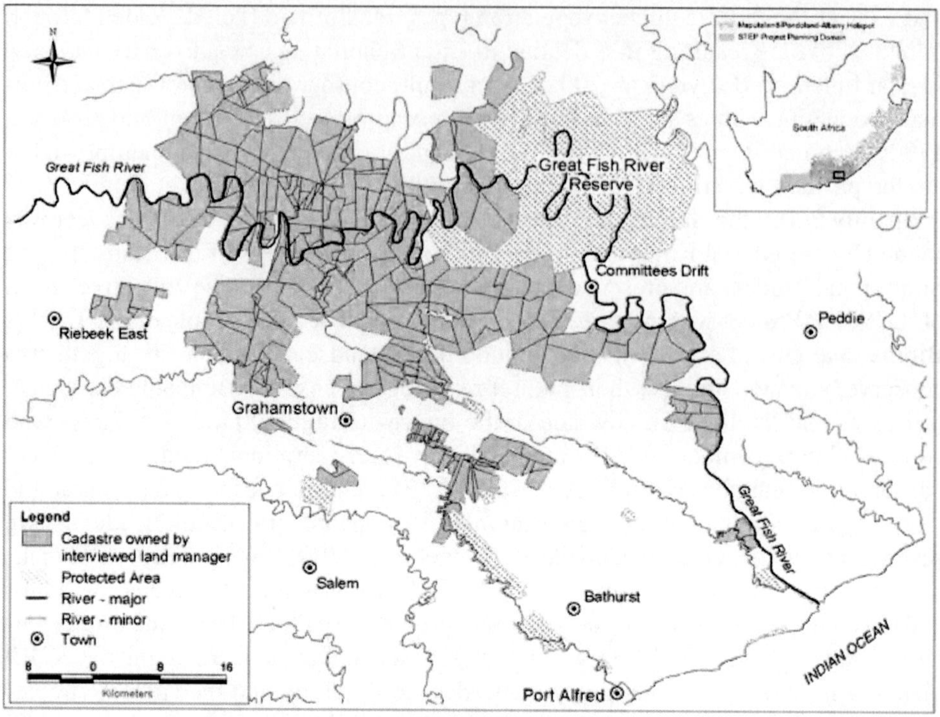

Fig. 11.1 The portion of the Fish-Kowie conservation corridor within the Eastern Cape province, South Africa, where 48 land managers were interviewed to assess factors influencing conservation opportunity for improving ecological and social connectivity

Fish-Kowie corridor has been variably compromised by fences. Social processes required for effective conservation and sustainable land management, such as communication and co-ordinated management between neighbours, has also been compromised. Fences have become social as well as ecological barriers.

Our aim was to facilitate fine-scale planning to direct the implementation of conservation action (for more details see Knight et al. 2010, 2011b). We mapped human and social capital defining conservation opportunity throughout a portion of the Fish-Kowie corridor (Fig. 11.1) using semi-structured face-to-face interviews, which were earlier reviewed by other academics and then piloted on nearby land managers. Of the 49 land managers approached, 48 agreed to be interviewed. Land managers were identified from the telephone directory or through other land managers during interviews ("snowballing", sensu Goodman 1961). The spatial location of cadastres was monitored as interviews proceeded, and land managers targeted who owned cadastres adjoining those of land managers already interviewed, improving the spatial contiguity of the final sample. Interviews lasted from 1 to 6 hours, and were conducted generally in the land manager's residence. Most land managers were small stock (i.e., goat, sheep) pastoralists who owned, and whose income was primarily generated from, their properties. A significant number supplement income with small-scale eco-tourism or hunting ventures, or exclusively through eco-tourism.

We defined indexes and scales for a suite of factors defining conservation opportunity, which provided an area-wide assessment of, and comparisons between, land managers. We calculated McDonald's $\acute{\omega}_h$ (Omega) for each factor of conservation opportunity (Table 11.1) as it is regarded the most accurate coefficient of internal consistency for questionnaire data (Zinbarg et al. 2005), and also coefficients of Cronbach's $\acute{\alpha}$ (Alpha) (Cronbach 1951) and Revelle's β (Beta) (Revelle 1979). Alternative combinations of subsets of items were trialled in exploring coefficient values in the search for internally consistent indexes and scales. The desired degree of internal consistency is a function of the purpose of the research, and we used 0.60 for McDonald's $\acute{\omega}_h$, and 0.80 and 0.70 for Revelle's β and Cronbach's $\acute{\alpha}$, respectively (Rossiter 2002). Our findings provide insights into the feasibility of restoring ecological connectivity through an understanding of factors defining land managers' motives, as they influence collective action and fencing (Ostrom 1990; Briggs 2001; Brunckhorst 2002). Analysis of motives and values is fundamental to implementing collective action (Ehrlich and Kennedy 2005), as choices made by individual people determine the effectiveness of conservation planning initiatives (Cowling and Pressey 2003).

Land managers in the Fish-Kowie corridor demonstrate a strong sense of belonging to their local area, which manifests itself in the way in which they identify themselves collectively as "Albany farmers" – small stock (i.e., goat, sheep) pastoralists who own, and whose income is primarily generated from, their properties, and whose ancestors were the first English settlers in the area. A small proportion supplement their income with small-scale eco-tourism or hunting ventures. Only about 20% of land managers are willing to consider selling their land to conservation organisations (Knight et al. 2011b). Accordingly, different types of private land conservation instruments, not land acquisition for formally protected areas, was identified as the most feasible Landscape Management Model for implementing the conservation corridor. This supported findings from the collaborative visioning process run earlier by the STEP project (Knight et al. 2003), and clearly indicates the importance of understanding land managers' motivations affecting fencing.

Willingness-to-Sell ranked highest for Alpha and Beta coefficients (0.90), and also had the highest RV-coefficient (0.79). It represented the most reliable factor, with the Willingness-to-Collaborate factor also scoring highly. The Confidence-in-Governance coefficient was also relative strong. Both Conservation Knowledge and Entrepreneurial Orientation had moderately low, but acceptable, coefficients. It is worth noting that the Entrepreneurial Orientation factor had a low RV-coefficient, which may have resulted because we selected and structured the items as distinct sub-scales (see Hermansen-Kobulnicky and Moss 2004). The measures of Local and Broader Networks ranked low (Omega 0.65). Local Sense of Belonging had variable Beta and Omega coefficients, but a high Omega coefficient. Willingness-to-Participate had relatively high values for Alpha, Beta and Omega coefficients, and a relatively high RV-coefficient. The Local Champion – Peers factor was not tested for internal consistency as it comprises only one item and so is, by definition, internally consistent. Conservation Behaviour and Local Champion – Personal both had relatively low Omega coefficients. No major sub-scales are apparent in the factors we used, which confirms their internal consistency, though there is some relationship exhibited between Entrepreneurial Orientation and Local champion – Peers.

Values for individual land manager's factors varied, often markedly. Conservation Knowledge was generally low. In contrast, Entrepreneurial Orientation was high, perhaps because most land managers run their own business. Willingness-to-Collaborate indexes varied markedly, from very low to very high, but were generally positive (mode=0.608; median=0.675). Willingness-to-Participate was assessed as three sub-scales – willingness to (1) adopt conservation instruments, (2) engage incentives, and (3) forgo production activities, with all having high values. Four land managers were identified by their peers as Local Champions. The factors of social capital were mixed, with Local Networks and Broader Network factors measuring low and very low, respectively, whilst Confidence in Governance was generally moderate (mode=0.5333; median=0.4667), notably with very poor confidence in local government, but moderate confidence in national government. Local Sense of Belonging was high, indicating land managers' attachment to the Subtropical Thicket biome.

Proposed implementation activities for the Fish-Kowie conservation corridor collaboratively sanctioned by stakeholders during Phase One of the STEP project (Knight et al. 2003) have stalled, primarily because little investment has been made in improving the human and social capital of land managers by conservation and land management organisations (Knight et al. 2011a). Generally, human and social capital are heterogeneous, sometimes markedly so, which necessitates a substantial investment by conservation organisations in promoting collaboration if collective action is to be effectively implemented. Presently, a common, united vision for improving connectivity and sustainable land management does not exist amongst land managers throughout the proposed Fish-Kowie conservation corridor. Few land managers have significant knowledge of the STEP Project, the conservation targets (Rouget et al. 2006), or of the global significance of the Subtropical Thicket Biome. A common vision is also lacking, in part because land managers have differing, multiple goals for their properties, and wish to pursue a diverse range of activities, some of which are compatible with conservation, whilst others are not. This diversity of activities means that some land managers require low numerous stock fences, others game fences and others fences sufficient to enclose elephants and rhinoceros. The removal of fences is therefore problematic, which is further complicated by land managers' general lack of knowledge about conservation approaches and activities and of the options that exist for improving the ecological functioning of the landscape and how this may improve their business ventures.

Declining returns on stock farming coupled with the expansion of private game reserves, often purchased by wealthy foreign nationals prepared to pay premium prices for land, has encouraged some land managers running small stock to sell their properties over the last 10 years. Private reserves have purchased multiple stock farms, which is beneficial for connectivity within the corridor as private reserves remove internal fences. This is done due to the higher financial returns of eco-tourism compared to stock farming (Langholz and Kerley 2006), and the need to "re-wild" their landscapes to ensure visitors a genuine African safari experience.

However, it has also resulted in the landscapes of the Fish-Kowie corridor slowly being emptied of people. For example, Kwandwe Private Game Reserve, the largest private reserve in the study area, comprises some 22,000 ha, which was previously multiple stock farms and their families. These families have now moved away. Traditional social institutions which historically maintained strong social capital, such as sporting clubs and Farmers Associations, are generally in decline or abandoned. Local and broader social networks are therefore poor, leaving land managers largely isolated.

Complicating the removal of fencing further is the strong sense of autonomy exhibited by land managers, a moderate and highly variable willingness-to-collaborate with each other and with conservation and land management organisations generally. Land managers are highly selective regarding whom they will collaborate with. Occasionally, relationships between neighbouring land managers can be extremely poor, decreasing the likelihood of fence removal. Levels of trust are generally low between land managers and land management organisations. Local government, which has significant land management influence, is generally neither respected nor trusted by land managers, with the South African national government commonly preferred. Eleven percent of land managers are very unwilling to collaborate with either the provincial nature conservation agency or nongovernment conservation organisations (NGOs), due to negative past experiences or concerns about their motives. Accordingly, voluntary non-binding agreements are more highly favoured than legally binding agreements. This has obvious potential implications for which specific organisations should lead initiatives that attempt to improve connectivity and that can run for long periods of time. A divide between managers of stock farms and formal protected areas and private game reserves is also apparent, with managers of stock farms viewing these "conservation areas" as refugia for "vermin" such as black-backed jackal, *Canis mesomelas*, and caracal, *Caracal caracal*, which predate domestic stock. The removal of fences between private reserves and stock farms is, in the mind of many stock farmers, likely to lead to increased predation of their stock by large carnivores. Following a substantial historical decline in support – and hence economic viability – from national government coinciding with the fall of Apartheid, the removal of fences is possibly also symbolic of relinquishing further control of their businesses at a time when land restitution is a primary government goal. The corridor-wide removal of fences in such a context will require a long-term perspective and an intensive investment of time and resources by conservation organisations.

There has been informal discussion for several years about the benefits to adjoining private game reserves within the Fish-Kowie corridor of removing their fences and being managed collectively, but no concrete action has so far manifest. The opportunity to enhance the ecological connectivity of the proposed corridor across private game reserves is contrasted by the challenges facing the managers of stock farms. Marginal returns from small stock farming (driven in part by the broad-scale and intensive degradation of these landscapes which has reduced productivity), coupled with the volatility of international markets and exchange rates, leaves few land managers who run stock able to afford fence removal, because they lack the financial resources to transition to alternative business models (such as five-star private game reserves). Many are also reticent because of the risk if a new venture should fail – the cost of re-establishing fences is now prohibitive. These impediments are reinforced by their need to rotate stock between camps to manage grazing pressure. Fences between stock farms and reserves (both private game reserves and formally protected areas) are perceived as essential by managers of stock farms because of the damage eventuating from incursions by mega-herbivores and large and medium-sized carnivores which predate stock. Persecution of black-backed jackal, caracal and leopard, *Panthera pardus*, is probably substantial (Lindsey et al. 2009). Removing fencing is probably also hampered by land managers self-image as "Albany farmers", which is founded (in part) upon running small stock (which require small fenced camps) in Subtropical Thicket. Given that only 20% of land managers are considering moving out of the region (i.e., are willing to sell), the fencing situation is unlikely to change through the influx of more sympathetic land managers, even through land reform and redistribution, as no land claims are pending the study area.

Given the decline in social networks and the generally low human capital which undermines the functionality of social processes, the opportunity to establish institutions to orchestrate the improved knowledge and refined behaviour required to initiate broad-scale removal of fences is minimal. The Association of Eastern Cape Private Nature Reserves & Frontier Country, otherwise known as the Indalo group, is perhaps the most obvious existing institution to promote landscape connectivity – it meets regularly, is an existing institutional structure with adequate financial capacity, its members are relatively spatially contiguous and have already removed the majority of internal fences within the borders of their individual private game reserves. A visioning process, and on-going negotiation, is required to focus and align the goals of its members. Such a collaborative institutional arrangement will need to address: (1) the need for autonomy expressed by individual private game reserves (which are mostly managed by entrepreneurs unused to sharing leadership); (2) the challenge of redefining between individual private game reserves to negate inter-reserve competition whilst promoting joint commercial gain; (3) a policy on exotic species; and (4) impact on tourist satisfaction with their ecotourism experience if neighbours are hunting.

Of institutions that promote connectivity among stock farms, conservancies are a well-recognised land management institution in South Africa and have been included in gap analyses and conservation plans (e.g. Driver et al. 2005). They have generally been assumed to provide a degree of protection for conservation-worthy land (e.g. Cowling et al. 2003; Gallo et al. 2009) and to represent areas where minimal institutional investment will promote achievement of conservation goals (e.g., Rouget et al. 2006). However, qualitative data from our surveys indicate that conservancies exhibit no more social connectivity than do non-conservancy groups of land managers, with several displaying severely fractured internal relationships. Furthermore, they do not provide secure long-term protection for valued nature (Driver et al. 2005; Downsborough et al. 2011) and may even be problematic for conservation, because land managers with game or stock farms (who commonly implement conservancies) may still persecute predators, over-stock with game or domestic species, introduce extra-limital species or artificially select hunted species for specific preferred traits, such as colour variations (Lindsey et al. 2009), all of which compromise ecosystem function. Conservancies are therefore no more likely to provide short-cuts for the establishment of land management institutions which promote ecological and social connectivity than other groups of land managers.

On the positive side, land managers exhibited high levels of Willingness-to-Participate in conservation initiatives, to engage incentives and to reduce production activities, which are all beneficial for achieving conservation goals. In the case of private game reserves, this may align with their existing conservation philosophy and/or provide a marketing advantage. For managers of stock farms, this Willingness-to-Participate is possibly due to wanting to conserve landscapes managed by their ancestors, their personal attachment to these landscapes, or the desire to improve the financial viability of their farming business through diversification. These motivations should be considered in the development of conservation initiatives promoting connectivity. Burnout in land managers is also low, meaning that the likelihood of them exiting a conservation initiative due to emotional or psychological stress is low. High levels of burnout have been linked to land managers exiting natural resource management programmes (Byron et al. 2001). Four land managers also present strongly as local champions, who might possibly provide the leadership required for initiating and maintaining effective collective land management. It is unlikely: however, that these land managers could drive effective collective action without a substantial investment of resources and external leadership.

Conclusions

When designing conservation corridors, connectivity is typically defined according to environmental processes. However, where ecological functioning of landscapes is compromised, the cause is generally human land management practices. The establishment of fences on multiple, privately owned properties disrupts species movement and migration (e.g., elephants), and can directly increase mortality rates and compromise species activities and habitat functioning. In this regard, ecological connectivity is a function of social connectivity, as people's interactions and networks are the mechanisms for developing the common vision and institutions essential for orchestrating the collective action necessary to motivate and co-ordinate broad-scale removal of fencing. Social connectivity can also be highly heterogeneous and fragmented, across landscapes, much as ecological connectivity may be. The mapping of human and social capital dimensions of conservation opportunity throughout a portion of the Fish-Kowie conservation corridor indicates that although the majority of land managers demonstrate significant Willingness-to-Participate in nature conservation initiatives, the generally low levels of human and social capital ensure that substantial investments of time and resources in long-term collaboration processes will be required before fence removal becomes possible. In a significant number of instances, removing fences will not be a feasible option. The landscapes and social networks within the Fish-Kowie conservation corridor are fragmented, and there appears to be little readiness by land managers to embrace the collective action required to holistically and effectively implement the corridor. The importance of developing a "true ecology" for addressing fencing fragmentation problems, one which integrates the human, social and ecological dimensions of connectivity, cannot be over-stated. Leadership and funding from all tiers of government, with support from non-government organisations, will be essential for building an institution capable of ensuring the collective action required to restore connectivity throughout the proposed Fish-Kowie conservation corridor.

References

Beier, P., Majka, D.R. & Spencer, W.D. (2008) Forks in the road: choices in procedures for designing wildland linkages. *Conservation Biology*, **22**, 836–851.

Bohner, G. & Wanke, M. (2002) *Attitudes and Attitude Change*. Psychology Press, East Sussex.

Briggs, S.V. (2001) Linking ecological scales and institutional frameworks for landscape rehabilitation. *Ecological Management and Restoration*, **2**, 28–35.

Brunckhorst, D.J. (2002) Institutions to sustain ecological and social systems. *Ecological Management and Restoration*, **3**,108–116.

Brunckhorst, D.J. & Coop, P.C. (2003) Tilbuster Commons: Synergies of theory and action in new agricultural commons on private land. *Ecological Management & Restoration*, **4**, 13–22.

Byron, I. & Curtis, A.L. (2002) Maintaining volunteer commitment to local watershed initiatives. *Environmental Management*, **30**, 59–67.

Byron, I., Curtis, A.L. & Lockwood, M. (2001) Exploring burnout in Australia's Landcare program: a case study in the Shepparton region. *Society and Natural Resources*, **14**, 901–910.

Carroll, C., Noss, R.F. & Paquet, P.C. (2001) Carnivores as focal species for conservation planning in the Rocky Mountain region. *Ecological Applications*, **11**, 961–980.

Conservation International. (2000) Designing sustainable landscapes. Center for Applied Biodiversity Science, Washington, D.C.

Cowling, R. M. & Pressey, R.L. (2003) Introduction to systematic conservation planning in the Cape Floristic Region. *Biological Conservation*, **112**, 1–14.

Cowling, R. M., Pressey, R.L., Lombard, A.T., Desmet, P.G. & Ellis, A.G. (1999) From representation to persistence: requirements for a sustainable system of conservation areas in the species-rich Mediterranean-climate deserts of southern Africa. *Diversity and Distributions*, **5**, 51–71.

Cowling, R.M., Pressey, R.L., Rouget, M. & Lombard, A.T. (2003) A conservation plan for a global biodiversity hotspot - the Cape Floristic Region, South Africa. *Biological Conservation*, **112**, 191–216.

Cowling, R.M., Knight, A.T., Faith, D.P. Ferrier, S., Lombard, A.T., Driver, A.T., Rouget, M., Maze, K. & Desmet, P.G. (2004) Nature conservation requires more than a passion for species. *Conservation Biology*, **18**, 1674–1677.

Cronbach, L.J. (1951) Coefficient alpha and the internal structure of tests. *Psychometrika*, **16**, 297–334.

Curtis, A.L. & de Lacy, T. (1998) Landcare, stewardship and sustainable agriculture in Australia. *Environmental Values*, **7**, 59–78.

Curtis, A.L., Lockwood, M. & MacKay, J. (2001) Exploring landholder willingness and capacity to manage dryland salinity in the Goulburn Broken catchment. *Australian Journal of Environmental Management*, **8**, 79–90.

de Tores, P.J., Marlow, N. (2012) The relative merits of predator-exclusion fencing and repeated fox baiting for protection of native fauna: five case studies from Western Australia. In *Fencing for Conservation* (eds. M.J. Somers and M.W. Hayward). Springer: New York.

Dinerstein, E., Powell, G., Olson, D., Wikramanayake, E., Abell, R., Louchs, C., Underwood, E., Allnutt, T., Wettengel, W., Ricketts, T., Strand, H., O'Connor, S. & Burgess, N. (2000) A workbook for conducting biological assessments and developing biodiversity visions for ecoregion-based conservation. Part 1: terrestrial ecoregions. Conservation Science Program, World Wildlife Fund–USA, Washington, D.C.

Dobson, A., Ralls, K., Foster, M., Soule, M.F., Simberloff, D., Doak, D., Estes, J.A., Mills, L.S., Mattson, D., Dirzo, R., Arita, H., Ryan, S., Norse, E.A., Noss, R.F. & Johns, D. (1999) *Connectivity: maintaining flows in fragmented landscapes*. Pages 129–170 in M. E. Soulé and J. Terborgh, ed. Continental conservation—scientific foundations of regional reserve networks. Island Press, Washington, D.C.

Downsborough, L., Shackleton, C.M. & Knight, A.T. (2011) The potential for voluntary instruments to achieve conservation planning goals: the case of conservancies in South Africa. *Oryx*, **45**, 357–364.

Driver, A., Maze, K., Rouget, M., Lombard, A.T., Nel, J., Turpie, J.K., Cowling, R.M., Desmet, P., Goodman, P., Harris, J., Jonas, Z., Reyers, B., Sink, K. & Strauss T. (2005) National Spatial Biodiversity Assessment 2004: Priorities for biodiversity conservation in South Africa. *Strelitzia* 17. South African National Biodiversity Institute, Pretoria.

Ehrlich, P.R. & Kennedy, D. (2005) Millennium assessment of human behaviour. *Science*, **309**, 562–563.

Goodman, L.A. (1961) Snowball sampling. *The Annals of Mathematical Statistics*, **32**, 148–170.

Gallo, J.A., Pasquini, L., Reyers, B. & Cowling, R.M. (2009) The role of private conservation areas in biodiversity representation and target achievement within the Little Karoo region, South Africa. *Biological Conservation*, **142**, 446–454.

Grootaert, C. & van Bastelaer, T. (2001) Understanding and Measuring Social Capital: A Synthesis of Findings and Recommendations from the Social Capital Initiative. Social Capital Initiative Working Paper No. 24. The World Bank, Washington, D.C.

Hayward, M.W. & Kerley, G.I.H. (2009) Fencing for conservation: restriction of evolutionary potential or a riposte to threatening processes? *Biological Conservation*, **142**, 1–13.

Hermansen-Kobulnicky, C.J. & Moss, C.L. (2004) Pharmacy student entrepreneurial orientation: a measure to identify potential pharmacist entrepreneurs. *American Journal of Pharmaceutical Education*, 68: Article 113.

Hobbs, R.J. (1992) The role of corridors in conservation: solution or bandwagon? *TREE*, **7**, 389–392.

Hoffman, M.T. & Cowling, R.M. (1990) Desertification in the lower Sundays River Valley, South Africa. *Journal of Arid Environments*, **19**, 105–117.

Holmes, C.M. (2003) The influence of protected area outreach on conservation attitudes and resource use patterns: a case study from western Tanzania. *Oryx*, **37**, 305–315.

Knight, A.T. & Cowling, R.M. (2007) Embracing opportunism in the selection of priority conservation areas. *Conservation Biology*, **21**, 1124–1126.

Knight, A.T., Boshoff, A.F., Cowling, R.M. & Wilson, S.L. (2003) *Keeping People on the Land in Living Landscapes: A Cooperative Strategy for Conserving Landscapes and Enhancing Livelihoods in the Subtropical Thicket Biome, South Africa.* TERU Report No.46. Terrestrial Ecology Research Unit, University of Port Elizabeth, Port Elizabeth.

Knight, A.T., Cowling, R.M. & Campbell, B.M. (2006a) An operational model for implementing conservation action. *Conservation Biology*, **20**, 408–419.

Knight, A.T., Driver, A., Cowling, R.M., Maze, K., Boshoff, A.F., Botha, M.A., Castley, J.G., Desmet, P.G., Goodman, P.S., Lombard, A.T. MacKinnon, K., Pierce, S.M., Rouget, M., Sims-Castley, R., Stewart, W.I. & von Hase, A. (2006b) Designing systematic conservation assessments that promote effective implementation: best practice from South Africa. *Conservation Biology*, **20**,739–750.

Knight, A.T., Cowling, R.M., Difford, M. & Campbell, B.M. (2010) Assessing and mapping human and social dimensions of conservation opportunity on private land. *Conservation Biology*, **24**, 1348–1358.

Knight, A.T., Cowling, R.M., Boshoff, A., Wilson, S. & Pierce, S.M. (2011a) Walking in STEP: lessons for linking spatial prioritisations to implementation strategies. *Biological Conservation*, **144**, 202–211.

Knight, A.T., Grantham, H.S., Smith, R.J., McGregor, G.K., Possingham, H.P. & Cowling, R.M. (2011b) Land managers' willingness-to-sell defines conservation opportunity for protected area expansion. *Biological Conservation*, **144**, 2623–2630.

Langholz, J.A. & Kerley, G.I.H. (2006) Combining Conservation and Development on Private Lands: An Assessment of Ecotourism-Based Private Game Reserves in the Eastern Cape. Centre for African Conservation Ecology (formerly TERU) Report No. 56, Nelson Mandela Metropolitan University, Port Elizabeth.

Lechmere-Oertel, R.G., Kerley, G.I.H. & Cowling, R.M. (2005) Patterns and implications of transformation in semi-arid succulent thicket, South Africa. *Journal of Arid Environments*, **62**, 459–474.

Lindsey, P.A., Romañach, S.S. & Davies-Mostert, H.T. (2009) The importance of conservancies for enhancing the value of game ranch land for large mammal conservation in southern Africa. *Journal of Zoology*, **277**, 99–105.

Lloyd, J.W., van den Berg, E.C. & Palmer, A.R. (2002) Patterns of Transformation and Degradation in the Thicket Biome, South Africa. Terrestrial Ecology Research Unit Report No. 39, University of Port Elizabeth, Port Elizabeth.

Lumpkin, G.T. & Dess, G.G. (1996) Clarifying the entrepreneurial orientation construct and linking it to performance. *Academy of Management*, **21**, 135–172.

Maslach, C., Jackson, S. & Leiter, M.P. (1996) *Maslach Burnout Inventory, 3rd edition*. Consulting Psychologists Press, Palo Alto.

Mittermeier, R.A., Hoffmann, M., Pilgrim, J.D., Brooks, T.M., Mittermeier, C.G., Lamoreux, J.L. & da Fonseca, G.A.B. (2004) *Hotspots Revisited: Earth's Biologically Richest and Most Endangered Ecoregions*. Cemex, Mexico City.

Muruthi, P. (2004) African heartlands: a science-based and pragmatic approach to landscape-level conservation in Africa. *Terrestrial ecoregions of Africa and Madagascar: a conservation assessment.* (eds. Burgess, N., D'Amicio Holes, J., Underwood, E., Dinerstein, E., Olson, D., Itona, I., Schipper, J. & Ricketts, T), pp 168–172. Island Press and World Wildlife Fund USA, Washington, D.C.

NACSA (National Association of Conservancies of South Africa). (2003) Conservancies Handbook. NACSA, Johannesburg.

Ostrom, E. (1990) *Governing the Commons: The Evolution of Institutions for Collective Action*. Cambridge University Press, Cambridge.

Pierce, S.M., Cowling, R.M., Knight, A.T., Lombard, A.T., Rouget, M. & Wolf, T. (2005) Systematic conservation assessment products for land-use planning: interpretation for implementation. *Biological Conservation*, **125**, 441–548.

Pressey, R.L. (1997) Priority conservation areas: towards an operational definition for regional assessments. *National parks and protected areas: selection, delimitation and management*. (eds. Pigram, J.J. & Sundell, R.C.), pp 337–357. University of New England, Centre for Water Policy Research, Armidale.

Pressey, R.L. (2002) The first reserve selection algorithm – a retrospective on Jamie Kirkpatrick's 1983 paper. *Progress in Physical Geography*, **26**, 434–441.

Pressey, R. L. & Taffs, R.L. (2001) Scheduling conservation action in production landscapes: priority areas in western New South Wales defined by irreplaceability and vulnerability to vegetation loss. *Biological Conservation*, **100**, 355–376.

Revelle, W. (1979) Hierarchical cluster analysis and the internal structure of tests. *Multivariate Behavioral Research*, **14**, 57–74.

Rhodes, H.M., Leland, L.S. & Niven, B.E. (2002) Farmers, streams, information and money: does informing farmers about riparian management have any effect? *Environmental Management*, **30**, 665–667.

Rossiter, J.R. (2002) The C-OAR-SE procedure for scale development in marketing. International *Journal of Research in Marketing*, **19**, 305–335.

Rouget, M., Cowling, R.M., Lombard, A.T., Knight, A.T. & Kerley, G.I.H. (2006) Designing large-scale conservation corridors for pattern and process. *Conservation Biology,* **20**, 549–561.

Sanderson, E.W., Redford, K.H., Vedder, A., Coppolillo, P.B. & Ward, S.E. (2002) A conceptual model for conservation planning based on landscape species requirements. *Landscape and Urban Planning*, **58**, 11–56.

Sanz, V. & Grajal, A. (1998) Successful reintroduction of captive-raised yellow-shouldered Amazon parrots on Margarita Island, Venezuela. *Conservation Biology*, **12**, 430–441.

Simberloff, D., Farr, J.A., Cox, J. & Mehlman, D.W. (1992) Movement corridors: conservation bargains or poor investments? *Conservation Biology*, **6**, 493–505.

Shanley, P. (2006) Science for the poor: how one woman challenged researchers, ranchers, and loggers in Amazonia. *Ecology and Society*, **11**, 28.

Smith, R.J., Easton, J., Nhancale, B.A., Armstrong, A.J., Culverwell, J., Diamini, S.D., Goodman, P.S., Loffler, L., Matthews, W.S., Monadjem, A., Mulqueeny, C.M., Ngwenya, P., Ntumi, C.P., Soto, B. & Leader-Williams, N. (2008) Designing a transfrontier conservation landscape for the Maputaland centre of endemism using biodiversity, economic and threat data. *Biological Conservation*, **141**, 2127–2138.

Steinmetz, R., Chutipong, W. & Seuaturien, N. (2006) Collaborating to conserve large mammals in southeast Asia. *Conservation Biology*, **20**, 1391–1401.

Tans, W. (1974) Priority ranking of biotic natural areas. *Michigan Botanist*, **13**, 31–39.

ten Kate, K., Bishop & Baryon, R. (2004) Biodiversity offsets: views, experience, and the business case. Executive summary. IUCN and Insight Investment, Gland, Switzerland and Cambridge.

Vlok J.H.J., Euston-Brown, D.I.W. & Cowling, R.M. (2003) Acocks' Valley Bushveld 50 years on: new perspectives on the delimitation, characterisation and origin of thicket vegetation. *South African Journal of Botany*, **69**, 27–51.

Wilson, E.O. (1998) *Consilience: The Unity of Knowledge*. Abacus, London.

Wilson, K. A., Pressey, R.L., Newton, P.W., Burgman, M.A., Possingham, H.P. & Weston, C. (2005) Measuring and incorporating vulnerability into conservation planning. *Environmental Management*, **35**, 527–543.

Yaffee, S.L. & Wondolleck, J.M. (2000) Making collaboration work: lessons from a comprehensive assessment of over 200 wide-ranging cases of collaboration in environmental management. *Conservation Biology in Practice*, **1**, 17–25.

Young, M.D., Gunningham, N., Elix, J., Lambert, J., Howard, B., Grabosky, P. & McCrone, E (1996) Reimbursing the future: an evaluation of motivational voluntary, price-based, property-right, and regulatory incentives for the conservation of biodiversity, parts 1 and 2. Biodiversity series, paper no. 9. Department of the Environment Sport and Territories, Canberra.

Zinbarg, R. E., Revelle, W., Yovel, I. & Li, W. (2005) Cronbach's α, Revelle's β, and McDonald's ω_h: Their relations with each other and two alternative conceptualizations of reliability. *Psychometrika*, **70**, 123–133.

Chapter 12
Ecological, Social and Financial Issues Related to Fencing as a Conservation Tool in Africa

Peter A. Lindsey, Chap L. Masterson, Andrew L. Beck, and Stephanie Romañach

Introduction

Fencing has taken on an increasingly important role in wildlife management in parts of Africa in recent years, particularly in southern Africa. Fencing is a legal requirement for ranchers to own wildlife in Botswana, Zambia and South Africa and facilitates the process of obtaining permits for some forms of consumptive utilization of wildlife on ranches in Namibia. Large areas of private land are broken up into small parcels by high game fencing across parts of the region as a result (Bond et al. 2004; Barnett and Patterson 2006). A number of protected areas in Africa are also partially or completely fenced, with the objective of limiting the movement of wildlife out of, and people into parks. While historically barriers were typically used to keep wild animals out, fencing developed as a tool in African conservation to keep wild animals inside protected areas. For example, Etosha National Park in Namibia, Matobo and Hwange national parks in Zimbabwe, and the Aberdare and Meru national parks in Kenya are partially or completely fenced (Boone and Hobbs 2004; Purchase 2008). Fencing of parks is particularly prevalent in South Africa, where virtually all parks are fenced, including most of the 20,000 km^2 Kruger National Park. Several lengthy veterinary fences have been constructed in parts of Africa

P.A. Lindsey (✉)
Mammal Research Institute, University of Pretoria, Pretoria 0002, South Africa
e-mail: palindsey@gmail.com

C.L. Masterson
Zvakanaka Wild Vets, Hluhluwe, South Africa

A.L. Beck
Animal, Plant and Environmental Sciences, University of the Witwatersrand, Johannesburg, South Africa

S. Romañach
African Wildlife Conservation Fund, Doral, FL, USA

M.J. Somers and M.W. Hayward (eds.), *Fencing for Conservation: Restriction of Evolutionary Potential or a Riposte to Threatening Processes?*,
DOI 10.1007/978-1-4614-0902-1_12,

with the objective of preventing disease transfer between wildlife and livestock, notably in Botswana, Namibia and Zimbabwe (Taylor and Martin 1987; Martin 2005; Mbaiwa and Mbaiwa 2006). In other parts of Africa, however, and particularly in West, Central Africa, and most of East Africa, the use of fencing in wildlife management is rare.

Despite the prevalence of fencing in southern Africa, and potentially significant ecological, financial and social impacts, relatively little has been written on the use of fencing as a wildlife management tool in the region. Hoare (1992) provided a review of the practical issues related to fencing wildlife in Africa. Several authors have assessed the conservation implications of veterinary fencing (Taylor and Martin 1987; Albertson 1998; Martin 2005; Mbaiwa and Mbaiwa 2006). Others have discussed the role of fencing in controlling human-wildlife conflict (Kassilly 2002; Nelson et al. 2003; Ogada et al. 2003), some have addressed the ecological impacts of fencing (Ben-Shahar 1993; Burger and Branch 1994; Boone and Hobbs 2004) and some coverage has been given to the social issues related to fencing (Wels 2000; Spierenburg and Wels 2006). Hayward and Kerley (2009) provided a recent global review of the issues relating to fencing. However, as yet, the literature lacks a comprehensive review of the ecological, social and financial issues relating to fencing as a conservation tool in Africa, which is what we set out to achieve in this chapter.

Types of Fencing

A variety of fencing types are employed to restrict the movement of wildlife, including galvanized steel wire (which may or may not be electrified) with metal or wooden poles, stone walls, live fences comprised of prickly pear *Opuntia littoralis* or other cactus species and thorn fences comprised of heaps of branches from trees such as *Acacia* spp. Throughout this review, unless otherwise specified, we restrict discussion to wire fencing, which represents the only type of fencing commonly used to create barriers over large distances. It is important to bear in mind that, no matter what their design, fences cannot be relied upon as being absolute barriers, especially with regard to the mega-herbivores, but their effectiveness is greatly enhanced by the appropriate use and maintenance of electrification.

Commonly used types of wire fencing include:

(a) Fencing designed to restrict the movements of mega-herbivores such as rhinoceroses *Diceros bicornis* and *Ceratotherium simum*, hippopotamuses *Hippopotamus amphibius*, elephants *Loxodonta africana* and buffaloes *Syncerus caffer*, allowing passage of most other species. Although in most cases more robust designs are used a single, electrified wire is often sufficient to discourage the passage of mega-herbivores, strung at a height of approximately 1.5 m for elephants, or 300–500 mm for a hippopotamus. Examples of the effective use of fencing for the containment and management of mega-herbivores include, inter alia: the containment of elephants and other mega-herbivores in reserves; the exclusion of elephants from sensitive habitat patches (e.g. from an area of sand forest in Phinda Resource Reserve, Zululand); the exclusion of mega-herbivores from tourist and staff camps; and, the confinement of hippopotamuses to single bodies of water during capture. Elephants frequently become adept at challenging fences, including those that have been electrified, by either avoiding electrified strands while pulling fence posts over or by targeted snapping of electrified wires using their tusks (which have low electrical conductivity). The subsequent material damage, need to recapture escaped animals, and possible crop-damage can be costly. In such cases, innovative fence designs are required to prevent elephants from challenging the fence. For example, at Ol Pejeta conservancy in Kenya, protruding wires (fixed to the fence about a metre above the ground) of ~1.5 m long, angled upwards at 45° and spaced about a metre apart (outriggers), serve to keep elephants away from the fence and prevent them from hooking their tusks under a fencing wire and snapping it (Graham et al. 2009). Relatively short (1–1.2 m) fencing is often used to contain buffaloes and rhinoceroses, comprised of four to six strands of thick (10 mm) cable wire.

(b) Fencing designed to restrict the movement of medium-sized wild ungulates: several antelope species, and notably eland *Taurotragus oryx*, waterbuck *Kobus ellipsiprymnus*, greater kudu *Tragelaphus strepsiceros* and impala *Aepyceros melampus* are prodigious jumpers, and so fences have to be constructed to a height of at least 2.4 m to contain them. Such fences are typically comprised of 12–21 strands of high tensile steel wire and/or barbed wire (80–170 mm apart) and may be electrified (van Rooyen et al. 2002). Where fences are electrified, electrified wires are typically offset from the main fence by a distance of ~300 mm at various heights above ground level to accommodate game of various sizes; for example at 250–300, 1,000, 1,500 and 2,000 mm above ground level. Meshed wire fencing (e.g. Bonnox™ or Veldspan™) of approximately 100–200 mm mesh size may also be used. To constrain "non-jumping" ungulates, a standard 1.5 m wire mesh fence with a strand of wire 150 mm above the mesh, and another strand 150 mm above the first strand is generally sufficient (van Rooyen et al. 2002). Non jumping ungulates include inter alia: bushbuck *Tragelaphus scriptus*; blesbok *Damaliscus dorcas*; grey duiker *Sylvicapra grimmia* red hartebeest *Alcelaphus bucelaphus*; oribi *Ourebia ourebi*; oryx *Oryx gazella*; roan antelope *Hippotragus equinus*; sable antelope *Hippotragus niger*; springbok *Antidorcas marsupialis*; and tsessebe *Damaliscus lunatus* (van Rooyen et al. 2002). The use of fencing to contain non-jumping wild ungulates is particularly common in Namibia where oryx, red hartebeest and springbok are among the most widespread and common species (Lindsey 2011).

(c) Predator-proof fencing. Predator-proof fencing generally involves similar materials as for the fences used to restrain wild ungulates, but with reinforcing at the base. Predator-proof fences are always electrified. The lower half of the fence is comprised of tight wire (pig) meshing, which may be dug into the soil to a depth of up to 50 cm, or alternatively be folded into an apron extending along the ground up to 1 m on the inside of the fence. Both arrangements may or may not be rock packed or incorporate a low off-set and electrified trip wire,

50–100 mm above ground level, to discourage digging under the fence. The installation of “swing-gates” permits passage of digging species such as aardvarks *Orycteropus afer* or warthogs *Phacochoerus africanus* (Schumann et al. 2006).

Advantages of Fencing as a Wildlife Management Tool

Ecological and Epidemiological Issues

Fencing and the Utilization of Small Habitat Fragments

With burgeoning human populations competing for land, natural habitat is becoming increasingly fragmented and patches of available habitat are shrinking in size (Norton-Griffiths 2007). Fencing permits the utilization of small habitat patches by reducing edge effects associated with wildlife moving out, or humans moving in to the area encompassed (edge effects are impacts imposed on organisms by the juxtaposition of human-modified landscapes on natural habitat (Murcia 1995)). Wildlife moving out of habitat patches is vulnerable to being hunted or being persecuted in response to human-wildlife conflict. Similarly, people moving into habitat patches can threaten wildlife through disturbance, hunting or by extracting crucial resources. By limiting movement of wildlife and people to and from small habitat patches, fencing can reduce the gradual loss of species from protected areas (Caro and Scholte 2007). Fencing has particular applicability to fertile, high rainfall areas which often have high densities of humans and fragmented natural habitat, and where the prevailing habitat types are under-represented in protected area networks (that were generally delineated on the basis of poor suitability for agriculture, Cumming 2004a). Fencing is also potentially useful for enhancing the prospects for the conservation of wide-ranging predators in habitat patches in which edge effects would otherwise cause local extinction (Woodroffe and Ginsberg 1998). In South Africa, for example, species such as wild dogs *Lycaon pictus*, cheetahs *Acinonyx jubatus*, lions *Panthera leo* and spotted hyenas *Crocuta crocuta* have been successfully reintroduced into a number of small, fenced reserves albeit with ongoing population and genetic management necessitated by virtue of restricted reserve size (Lindsey et al. 2005a; Hunter et al. 2007; Hayward and Somers 2009). However, while fences may impart benefits through reducing edge-effects on large-bodied animals, they may actually impose edge-effects on smaller species through mortality associated with entanglement and electrocution (discussed in detail below, Beck 2007). Counter-intuitively, fences may be to the disadvantage of certain species within protected areas by obliging them to compete for resources with other wild species within a restricted fenced area. An example of this is provided by the thriving population of suni antelope *Neotragus moschatus*, found in the KwaNgwenya Communal Area (in northern Zululand, South Africa) whereas the same species is markedly under-represented in the adjacent Kube Yini Private Game Reserve.

This discrepancy is plausibly due to the competition between suni and nyala, *Tragelaphus angasi* which are abundant within the reserve but rare or absent outside.

Fencing as a Wildlife Management Intervention

By enabling manipulation of the movement of wildlife, fencing is an important tool in the management of endangered species and of habitat. For example, fencing can be used to separate black rhinoceros bulls, which may otherwise suffer high mortality rates due to fighting (Hrabar and du Toit 2005). Fencing is sometimes used to create enclosures to protect locally rare ungulates such as roan antelope or tsessebe from predation and competition, as has been attempted in Kruger National Park (Grant et al. 2002). Fencing is used as an important tool for wildlife reintroductions, by creating release paddocks (or "bomas") which allow for an acclimatization and familiarization period for wildlife prior to reintroduction. Bomas are of particular importance for reintroductions involving wide-ranging species such as wild dogs by reducing post-release dispersal (Hofmeyr 2001).

Fencing can be used to protect sensitive habitats within wildlife areas from pressure from herbivores. In Addo Elephant National Park and Phinda Resource Reserve in South Africa, for example, fencing has been used to protect endemic plant species from impacts from elephants (Grant et al. 2007). Fencing is also used to protect individual trees and infrastructure such as windmills, taps or pipes from damage by elephants (Grant et al. 2007).

Fencing in Disease Control

Fencing is widely employed in southern Africa as a tool for limiting the spread of diseases between livestock and wildlife (Thomson et al. 2004). For example, fences have been constructed in several locations in Botswana, Namibia, around Kruger National Park in South Africa, and widely in Zimbabwe, including in the South, Zambezi Valley and escarpment areas (Albertson 1998; du Toit 2005; Martin 2005). Fences are most commonly used to prevent the transfer of foot-and-mouth disease from buffaloes and other wildlife species to cattle, to retain access to European Union export-markets for beef and meat from wildlife (Thomson et al. 2004). Fences have also been used to control transfer of other diseases, including corridor disease, rinderpest, African swine fever, contagious bovine pleuropneumonia, malignant catarrhal fever and trypanosomiasis (Taylor and Martin 1987; Grant et al. 2007). The assumed ability of fencing to limit disease transfer was significant in the development of the wildlife ranching industry, which may otherwise have been subject to greater resistance. Conversely, with the formation of trans-frontier parks, wild animals are free to move across international boundaries, thus facilitating the cross-border spread of important restricted trans-boundary animal diseases. For example bovine tuberculosis is now known to have spread from buffaloes in Kruger National Park in South Africa to populations in Gonarezhou National park in Zimbabwe (Foggin 2008).

Social Issues

Fencing and Human-Wildlife Conflict

Well constructed and maintained electrified fencing represents an effective tool in limiting several forms of human-wildlife conflict. Most commonly, fencing is used to prevent the movement of animals from protected areas into adjacent agricultural lands. Fencing can permit the retention of wildlife or establishment of wildlife-based land uses in areas dominated by agriculture. For example, the 766 km^2 Aberdare National Park in Kenya is surrounded by high densities of humans practising small-scale agriculture, and fencing is crucial for controlling the movement of elephants from the park and permitting coexistence with neighbouring communities. In Savé Valley Conservancy in Zimbabwe, the removal of electrified fencing by settlers during land "reform" resulted in an upsurge in conflict and human fatalities caused by elephants due to movement of the species into adjacent agricultural areas (Lindsey et al. 2008).

In some cases, fencing is used to reduce human-wildlife conflict by creating enclaves of human activity within wildlife areas. For example, some villages in Niassa Game Reserve in Mozambique are fenced in to reduce crop raiding by elephants (Osborn and Anstey 2002). Similarly, the Masoka community in the Guruve area of Zimbabwe used donor funds to fence settlements, leaving the rest of their land for wildlife production in response to potential benefits from the CAMPFIRE programme (Taylor 2009). Similarly, in Kenyan pastoral lands, thorn "fences" enclosing livestock corral are effective at reducing losses of cattle and sheep to lions and spotted hyaenas (Ogada et al. 2003). In Namibia, electric fencing is used in some communal land conservancies to protect cattle from crocodiles *Crocodilus niloticus* at drinking sites on rivers (Lamarque et al. 2008).

Fencing and the Security of Wildlife Areas

Electrified fencing is used as a tool to restrict access to wildlife areas by people to reduce illegal extraction of natural resources and for security where crime is prevalent (such as in South African wildlife ranching areas). Fencing as an anti-poaching strategy is particularly important for areas containing high-risk, valuable wildlife such as black and white rhinoceroses. In addition to reducing access, cleared areas adjacent to fences enable fence patrols to scan for human and animal footprints crossing the 4–10 m open patch of ground generally maintained on the inside of fence lines. The same patch of open ground also plays a role in controlling fires by acting as a firebreak and may reduce the cross-fence transference of arthropod disease vectors such as the brown ear tick *Rhipicephalus appendiculatus*, the vector responsible for the transmission of Corridor Disease *Theilleria parva lawrencii* from buffalo to cattle. The brown ear tick is an "ambush tick" (as opposed to a "hunting tick") and thus is less likely to cross a fence with an adjacent cleared area than if vegetation grew right up to the fence on both sides.

The majority of fences present little hindrance to human's intent on entering an area. However, fences nonetheless clearly delineate legal boundaries to persons illegally entering wildlife reserves and if such people are apprehended within the fenced area they may then be liable to prosecution. In South Africa, for example, persons who cross into reserves at sites other than at designated entry points will be deemed to have read the indemnities, terms and conditions of entry displayed at such designated entry points in the case of such persons making claim for damages against landowners.

Financial Issues

Allocation of User Rights over Wildlife

Wildlife has the status of being *res nullius* in most southern African countries, or without ownership until it has been captured, killed or enclosed. This status effectively means that wildlife belongs to the person whose land it is on, but if it leaves the property, ownership is lost. In South Africa, the Game Theft Act (1991) was promulgated to prevent the theft of wildlife: if a property is adequately fenced, and if the wildlife is positively identifiable (e.g. with micro-chips, brands, ear notches or ear tags), wildlife remains the property of the owner even if it escapes or is lured or otherwise removed from his/her property, or is killed by a poacher (Boshoff 2008). Legislation requires that fences are present for landowners to effectively own wildlife in South Africa, Botswana, and Namibia (in the case of huntable and exotic wildlife) (Bond et al. 2004; Barnett and Patterson 2006). Consequently, fencing was an important component in the decision to allocate user rights to wildlife, a decision which resulted in the development of the wildlife ranching industry across large areas of southern Africa by enabling landowners to benefit financially from wildlife. In Namibia and South Africa, for example, wildlife ranches comprise at least 288,000 km^2 (with 32,000 km^2 used exclusively for wildlife production [i.e. lacking livestock], Lindsey, 2011) and approximately 205,000 km^2 respectively (National Agricultural Marketing Council 2006).

Disadvantages of Fencing as a Wildlife Management Tool

There are a number of potential social, ecological and financial problems associated with the use of fencing as a conservation tool which must be considered by wildlife managers. They include the following issues.

Ecological and Epidemiological Issues

Fencing Inhibits a Variety of Ecological Processes

Fencing can inhibit or prevent natural ecological processes such as immigration, emigration and migration. By limiting movement, fenced reserves are vulnerable to problems associated with small populations and islands, and render populations more susceptible to environmental, demographic and genetic stochasticity (Macarthur and Wilson 1967; Caughley 1994). Fencing interrupts gene flow between populations, introducing a risk of inbreeding and enhancing the prevalence and impacts of founder effects and genetic drift (Caughley 1994; Hayward et al. 2007). Managing fenced populations as components of a meta-population can help to avoid many of these problems (Hayward et al. 2008). For example, meta-population management of wild dogs in a series of small reserves in South Africa has enabled the re-establishment of a viable population comprising several sub-populations in a network of small fenced reserves, and similar plans are in place for cheetahs (Davies-Mostert et al. 2009; Lindsey and Davies-Mostert 2009).

Fencing also limits the extent to which wildlife populations can move to utilize patches of primary productivity, and can reduce the ecological capacity of land as a result (Ben-Shahar 1993; du Toit 1998; Boone and Hobbs 2004). The veterinary fences of Botswana for example, reduced access of Burchell's zebra *Equus quagga*, blue wildebeest *Connochaetes taurinus* and other wildlife to water and dry-season grazing and caused massive die-offs as a result (Williamson and Mbano 1988; Albertson 1998; Mbaiwa and Mbaiwa 2006). Fences in northern Botswana have been implicated in the decline of populations of species of conservation significance in the Caprivi Strip of Namibia, including tsessebe, sable antelope and roan antelope (Martin 2005). In Kenya, the construction of fencing is exacerbating the impacts of habitat fragmentation associated with the sub-division of communal rangelands into private smallholdings (Western and Nightingale 2005).Similarly, in South Africa, wildlife ranchers are increasingly constructing predator-proof fencing to protect valuable antelopes, thus reducing habitat availability for free-ranging populations of threatened species such as wild dogs and cheetahs (Lindsey et al. 2009). Perhaps most seriously, fences may exacerbate the impacts of global warming on wildlife conservation by constraining adaptive responses of wildlife to climate change through adjustments in their spatial distribution (Cumming 2004b).

The confinement of wildlife populations with fencing appears to affect density dependent population regulation, and fenced areas are susceptible to unnaturally high densities of some wildlife, resulting in environmental degradation and the risk of population crashes (Boone and Hobbs 2004; Hayward and Kerley 2009). For some wildlife, such as the white rhinoceros, unusually high densities may increase the frequency of aggressive social interactions, resulting in elevated incidence of infanticide and reduced reproductive rates (Masterson, unpublished data). As a result of these phenomena, small fenced reserves require frequent management intervention to prevent over-population or local extinction, which can impose additional financial costs on the land manager. Predators reintroduced into small reserves in South Africa, for example, commonly reach unusually high densities, and management intervention is sometimes required to prevent precipitous prey population declines (Davies-Mostert et al. 2009; Hayward and Kerley 2009). Furthermore, fencing can modify the behaviour of predators and impose additional financial impacts in small reserves. Wild dogs learn to chase animals against fences during hunting, enabling them to kill large species such as waterbuck and kudu more frequently than is typical in the absence of fencing (van Dyk and Slotow 2003).

Similarly, fencing may increase the local impacts of wild dogs on prey populations during the time that they have puppies at a den site by preventing the return of prey to an area following departure of the pack (Romañach and Lindsey 2008).

Fencing and Disease Control Objectives

Fencing is effective at controlling diseases (e.g. corridor disease) that are transferred by non-flying arthropods. However, there is evidence that fences are not entirely effective at controlling diseases transferred by aerosol, such as foot-and-mouth disease and bonvine tuberculosis (Boone and Hobbs 2004; du Toit 2005). Though buffaloes, which are the primary maintenance host for foot-and-mouth disease are constrained by fences, other ungulates such as the impala, greater kudu and warthog frequently traverse fences. Data from Kruger National Park and from Zimbabwe indicate that 18.4% of antelopes and 7.8% of small wildlife species have antibodies to foot-and-mouth disease, indicating that there is risk of it being transferred to cattle outside of fenced wildlife areas when those species escape (Stutmoller 2002). Correspondingly, the foot-and-mouth disease fence in Zimbabwe has failed to prevent multiple outbreaks of the disease (du Toit 2005). The use of fencing to control tsetse fly and trypanosomiasis has become obsolete due to the development of the technique of using pesticides to control tsetse flies in the presence of wildlife populations (Taylor and Martin 1987).

Fencing and Animal Welfare Implications

Electrified fencing has implications for animal welfare due to the frequency with which small animals are entangled in fence lines and are killed either through electrocution or dehydration. Fencing may also prevent escape of animals from bush fires. In South Africa, a recent study conducted in a number of study areas (Fig. 12.1) over 2 years recorded 33 species (16 Mammals, 14 Reptiles and 3 Amphibians) killed due to entanglement or electrocution on fence lines. The mean mortality rate for reptiles was calculated to be 0.48 reptiles/km/year with the bulk of deaths occurring on electrified strands at heights of 50–200 mm (Beck 2007; Table 12.1). Chelonians appear to be particularly affected: one study recorded 52 tortoises being killed along a fence of 8.4 km during 132 days (Burger and Branch 1994), with 91% of all reptile mortalities involving the leopard tortoise *Geochelone pardalis*.

Social Issues

Negative Community Perceptions

The development of preservationist conservation strategies by colonial and post-colonial governments excluded rural people from the natural resources that they had

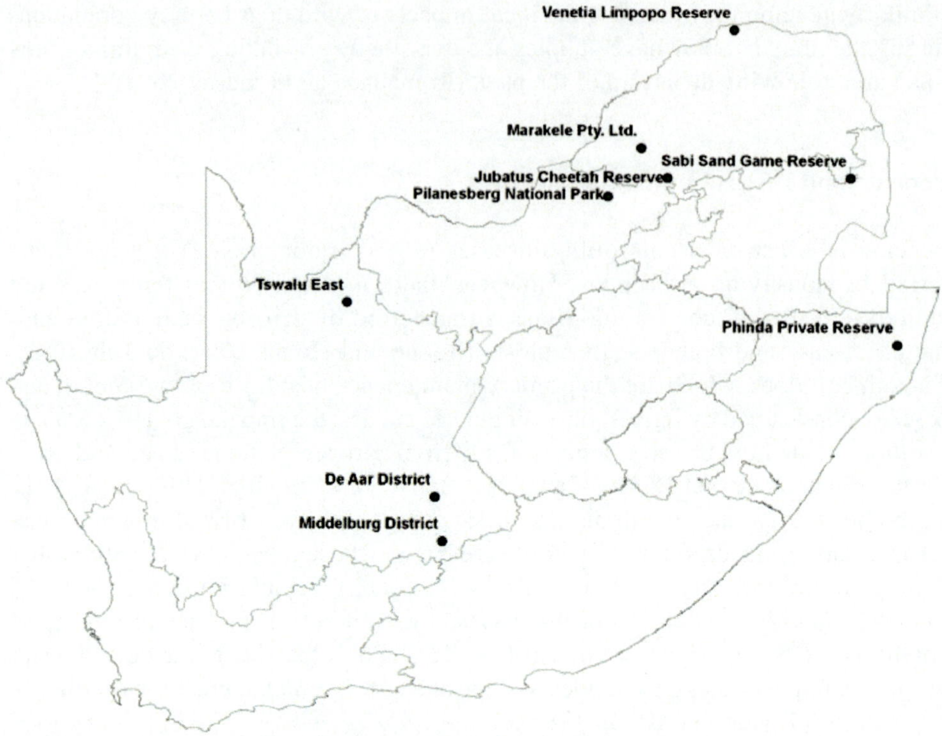

Fig. 12.1 Geographic location of the study areas in South Africa (Beck, unpublished data)

historically utilized (Els 2002). Wildlife fencing represented both a functional and symbolic barrier to local communities and became perceived by them as representing a negative, exclusive, imposition designed by wildlife authorities to protect wildlife with little consideration for human welfare (Spierenburg and Wels 2006). In Botswana, additional complaints of communities regarding fencing include the fact that it prevents the movement of wildlife into traditional hunting areas, and that fences have divided communities spatially, preventing interaction among family members (Albertson 1998). Though more recently, conservation efforts have aimed to "move beyond fences" and extend benefits from protected area management to neighbouring communities, negative perceptions persist in some areas (Spierenburg and Wels 2006). Antagonism created by the presence of a fence can encourage vandalism of or theft of parts of the fence, and contribute to the illegal extraction of resources from a wildlife area.

Theft of Fencing Materials to Make Snares for the Bushmeat Trade

The use of wire snares to capture animals for illegal bush-meat is a major threat to wildlife populations in Africa (Noss 1998). Steel and barbed wire of the kind used to construct wildlife-fencing are ideal material for making snares, and thus, wildlife-

Table 12.1 Indigenous South African fauna killed or injured by electrified fences in all study areas between July 2007 and June 2008 (Beck, unpublished data)

		Rate of recurrence		
Species	Common name	Frequent	Infrequent	Occasional
Mammals				
Atelerix frontalis	South African hedgehog			*x*
Canis mesomelas	Black-backed jackal			*x*
Cephalopus natalensis	Red duiker			*x*
Crocuta crocuta	Spotted hyaena			*x*
Galago moholi	Lesser Bush-baby		*x*	
Genetta genetta	Small spotted genet		*x*	
Hystrix africaeaustralis	Porcupine	*x*		
Manis temminckii	Pangolin	*x*		
Mellivora capensis	Honey badger		*x*	
Oreotragus oreotragus	Klipspringer			*x*
Orycteropus afer	Aardvark		*x*	
Oryx gazella	Gemsbok			*x*
Otolemur crassicaudatus	Thick tailed bushbaby		*x*	
Phacochoerus africanus	Warthog		*x*	
Potamochoerus larvatus	Bushpig			*x*
Simia aethiops	Vervet monkey		*x*	
Reptilians				
Chameleo dilepis	Flap necked chameleon		*x*	
Dendroaspis polylepis	Black mamba			*x*
Dispholidus typus	Boomslang			*x*
Kinixys belliana	Bells hinged tortoise		*x*	
Kinixys lobatsiana	Lobatse hinged tortoise	*x*		
Pelomedusa subrufa	Marsh terrapin		*x*	
Philothamnus semivariegatus	Spotted bush snake			*x*
Psammobates oculiferus	Kalahari tent tortoise			*x*
Psammophis mossambicus	Olive grass snake			*x*
Psammophis subtaeniatus	Stripe-bellied sand snake Snake			*x*
Python natalensis	Southern African python	*x*		
Geochelone pardalis	Leopard tortoise	*x*		
Thelotornis capensis	Southern vine snake			*x*
Varanus albigularis	Rock monitor	*x*		
Amphibians				
Bufo pantherinus	Leopard toad			*x*
Pyxicephalus adspersus	Giant bullfrog		*x*	
Bufo rangeri	Raucous toad		*x*	

proof fences provide enormous quantities of potential snare material. In Savé Valley Conservancy for example, >84,000 snares were removed by anti-poaching teams during an 8-year period, most of which were comprised of wire stolen from the fence (Lindsey et al. 2011). In parts of Africa where wire is less readily available, poachers are forced to use alternative methods which may be easier to control. In central Mozambique for example, poachers use steel gin traps which if confiscated by anti-poaching scouts are costly to replace (Lindsey et al. 2011). In the Zululand region of KwaZulu-Natal in South Africa, most ranchers use mesh Bonnox fencing for various practical reasons including the fact that meshed fencing is less conducive to snare construction than steel wire (Lindsey et al. 2005b).

Financial Issues

Fencing and Wildlife-Based Land Use Options

The fragmentation of ranching areas with perimeter fencing along individual properties limits land use options which has implications for conservation. Ranches in South Africa are generally small (8.2–49.2 km^2 depending on the province, Bothma 2002). Large charismatic species cannot be reintroduced to small properties with perimeter wildlife fencing, which precludes high-end ecotourism, and limits land use to high off-take, low income meat and trophy hunting of antelopes (Lindsey et al. 2009). Under such conditions, predators are perceived to impose direct financial costs by removing animals that could otherwise have been hunted, and are widely persecuted (Marker et al. 2003; Lindsey et al. 2005b). By contrast, where neighbouring wildlife ranchers have removed internal fencing to create conservancies, the full complement of mammal fauna can be reintroduced, and more profitable land uses such as high-end ecotourism, and "big game" trophy hunting can be practised. In conservancies, several of the conservation problems inherent with small fenced ranches (e.g. persecution of predators, over-stocking of ungulates, unethical hunting practices) generally fall away and natural ecological processes are allowed to recover (Lindsey et al. 2009).

The Cost of Constructing and Maintaining Wildlife-Proof Fences

In Kenya, the 3.3 m tall electrified perimeter fence surrounding Aberdare National Park cost ~US $20,000/km to construct (Lamarque et al. 2008). Wildlife-proof fencing in South Africa currently costs approximately US $4,500–6,500/km for mesh fencing designed to contain antelopes and US $5,250–7,250/km for predator-proof fencing (Masterson, unpublished data). Costs are significantly influenced by the terrain (costs may be double for mountainous vs. flat terrain), the shape of the ranch/reserve and whether or not public roads are present which may require fencing.

Fence maintenance is also costly: in South Africa, on a reserve with 100 km of fencing, fence maintenance costs are likely to be ~US $32,000/year (Masterson, unpublished data). Following rainfall, gaps often appear under fences along erosion gulleys or stream beds and fences are frequently bridged by species such as warthogs, porcupines and aardvarks which burrow under the bottom wire creating gaps for other animals to pass through. These breaches in security impose intensive and costly maintenance requirements. Maintenance costs can be reduced by installing swing-gates, concrete drainpipes or old vehicle tyres to allow passage of warthogs, aardvarks and other "fence-challengers" (van Rooyen et al. 2002) provided that no large predators are meant to be contained by the fence. The costs per km of maintaining fencing are likely to be higher on smaller reserves, due to the loss of economies of scale associated with fixed costs such as vehicle purchase.

The components of electrified fencing (such as solar panels, batteries, chargers and fence wire) are valuable and vulnerable to theft, which affects fence integrity and efficacy. In Kruger National Park, for example, break outs of elephants through the fence have been attributed to vandalism and theft from the fence (Grant et al. 2007). Due to the expenditure and diligence required for maintenance, fencing can be quickly compromised under conditions of political instability as has occurred in Zimbabwe recently (Lindsey et al. 2009).

In southern Africa, the expense of fencing is justified by the fact that wildlife is a valuable, tradable asset. In most of the rest of Africa, ownership of wildlife is retained by the state and such value does not exist. Furthermore, most African countries are grappling with severe funding shortages for protected area networks, and many lack the finance to provide anything more than nominal protection for parks (Wilkie et al. 2001; Child et al. 2004). As a result, outside southern Africa, the use of fencing is restricted to a handful of smaller parks and areas experiencing severe human-wildlife conflict.

Fencing as a Conservation Tool

In light of the problems and limitations that are commonly associated with fencing, we outline several issues that should be considered when designing conservation strategies involving fences:

1. Using the minimum possible amount of fencing

   The key ecological problems associated with fencing stem from the isolation of small populations and inhibition of the processes of immigration, emigration, and movement to exploit patchy primary productivity. Most of these problems can be resolved by increasing the size of an area encompassed by a fence and thus reducing the length of fencing employed per unit area. A practical means of achieving this is through the creation of transfrontier conservation areas whereby fences are removed between protected areas occurring on the borders of adjacent countries. For example, fences have been removed between the Kruger National Park in South Africa and Limpopo National Park in Mozambique, which will ultimately be linked with Gonarezhou National Park in Zimbabwe to create a single large (35,000 km^2) protected area, the Great Limpopo Transfrontier Park. Even fenced parks as large as Kruger (20,000 km^2) have suffered population declines related to isolation (Hayward and Kerley 2009). Creating linkages with adjacent parks may limit the occurrence of such trends in future. On private land, similar ecological benefits can be achieved through the creation of conservancies. Conservancies are created when adjacent landowners remove internal fencing to create a larger area, usually encompassed by a single perimeter fence. Conservancies can support larger populations of wildlife than isolated ranches, and contain larger areas which are more resilient to stochastic events. Consequently, they are less susceptible to localized variation in rainfall (du Toit 1998; Lindsey et al. 2009). Furthermore, the larger areas encompassed by conservancies permit the re-establishment of the full complement of mammal fauna and encourage land uses such as ecotourism which are more closely aligned with conservation objectives than the high off-take meat hunting typical of small, fenced ranches (Lindsey et al. 2009). National governments should consider repealing legislation which requires perimeter fencing to be present for landowners to own/utilize wildlife. In Zimbabwe, land owners are not required by law to have perimeter fencing on their property, and consequently, fencing was less prevalent in wildlife ranching areas than in South Africa, creating more open ecological systems with a lower prevalence of the problems associated with fenced wildlife ranching. Conservancies containing charismatic mega fauna attract ecotourists and have potential to generate significant foreign currency income, and so governments should consider introducing incentives such as tax breaks to promote their development (Lindsey et al. 2009).

A key barrier to the formation of conservancies is fear among ranchers of the loss of wildlife to neighbouring land owners. However, effective models for collaborative management of wildlife exist which enable equitable access among neighbouring ranchers to the mobile resource, and which take into account and address differential initial investments in wildlife. The development of such agreements can effectively remove the need for physical barriers between adjacent wildlife ranches.

2. Re-thinking fences for veterinary purposes

A re-think of the strategy of fencing for veterinary purposes is required. Veterinary fences in Botswana have resulted in high mortalities of wildlife due to the disruption of migration routes. Moreover, in several southern African countries, restrictions imposed by veterinary fencing severely limit the profitability of wildlife-based land uses (Albertson 1998; du Toit 2005; Mbaiwa and Mbaiwa 2006). Efforts of southern African countries to develop foot-and-mouth disease-free zones preclude the reintroduction of buffalo over large areas. Buffaloes are the most important species for generating income from trophy hunting as they can be hunted in relatively large numbers, they command high trophy fees and daily hunting rates, and are used to sell hunting packages (du Toit 2005; Lindsey et al. 2007). On Namibian wildlife ranches, and most communal land conservancies, the reintroduction of buffalo is precluded by veterinary restrictions (Lindsey et al. 2011). As a result, hunting safaris on private land in Namibia target almost entirely of low-value "plains game" hunts (Humavindu and Barnes 2003; Lindsey et al. 2007) reducing income significantly relative to what could be achieved if buffalo reintroductions were permitted. Veterinary fencing is typically supported by state or international funding, whereas the benefits accrue to private beef producers. Such subsidies artificially inflate the profitability of livestock-based land uses (Scoones & Wolmer 2008), while undermining wildlife-based land uses through associated restrictions on buffalo reintroductions and translocation of other species (Lindsey 2011). The advisability of these subsidies is declining in some cases as the profitability and productivity of livestock production appears to be waning in some areas. For example, the livestock industry in Botswana now acts as a net drain to the treasury (McLaughlin 2010). Similarly, in Botswana, the beef export industry is supported by subsidies from the European Union, providing livestock production with an unfair advantage over wildlife-based land uses and promoting continued erection and maintenance of veterinary fences. Furthermore, tourism (relying largely on wildlife) has already exceeded beef production in its contribution to Gross Domestic Product, and would increase further in profitability if veterinary fences were removed and wildlife populations allowed to recover (Mbaiwa and Mbaiwa 2006). Research is required to compare the profitability of livestock farming vs. potential returns from wildlife-based land uses unhindered by veterinary restrictions to assess the appropriateness of current foot-and-mouth disease control policies. Such a review would determine whether the promotion of wildlife-based land uses through an expansion of foot-and-mouth disease endemic zones was justified in economic terms. In Zimbabwe, for example, several authors have suggested that the foot-and-mouth disease fence in the South East of the country could be realigned to create a larger foot-and-mouth disease zone without adversely affecting potential for beef exports (Taylor and Martin 1987; du Toit 2005).

Ideally, alternative strategies for the control of wildlife-borne diseases are required which do not impose limitations on the development of wildlife-based land uses. One option would be to create protected enclaves for livestock from which wildlife is excluded and where wildlife diseases are strictly controlled, while allowing unfettered development of wildlife-based land uses elsewhere (Kock 2005). Perhaps the most promising potential strategy is through lobbying for acceptance of commodity based trading of processed meat products (Thomson et al. 2004). Under commodity based trading, meat processed in a manner proven to provide minimal risk of transmitting foot-and-mouth disease virus (e.g. through removal of bones and lymph nodes) could be considered acceptable for export (Thomson et al. 2004). The most pressing issue for gaining acceptance of commodity-based trade is obtaining support of the concept from the World Organisation for Animal Health (AHEAD 2008).

Where fences will continue to be used to control diseases, it is important that the design of fence employed in disease control be appropriate to contain the applicable animal hosts of that disease. For example, a fence aimed at restriction of African Swine Fever should be constructed in such a way as to be pig-proof and would not need to be higher than 1 m whereas a fence to contain foot-and-mouth disease within a wildlife area should incorporate a corridor to prevent direct contact between animals on either side of the fence as well as being appropriate to contain all cloven-hoofed wildlife (i.e. being pig and "digger" proof at the bottom while being high enough to contain "jumpers").

3. Adequate environmental and social impact assessments

Environmental and social impact-assessments should be a legal prerequisite for the development of fences for conservation purposes in all African nations (as is already the case in South Africa), a process that would ensure that the more obvious problems associated with fences are avoided. Had environmental impact assessments been conducted prior to the construction of veterinary control fences in Botswana, for example, wildlife die-offs may have been avoided by employing different fence designs or alignments, or by advising alternative veterinary control strategies (Albertson 1998).

The erection of fences around wildlife areas adjacent to rural communities should only occur following consultation with and acceptance from those communities. Fences are more likely to be accepted if efforts are made to provide communities with a stake in the management and use of wildlife and of the fence line. Employment opportunities for communities can be created in this way. Similarly, allocating shareholdings in protected areas to communities have the potential to increase tolerance of fences. In Kenya, wildlife populations in protected areas with outreach programmes involving adjacent rural communities performed better than those in parks lacking such efforts (Western et al. 2009). Community stake holdings in protected areas could be achieved by moving fences out to incorporate community land, as was recently done along the southern border of Savé Valley Conservancy in Zimbabwe (Lindsey et al. 2009). Alternatively, by engaging in co-management agreements with communities, park agencies could improve neighbour relations and effectively reduce the negative symbolism of fences. For example, in South Africa, South African National Parks have engaged in agreements with several communities to form contractual parks within existing national parks (e.g. Kruger and Kgalagadi) following successful land claims from communities evicted from by the apartheid government (Grossman and Holden 2009). Similar models have potential to extend benefits "beyond fences" in other countries where rural communities were dispossessed of land during the creation of parks and game reserves.

4. Re-designing of wildlife-proof fences

All wildlife proof fences should be designed to minimize the loss of wild animals due to entanglement or electrocution. For example, Burger and Branch (1994) recommend that the lowest electrified wire should be positioned at ≥250 mm above the ground to minimize tortoise mortality. Such a placement may also minimize the extent to which the fences are damaged by warthogs, aardvarks and porcupines, but would mean that predators and smaller species would not be effectively contained. Beck (2007) suggests the following additional modifications:

- The lowest electrified strand be offset by at least 400 mm to reduce the likelihood of tortoises being trapped between the mesh fence and an electrified strand.
- Where possible, rock-packed aprons should be used instead of low-level trip wires to reduce tortoise mortality.
- If low-level trip wires must be used, then a paired earth wire should be offset by an additional 100 mm to prevent direct contact with the live wire.

The frequency with which fences are challenged (and that entanglements occur) can be reduced by increasing the visibility of the fence by using wire of at least

2.5 mm diameter, and attaching objects such as cans or sheet material to the fence (du Toit 2005) as well as by avoiding construction of acute corners in fences. Finally, care is required when constructing fences to ensure that the material used cannot readily be made into snares, especially in areas where the supply of metallic wire is limited.

Conclusions

Fencing is an important tool in the management of wildlife and one that has played a significant role in the development of wildlife-based land uses in southern Africa. Furthermore, fencing is likely to become increasingly important as human populations continue to encroach on wildlife areas and habitats become more fragmented. However, due to financial constraints, fencing is likely to be of limited applicability in many countries in Africa. In addition, there are a number of conservation problems that can result from the improper use of fencing. Cognizance of all ecological, financial and social issues related to fences during planning can reduce problems and optimize conservation and economic gains achieved. Conservation planners should aim to achieve maximum gain with the minimum use of fencing, conduct thorough environmental and social impact assessments before establishing a fence, obtain support from local people, and give careful consideration to the alignment and type of fencing to be used.

References

AHEAD (2008) Report of the workshop held at Chobe Marina Lodge, Kasane, Botswana 11 to 14 November 2008. Available from: http://www.wcsahead.org/documents/sadc_tfcas_tads_nov08.pdf.

Albertson, A. (1998) *Northern Botswana veterinary fences: critical ecological impacts*. Okavango Peoples Trust report, Gabarone.

Barnett, R. & Patterson, C. (2006) *Sport hunting in the SADC Region: an overview*. TRAFFIC East/Southern Africa, Johannesburg.

Beck, A. (2007) *Shocking impacts of electric fences*. Endangered Wildlife Trust report, available at http://www.ewt.org.za/news_fullstory.aspx?status=0&newsID=445.

Ben-Shahar, R. (1993) Does fencing reduce carrying capacity for populations of large herbivores? *Journal of Tropical Ecology*, **9**, 249–253.

Bond, I., Child, B., de la Harpe, D., Jones, B., Barnes, J. & Anderson, H. (2004). Private-land contribution to conservation in South Africa. *Parks in transition*. (ed Child, B.) pp. 29–62. Earthscan, London.

Boone, R. & Hobbs, N. (2004) Lines around fragments: effects of fencing on large herbivores. *African Journal of Range & Forage Science*, **21**,147–158.

Bothma, J. du P. (2002) Some economics of game ranching. In B. Penzhorn (ed), *Proceedings of a symposium on game ranch planning and management* Onderstepoort: Wildlife Group, SA Veterinary Association, 23–40.

Burger, M. & Branch, W. (1994) Tortoise mortality caused by electrified fences in the Thomas Baines Nature Reserve. *South African Journal of Wildlife Research*, **24**, 32–37.

Caro, T.M. & Scholte, P. (2007) When protection falters. *African Journal of Ecology*, **45**, 233–235.
Caughley, G. (1994) Directions in conservation biology. *Journal of Animal Ecology*, **63**, 215–244.
Child, B., McKean, S., Kiss, A., Munthali, S., Jones, B., Mtsambiwa, M., Castley, G., Patton, C., Magome, H., Pangeti, G. & Fearnhead, P. (2004) Park agencies, performance and society in southern Africa. *Parks in transition*. (ed. Child, B) pp. 125–164. Earthscan.
Cumming, D. (2004a) Performance of parks in a century of change. *Parks in transition*. (ed. Child, B) pp. 105–204. Earthscan, London.
Cumming, D. (2004b) Movements and Seasonal migrations of large mammals in the four corners area. *Biodiversity of the Four Corners* area. (eds. Timberlake, J. & Childes, S.) pp. 444–457. Technical Reviews. Biodiversity Foundation, Bulawayo.
Davies-Mostert, H.T., Mills, M.G.L. & MacDonald, D.W. (2009) South Africa's wild dog *Lycaon pictus* meta-population management programme. *The re-introduction of top-order predators*. (eds. Hayword, M.W. & Somers, M.J.). pp. 10–42. Wiley-Blackwell, Oxford.
du Toit, R. (1998) Case study of policies that support sustainable development in Africa: the Savé Valley Conservancy, Zimbabwe. Paper presented at: *A workshop on African perspective of policies which support sustainable development*. The Scandinavian Seminar College, Harare.
du Toit, R. (2005) Foot-and- mouth disease management and land-use implications in the Zimbabwean Lowveld: the rationale for creating a biosphere reserve. In: S. Osofsky (ed.), *Conservation and development interventions at the wildlife/livestock interface: implications for wildlife, livestock and human health*. (pp. 109–112). IUCN, Gland.
Els, H. (2002) Rural development and game ranch management. *Game Ranch Management*. (ed. Bothma, J & du P.) pp. 677–686. van Schaik, Pretoria.
Graham, M.D., Gichohi, N., Kamau, F., Aike, G., Craig, B., Douglas-Hamilton, I. & Adams, W.M. (2009) The Use of Electrified Fences to Reduce Human Elephant Conflict: A Case Study of the Ol Pejeta Conservancy, Laikipia District, Kenya, Working Paper 1, Laikipia Elephant Project, Nanyuki, Kenya.
Grant, C., Bengis, R., Balfour, D. & Peel, M. (2007) Controlling the distribution of elephants. *The 2007 scientific assessment of elephant management in South Africa*. (eds. Mennell, K. & Scholes, R.) Witwatersrand University Press, Johannesburg.
Grant, C., Davidson, T., Funston, P. & Pienaar, D. (2002) Challenges faced in the conservation of rare antelope: a case study on the northern basalt plains of the Kruger National Park. *Koedoe*, **45**, 45–62.
Grossman. D. & Holden, P. (2009) Towards transformation: contractual national parks in South Africa. Evolution and innovation in wildlife conservation. (eds. Such, H., Child, B. & Spenceley, A.), pp. 357–372. Earthscan, London.
Hayward, M.W. & Kerley, G.I.H. (2009) Fencing for conservation: restriction of evolutionary potential or a riposte to threatening processes? *Biological Conservation*, **142**, 1–13.
Hayward, M.W. & Somers, M. J. (eds.) (2009) Reintroduction of top-order predators. Wiley-Blackwell, Oxford.
Hayward, M.W., O' Brien, J. & Kerley, G.I.H. (2007) Carrying capacity of large African predators: predictions and tests. *Biological Conservation*, **139**, 219–229.
Hayward, M.W., Hayward, G.J., Druce, D. & Kerley, G.I.H. (2008) Do fences constrain predator movements on an evolutionary scale? Home range, food intake and movement patterns of large predators reintroduced to Addo Elephant National Park, South Africa. *Biodiversity and Conservation*, **18**, 887–904.
Humavindu, M.N. & Barnes, J.L. (2003) Trophy hunting in the Namibian economy: an assessment. *South African Journal of Wildlife Research*, **33**, 65–70.
Hoare, R.E. (1992) Present and future use of fencing in the management of larger African mammals. *Environmental Conservation*, **19**, 160–164.
Hofmeyr, M. (2001) Introduction of the African wild dog *Lycaon pictus* to game reserves. Proceedings of a Symposium on the re-introduction of larger mammals to game reserves. Faculty of Veterinary Sciences report, University of Pretoria, Pretoria.
Hrabar, H. & du Toit, J.G. (2005) Dynamics of a protected black rhino (*Diceros bicornis*) population: Pilanesberg National Park, South Africa. *Animal Conservation*, **8**, 259–267.

Hunter, L., Pistorius, K., Carlisle, M., Rickleton, C., Walker, C., Slotow, R. & Skinner, J.D. (2007) Restoring lions *Panthera leo* to northern KwaZulu-Natal, South Africashort-term biological and technical success but equivocal long-term conservation. *Oryx*, **41**, 196–204.

Kassilly, F. (2002) The fence as a moderator of the wildlife menace in Kenya. *African Journal of Ecology*, **40**, 407–409.

Kock, R. (2005) What is this infamous "wildlife/livestock disease interface?" A review of current knowledge for the African continent. *Conservation and development interventions at the wildlife/livestock interface implications for wildlife, livestock and human health.* (ed. Osofsky, S.), pp. 1–13. IUCN, Gland.

Krug, W. (2001) Private supply of protected land in southern Africa. Workshop paper on market creation for biodiversity products and services Paris: World Bank/OECD.

Lamarque, F., Anderson, J., Chardonnet, P., Fergusson, R., la Grange, M., Osei-Owusu, Y., Bakker, L., Belemsobgo, U., Beytell, B., Boulet, H., Soto, B. & Tabi Tako-Eta, P. (2008) Human-wildlife conflict in Africa: an overview of causes, consequences and management strategies. FAO internet report: http://www.wildlife-conservation.org/en/content/download/1034/4568/file/HumanWildlifeConflict2008.pdf, Washington D.C.

Lindsey, P. & Davies-Mostert, H. (2009) South African Action Plan for the Conservation of cheetahs and African wild dogs. Report from a National Conservation Action Planning Workshop, Bela Bela, Limpopo Province, South Africa, 17–19 June 2009.

Lindsey, P. (2011) An analysis of game meat production and wildlife-based land uses on freehold land in Namibia: linkages with food security. TRAFFIC report, pp. 77, available from: http://www.traffic.org/general-reports/traffic_pub_gen43.pdf. Accessed September 2011.

Lindsey, P., du Toit, J. & Mills, M.G.L. (2005a) The cost efficiency of wild dog conservation in South Africa. *Conservation Biology*, **19**, 1205–1214.

Lindsey, P., du Toit, J. & Mills, M.G.L. (2005b) Attitudes of ranchers towards African wild dogs *Lycaon pictus*: conservation implications for wild dogs on private land. *Biological Conservation*, **125**, 113–121.

Lindsey, P., Roulet, P. & Romañach, S. (2007) Economic and conservation significance of the trophy hunting industry in sub-Saharan Africa. *Biological Conservation*, **134**, 455–469.

Lindsey, P., du Toit, J., Pole, A. & Romañach, S. (2008) Savé Valley Conservancy: a large scale experiment in cooperative wildlife management. In *Evolution and innovation in wildlife conservation*. (eds. Such, H., Child, B. & Spenceley, A.), pp. 163–186. Earthscan, London.

Lindsey, P., Romanach, S. & Davies-Mostert, H. (2009) The importance of conservancies for enhancing the value of game ranch land for large mammal conservation in southern Africa. *Journal of Zoology*, **277**, 99–105.

Lindsey, P., Romañach, S., Tambling, C. & Chartier, K. (2011) Ecological and financial impacts of illegal bushmeat trade in Zimbabwe. *Oryx*, **45**, 96–111.

MacArthur, R.H. & Wilson, E.O. (1967) *The theory of island biogeography*. Princeton University Press, New Jersey.

Marker, L.L., Mills, M.G.L. & Macdonald, D.W. (2003) Factors influencing perceptions of conflict and tolerance toward cheetahs on Namibian farmlands. *Conservation Biology*, **17**, 1290–1298.

Martin, R. (2005) The influence of veterinary control fences on certain wild large mammal species in the Caprivi, Namibia. *Conservation and development interventions at the wildlife/livestock interface implications for wildlife, livestock and human health*. (ed. Osofsky, S.), pp 27–39. IUCN, Gland.

Mbaiwa, J.E. & Mbaiwa, O.I. (2006) The effects of veterinary fences on wildlife populations in Okavango Delta, Botswana. *International Journal of Wilderness*, **12**, 17–41.

McLaughlin, D. (2010). *Botswana Beef Industry and EU Trade Policy*. Unpublished report, pp. 33.

Murcia, C. (1995) Edge effects in fragmented forests: implications for conservation. *Trends in Ecology and Evolution*, **10**, 58–62.

National Agricultural Marketing Council (2006) Report on the investigation to identify problems for sustainable growth and development in South African wildlife ranching. National Agricultural Marketing Council Report, 2006 number 3, available from: http://www.namc.co.za/ASSETS/PDF/Report%20on%20Wildlife%20ranching.pdf.

Nelson, A., Bidwell, P. & Sillero-Zubiri, C. (2003) A review of human elephant conflict management strategies. People and Wildlife Initiative. Wildlife Conservation Research Unit, Oxford University. Available from: http://www.peopleandwildlife.org.uk/crmanuals/HumanElephantConflictP&WManual.

Norton-Griffiths, M. (2007) How many wildebeest do you need? *World economics*, **8**, 41–64.

Noss, A. (1998) The impacts of cable snare hunting on wildlife populations of the Central African Republic. *Conservation Biology*, **12**, 390–398.

Ogada, M.O., Woodroffe, R., Oguge, N. & Frank, L.G. (2003) Limiting depredation by African carnivores: the role of livestock husbandry. *Conservation Biology*, **17**, 1521–1530.

Osborn, L. & Anstey, S. (2002) Elephant/human conflict and community development around the Niassa Reserve, Mozambique. WWF-SARPO Report. Internet http://www.elephantpepper.org/downloads/Niassa%20ele%20report.pdf.

Romañach, S. & Lindsey, P. (2008) Conservation implications of prey responses to wild dogs *Lycaon pictus* during the denning season on wildlife ranches. *Animal Conservation*, **11**, 111–117.

Schumann, M., Schumann, B., Dickman, A., Watson, L. & Marker, L. (2006) Assessing the use of swing gates in game fences as a potential non-lethal predator exclusion technique. *South African Journal of Wildlife Research*, **36**, 173–181.

Scoones, I. & Wolmer, W. 2008. Foot-and-mouth disease and market access: challenges for the beef industry in southern Africa. Transboundary animal disease and market access: future options for the beef industry in southern Africa *Working Paper 1, Institute for Development Studies*, Brighton.

Spierenburg, M. & Wels, H. (2006) Securing space. *Space and Culture,* **9**, 294–312.

Stutmoller, P. (2002) The fencing issue relative to the control of foot-and-mouth disease. *Annals of the New York Academy of Science*, **969**, 191–200.

Taylor, R. (2009) The performance of CAMPFIRE in Zimbabwe: 1989–2006. *Evolution and innovation in wildlife conservation*. (eds. Such, H., Child, B. & Spenceley, A.). pp. 201–222. Earthscan, London.

Taylor, R. & Martin, R. (1987) Effects of veterinary fences on wildlife conservation in Zimbabwe. *Environmental Management*, **11**, 327–334.

Thomson, G., Tambi, E., Hargreaves, S., Leyland, T., Catley, A., van't Klooster, G. & Penrith, M. (2004) International trade in livestock and livestock products. *Veterinary Record,* **155**, 429–433.

van Dyk, G. & Slotow, R. (2003) The effects of fences and lions on the ecology of African wild dogs reintroduced into Pilanesberg National Park, South Africa. *African Zoology*, **38**, 79–94.

van Rooyen, N., du Toit, J.G. & van Rooyen, J. (2002) Fences. *Game ranch management*. Fourth edition. (ed. Bothma, J. du P.), pp. 79–94. JL van Shaik, Pretoria.

Wels, H. (2000) *Fighting over fences: organisational co-operation and reciprocal exchange between the Savé Valley Conservancy and its neighbouring communities*. PhD thesis, Free University, Amsterdam.

Western, D. & Nightingale, M. (2005) Keeping the East African rangelands open and productive. *Conservation and People*, **1**, 1–8.

Western, D., Russell, S. & Cuthill, I. (2009). The status of wildlife in protected areas compared to non-protected areas of Kenya. *Plos One* **4**, e6140.

Wilkie, D., Carpenter, J. & Zhang, Q. (2001) The under-financing of protected areas in the Congo Basin: so many parks and so little willingness-to-pay. *Biodiversity and Conservation*, **10**, 691–709.

Williamson, D. & Mbano, M. (1988) Wildebeest mortality during 1983 at Lake Xau, Botswana. *African Journal of Ecology,* **26**, 341–344.

Woodroffe, R. & Ginsberg, J. (1998) Edge effects and the extinction of populations inside protected areas. *Science*, **280**, 2126–2128.

Chapter 13
Do Fences or Humans Inhibit the Movements of Large Mammals in Białowieża Primeval Forest?

R. Kowalczyk, K. Schmidt, and W. Jędrzejewski

Introduction

Large areas of continuous forests and marshlands, which covered most of the European continent in the past, have been largely wiped out and its remnants fragmented during recent centuries as human habitation and cities developed. Along with habitat deterioration most of large animals disappeared or have been pushed into isolated pockets of habitat (Bibikov 1985; Pucek et al. 2004; Von Arx et al. 2004). Moreover, further fragmentation of habitats has been caused by anthropogenic barriers including roads and human settlements. This gives rise to a question – how does it affect the communities of animals and their genetic diversity? This is particularly important as the large mammals require vast spaces to exist. For instance, wolves (*Canis lupus*) with their territories covering from 80 to 4,300 km^2 (Jędrzejewski et al. 2007) or Eurasian lynx (*Lynx lynx*) utilising ranges up to 1,900 km^2 (Schmidt et al. 1997; Linnell et al. 2001) are often far above the extent of habitat available for sustainable populations of these carnivores. That is most likely why the distribution of continuous ranges of those species covering large areas of Asia ends up at the edges of uninterrupted forest cover in Eastern Europe. Indeed, the quantitative analysis of the wolf and lynx distribution in Poland has shown that woodlands are strongly preferred by both carnivores (Jędrzejewski et al. 2004; Niedziałkowska et al. 2006).

One of the last areas in Europe, which supports populations of keystone species, such as European bison, Eurasian lynx and wolf is the Białowieża Primeval Forest (BPF), located on the Polish-Belarussian border. The BPF constitutes the last remnants of the natural forests that once covered much of continental Europe. It covers 1,500 km^2 of continuous woodland (600 km^2 in Poland and 900 km^2 in Belarus),

R. Kowalczyk (✉) • K. Schmidt • W. Jędrzejewski
Mammal Research Institute, Polish Academy of Sciences, 17-230, Białowieża, Poland
e-mail: rkowal@zbs.bialowieza.pl

M.J. Somers and M.W. Hayward (eds.), *Fencing for Conservation: Restriction of Evolutionary Potential or a Riposte to Threatening Processes?*,
DOI 10.1007/978-1-4614-0902-1_13, © Springer Science+Business Media, LLC 2012

consisting mainly of unique, at a European scale, habitats such as oak-lime-hornbeam forests and ash-alder wet bog forests (Jędrzejewska and Jędrzejewski 1998). The BPF has been protected since the fifteenth century by Polish kings and later on by Russian tsars for hunting purposes (Samojlik 2005). Thanks to this protection, the forest and communities of animals, survived almost untouched until the twentieth century.

The BPF was an important refuge for the European bison and large carnivores in the past. It was one of two locations where the European bison survived into the twentieth century (Pucek et al. 2004), whereas the populations of wolves and lynx were able to quickly recover in BPF after periods of nearly total exterminations, due to its connectivity with other woodlands and marshland in the east (Jędrzejewska et al. 1996; Jędrzejewski et al. 1996). At present, however, these highly space-demanding species seem to face several constraints that may create barriers for their movements and contribute to the isolation of their populations. The BPF borders large open areas consisting of the agricultural lands and human settlements at its western edges that form inhospitable terrain for wildlife. Animal movements within the BPF have been yet more limited since 1981 due to the construction of a barbed-wire fence (2.5 m high) along the border of the former Soviet Union, which now separates its Polish and Belarussian parts. In this chapter, we discuss the costs and benefits of physical and metaphorical barriers, and we demonstrate how these barriers are influencing the spatial organisation and movements of large mammals, namely the European bison, the wolf and the Eurasian lynx, inhabiting the BPF.

European Bison: The Beast in the Cage

European bison is the largest extant terrestrial mammal in Europe and is a flagship species for nature conservation. After its extinction in the wild at the beginning of the twentieth century, the species was resurrected from only seven individuals and brought back to the wild (Pucek et al. 2004; Krasińska and Krasiński 2007). Bison are distributed in nearly 30 populations over large areas mainly in Belarus, Poland, Ukraine and Russia (Pucek et al. 2004; Krasińska and Krasiński 2007). Exchange of individuals among those scattered populations is limited mainly due to their high geographical isolation; however, its numbers are systematically growing, but most of the bison populations are small (below 100 individuals) (Pucek et al. 2004).

BPF protects the largest free ranging bison population, which numbers nearly 800 individuals (Bison Pedigree Book 2007). Yet this population has had a history of restriction in the forest, through land use and management factors. Since the beginning of the 1980s, the population has been divided by the Polish-Belarussian border fence into two separate populations. Before construction of the fence, European bison crossed the border at a rate of 5.8 bison/1 km/year (Bunevich 2004). These movements were more intensive during the rutting season (August-October), when bulls roamed in search of breeding cows. For nearly 30 years the two populations have remained isolated.

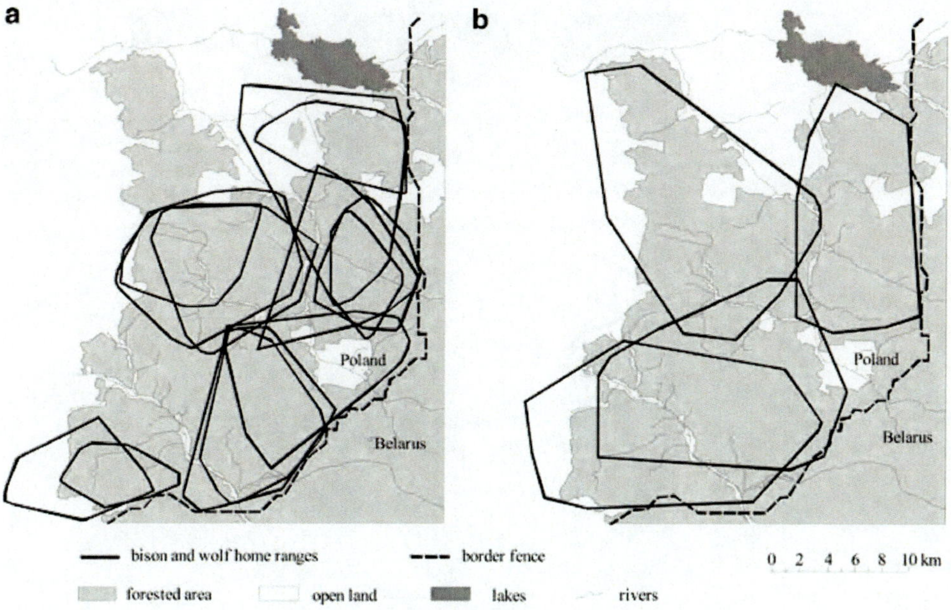

Fig. 13.1 Spatial organisation of large mammals studied by radio tracking in Polish part of the Białowieża Primeval Forest: (**a**) bison; (**b**) wolves

As estimated by many researchers, minimum viable populations of large mammals should number hundreds, or even thousands of individuals (see review by Traill et al. 2007). It is especially important for the species such as the European bison, which suffered a severe bottleneck effect (Pucek et al. 2004). Genetic variation of Białowieża's population of bison is very low ($H_e = 0.26$, $N_{alleles} = 2.15$; Tokarska et al. 2009; Wójcik et al. 2009). Thus, the population is under threat with regard to factors that are potentially able to decrease its genetic variability. First, it is the influence of the border fence that prevented any exchange of individuals between the two parts of the population. It is clear from radio-tracking data that movements or shape of the bison home-ranges adjacent to the border area are affected by the border fence (Figs. 13.1 and 13.2), indicating that the fence is a real barrier for them. Second, it is the agricultural land on the western side of the forest that prevents expansion of the population. Although bison are not so sensitive to habitat fragmentation, and migrations through open areas were often observed (Krasińska and Krasiński 2007), they are then exposed to conflicts with humans. That is due to the damage they cause to crops or the risk of attacks on people, when bothered while roaming close to human settlements. One of the management strategies in such cases was trapping or culling dispersing animals. Between 1985 and 2005, 32 dispersing bison were culled and 15 were captured and translocated (Krasińska and Krasiński 2007; Krasiński Z, unpublished data).

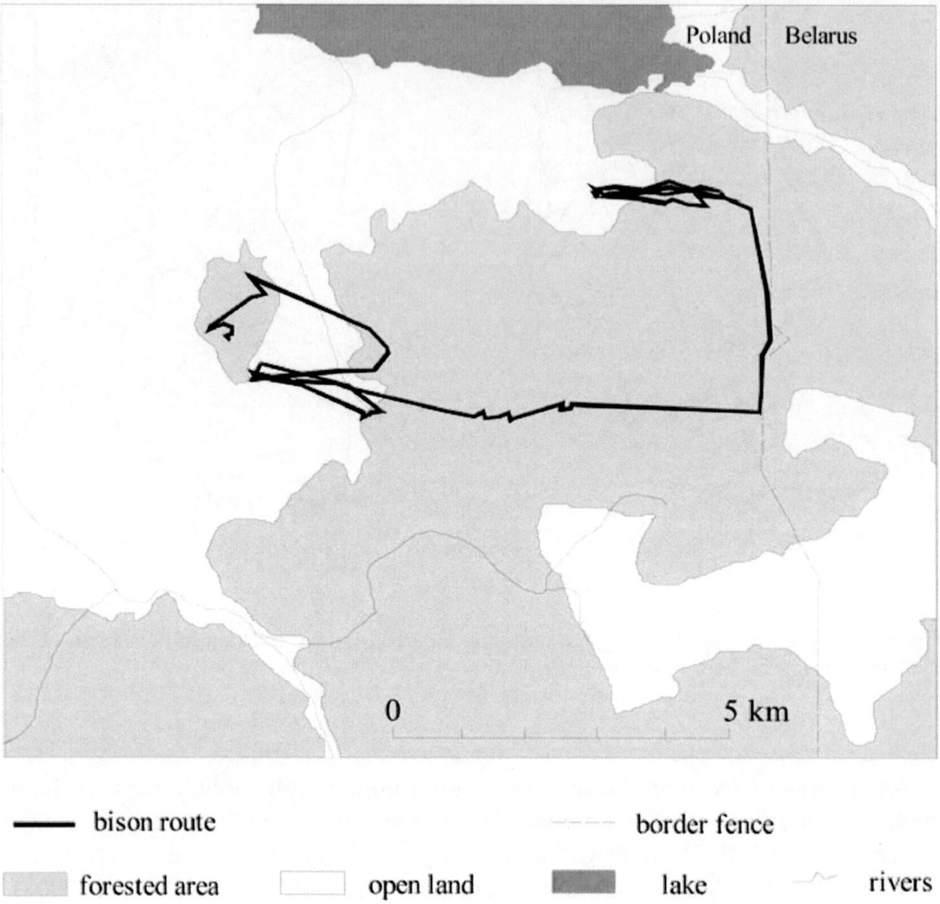

Fig. 13.2 Example of movements of bison cow fitted with GPS collar near Polish-Belarussian border (5–15 January 2008)

Not only are border fences and distribution of woodlands affecting the ranging pattern of bison, but also management practices strongly influence their spatial organisation and movements. Supplementary feeding is one such management strategy and is intended to increase bison survival, control their migration and decrease the potential of damage to trees. Bison in winter gather in 5 or 6 main feeding sites and create winter aggregations numbering up to 100 individuals (Krasińska and Krasiński 2007). In effect, the population is divided into a few sub-populations associated with different feeding sites, with no or limited exchange of individuals between them (Kowalczyk et al. 2010). It also influences space use by bison in the spring-autumn period, as the home ranges of individuals originating from the same feeding site largely overlap (Krasińska et al. 2000). Additionally, most males mate with females from the same sub-population (Kowalczyk et al. 2010). Of 18 males radio-collared in 2005–2008, only 11% roamed between herds of cows associated with different sub-populations. Supplementary feeding most strongly affects space use and movements of bison in winter (Rouys and Krasińska 2001). Ranges of individuals from the herds fed intensively (delivery of hay and silage 3–5 times a week) occupied very small ranges (mean 2.8 km^2), comparing to less-intensively (16.2 km^2) or non-fed bison (29.5 km^2) (Schneider 2008; Kowalczyk et al. 2010). Daily movement distances of bison are negatively correlated with intensity of supplementary feeding (Schneider 2008), and it declines from 0.9 km/day in non-fed bison to 0.5 km/day for individuals that are intensively fed.

Lynx and Wolves: Surviving on Restricted Hunting Grounds

Large predators, such as lynx and wolves, are highly mobile animals, whose long-distance movements are expected to ensure effective gene flow among sub-populations. However, because they utilise large territories, their densities are usually very low; therefore, if they occupy fragmented landscapes the sub-populations are much below minimum viable population size (Traill et al. 2007). Such sub-populations may exist in the long term only if an exchange of individuals among them is possible, otherwise genetic deterioration and/or stochastic events may drive them to extinction (Gilpin and Soule 1986). In the case of the Eurasian lynx, habitat fragmentation may particularly affect its survival, because of its inability to traverse large open areas (Schmidt 1998). Additionally, open landscapes inhabited by humans constitute a barrier due to higher mortality rates related to anthropogenic reasons (vehicle collision, illegal killing and predation from dogs). Among the lynx studied in the Białowieża Forest, nearly every radio-collared lynx whose home ranges included fragmented woodlands died due to human-related sources of mortality ($n=4$, i.e. 15% of all radio-collared lynx) (K. Schmidt, R. Kowalczyk, and W. Jędrzejewski unpublished data). Radio-tracking studies showed that the lynx population in BPF is maintained by exchange with other neighbouring populations in Poland and Belarus; however, the range and directions of their dispersal is strongly affected by the distribution and availability of woodlands and forest corridors (Schmidt 1998). This was especially apparent with two radio-collared siblings radio tracked during their dispersal, as they altered the directions of their movements after reaching the western limits of the BPF (Fig. 13.3). On the other hand, genetic research of lynx inhabiting the BPF and its vicinity suggested low gene flow between this population and the lynx from the more contiguous population in the Baltic countries (Schmidt et al. 2009). Thus, although not entirely physically fenced, the lynx are affected by metaphorical fences by way of habitat barriers.

In contrast, presence of real fences, including the fencing along the Polish-Belarussian border does not seem to play any role in the lynx movements. Radio-tracked lynx whose home ranges encompassed both the Polish and Belarussian parts of the Białowieża Forest, frequently crossed the border (Schmidt et al. 1997). Their ability to pass the fenced border is not surprising, as 31% of lynx resting sites in BPF is localised within fenced plots of plantations and thickets (Podgórski 2006; Podgórski et al. 2008). Selection for such plots is due to the fact that most of the

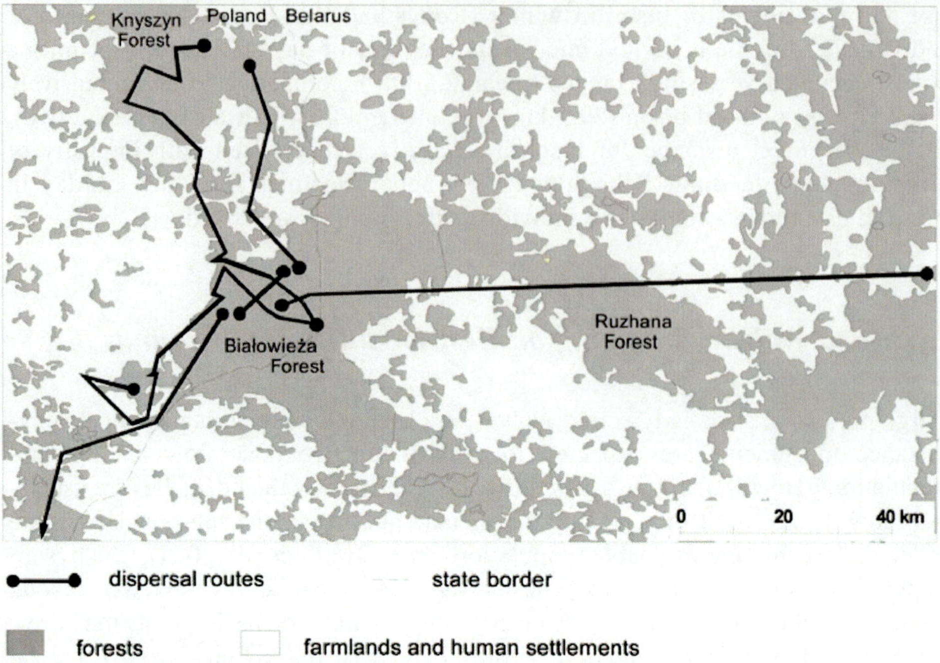

Fig. 13.3 Dispersal of radio-collared lynx in Białowieża Primeval Forest (based on Schmidt 1998; modified)

younger tree stands used by lynx for resting (Podgórski et al. 2008) are fenced against damages by ungulates.

The border fence is not a barrier to wolves either, as they are able to cross it most likely along the rivers. Similarly, the open spaces do not seem to prevent the wolves' movements as they were observed to wander out of the forest area more readily than lynx in the BPF (Schmidt 2008). Wolves have been found to be typically less dependent on habitat characteristics than felids (Husseman et al. 2003). In spite of that, territories of wolf packs in the Polish part of the BPF were distributed in a way that they covered mainly the continuous forested area – between the border fence in the east, and open agricultural and urban land in the west (Fig. 13.1) (Okarma et al. 1998; Theuerkauf et al. 2003a, b). The effect of the border zone can be a result of the fact that wolves are strongly persecuted in the Belarussian part of the BPF (62 wolves shot in 1995–2000; Jędrzejewski et al. 2005). A similar reason can determine the positioning and shape of the western territories, indicating the association of these predators with the forest habitat. Wolves settling in areas with little forest cover are more prone to conflicts with humans due to their depredation on livestock. Indeed, conflict areas may constitute a significant barrier for wolves' dispersal as the predators are more often removed from such areas (Kojola et al. 2006).

Human activity was found to directly influence the movement pattern of wolves in BPF (Theuerkauf et al. 2001, 2003a, b). The movement patterns of wolves occupying commercial parts of the forest (human activity and traffic during the daylight) were negatively correlated with the activity patterns of humans, whereas those inhabiting the strictly protected area of the BPF (Białowieża National Park: reduced human activity, no traffic) were not. Moreover, although the wolves did not reduce their activity during the daylight, they avoided being in the same place at the same time as humans (Theuerkauf et al. 2001, 2003a, b).

Conclusions

Our review showed that spatial organisation, movements and activity of large mammals are affected by past (habitat fragmentation) or present (management, traffic) human activity even in the area considered to be the best-preserved forest in Europe. These involve a real fence occurring along the state border that is physically impassable for big ungulates as well as virtual barriers in the form of habitat discontinuity that does not permit a sufficient number of animals to traverse. Both types of barriers increase mortality of animals on the one hand and decrease the exchange of individuals on the other. Therefore, we believe that fencing may have a largely negative impact on large mammal population viability.

Although fencing wildlife is often considered as a conservation tool aiming at preventing damage to human economy (Hayward and Kerley 2009), it should not be encouraged in Europe when large ungulates and carnivores are concerned. In contrast, measures directed at extending the availability of habitat and alleviating its fragmentation through connecting it via the network of ecological corridors should be strongly supported. An important task should be to remove the fence along the state border; however, this is a political decision, especially considering that the eastern border of Poland is also the border of the European Union. Nevertheless, given the growing pressure of human activity and increasing fragmentation and loss of habitats in Europe, future conservation efforts concerning large mammals should aim to mitigate all negative processes.

Acknowledgements We wish to thank Mathew Hayward for valuable comments and English revision.

References

Bibikov, D.I. (1985) *The wolf.* Nauka Publishers, Moscow [in Russian].

Bison Pedigree Book. (2007) J. Raczyński (ed.). Białowieski Park Narodowy, Białowieża.

Bunevich, A. (2004) Spatial structure and movements of European bison in the Belarusian part of Białowieża Forest. *Proceedings of the Conference "European Bison Conservation"*, (eds. Krasińska, M. and Daleszczyk, K.). pp 23–26. Mammal Research Institute PAS, Białowieża, Poland.

Gilpin, M.E. & Soule, M.E. (1986) Minimum viable population: processes of species extinction. *Conservation Biology: the science of scarcity and diversity*, (eds. Soule, M.E.). pp 19–34. Sinauer, Sunderland.

Hayward, M.W. & Kerley, G.I.H. (2009) Fencing for conservation: Restriction of evolutionary potential or a riposte to threatening processes? *Biological Conservation,* **142,** 1–13.

Husseman, J.S., Murray, D.L., Power, G., Mack, K., Wenger, C.R. & Quigley, H. (2003) Assessing differential prey selection patterns between two sympatric large carnivores. *Oikos,* **101,** 591–601.

Jędrzejewska, B. & Jędrzejewski, W. (1998) Predation in vertebrate communities: the Białowieża Primeval Forest as a case study. Ecological Studies 135, p. 450. Springer-Verlag, Berlin – Heidelberg – New York.

Jędrzejewska, B., Jędrzejewski, W., Bunevich, A.N., Miłkowski, L. & Okarma, H. (1996) Population dynamics of wolves *Canis lupus* in Białowieża Primeval Forest (Poland and Belarus) in relation to hunting by humans, 1847–1993. *Mammal Review*, **26**, 103–126.

Jędrzejewski, W., Jędrzejewska, B., Okarma, H., Schmidt, K., Bunevich, A.N. & Miłkowski, L. (1996) Population dynamics (1869–1994), demography, and home ranges of the lynx in Białowieża Primeval Forest (Poland and Belarus). *Ecography*, **19**, 122–138.

Jędrzejewski, W., Niedziałkowska, M., Nowak, S. & Jędrzejewska, B. (2004) Habitat variables associated with wolf (*Canis lupus*) distribution and abundance in northern Poland. *Diversity and Distribution*, **10**, 225–233.

Jędrzejewski, W., Branicki, W., Veit, C., Medugorac, I., Pilot, M., Bunevich, A.N., Jędrzejewska, B., Schmidt, K., Theuerkauf, J., Okarma, H., Gula, R., Szymura, L. & Förster, M. (2005) Genetic diversity and relatedness within packs in an intensely hunted population of wolves Canis lupus. *Acta Theriologica*, **50**, 1–22.

Jędrzejewski, W., Schmidt, K., Theuerkauf, J., Jędrzejewska, B. & Kowalczyk, R. (2007) Territory size of wolves *Canis lupus*: linking local (Białowieża Primeval Forest, Poland) and Holarctic-scale patterns. *Ecography*, **30**, 66–76.

Kojola, I., Aspi, J., Hakala, A., Heikkinen, S., Ilmoni, C. & Ronkainen, S. (2006) Dispersal in an expanding wolf population in Finland. *Journal of Mammalogy*, **87**, 281–286.

Kowalczyk, R., Kamiński, T. & Schneider, T.C. (2010) Influence of supplementary feeding on bison in Białowieża Primeval Forest. *European bison conservation in the Białowieża Forest. Threats and prospects of the population development*, (eds. Kowalczyk, R., Ławreszuk, D. & Wójcik, J. M.). pp 147–160. Mammal Research Institute PAS, Białowieża, Poland.

Krasińska, M., Krasiński, Z. A. & Bunevich, A. N. (2000) Factors affecting the variability in home range size and distribution in European bison in the Polish and Belarussian parts of the Białowieża Forest. *Acta Theriologica*, **45**, 321–334.

Krasińska, M. & Krasiński, Z. (2007) European bison. *The Nature Monograph.* Mammal Research Institute, Białowieża.

Linnell, J.D.C., Andersen, R., Kvam, T., Andren, H., Liberg, O., Odden, J. & Moa, P.F. (2001) Home range size and choice of management strategy for lynx in Scandinavia. *Environmental Management*, **27**, 869–879.

Niedziałkowska, M., Jędrzejewski, W., Mysłajek, R.W., Nowak, S., Jędrzejewska, B. & Schmidt, K. (2006) Environmental correlates of Eurasian lynx occurrence in Poland - Large scale census and GIS mapping. *Biological Conservation*, **133**, 63–69.

Okarma, H., Jędrzejewski, W., Schmidt, K., Śnieżko, S., Bunevich, A.N. & Jędrzejewska, B. (1998) Home ranges of wolves in Białowieża Primeval Forest, Poland, compared with other Eurasian populations. *Journal of Mammalogy*, **79**, 842–852.

Podgórski, T. (2006) *Characteristics and selectivity of resting and kill sites of Eurasian lynx in the Białowieża Primeval Forest*. M.Sc. Thesis. Wrocław University. Wrocław.

Podgórski, T., Schmidt, K., Kowalczyk, R., & Gulczyńska, A. (2008) Microhabitat selection by Eurasian lynx and its implications for species conservation. *Acta Theriologica*, **53**, 97–110.

Pucek, Z. (ed.); Pucek, Z., Belousova, I.P., Krasińska, M., Krasiński, Z.A. & Olech, W. (2004) *European biosn. Status survey and conservation action plan.* IUCN/SSC Bison Specialist Group. IUCN, Gland.

Rouys, S. & Krasińska, M. (2001) *Influence of supplementary winter feeding in the daily travel distance of European bison in the Białowieża Forest, Poland*. The Wildlife Society, Maryland.

Samojlik, T. (ed). (2005) *Conservation and hunting. Białowieża Forest in the time of Kings.* Mammal Research Institute, Białowieża.

Schmidt, K., Jędrzejewski, W. & Okarma, H. (1997) Spatial organization and social relations in the Eurasian lynx population in Białowieża Primeval Forest, Poland. *Acta Theriologica*, **42**, 289–312.

Schmidt, K. (1998) Maternal behaviour and juvenile dispersal in the Eurasian lynx. *Acta Theriologica*, **43**, 391–401.

Schmidt, K. (2008) Behavioural and spatial adaptation of the Eurasian lynx to a decline in prey availability. *Acta Theriologica*, **53**, 1–16.

Schmidt, K., Kowalczyk, R., Ozolins, J., Männil, P., & Fickel, J. (2009) Genetic structure of the Eurasian lynx population in north-eastern Poland and the Baltic states. *Conservation Genetics*, **10**, 497–501.

Schneider T. C. (2008) Microhabitat selection by European bison (Bison bonasus) in Białowieża Primeval Forest, Poland. MSC thesis, Göettingen University, Germany.

Theuerkauf, J., Jędrzejewski, W., Schmidt, K. & Gula, R. (2001) *Impact of human activity on daily movement patterns of wolves,* Canis lupus, *in the Białowieża Forest, Poland.* The Wildlife Society, Maryland.

Theuerkauf, J., Jędrzejewski, W., Schmidt, K. & Gula, R. (2003) Spatiotemporal segregation of wolves from humans in the Białowieża Forest (Poland). *The Journal of Wildlife Management*, **67**, 706–716.

Theuerkauf, J., Jędrzejewski, W., Schmidt, K., Okarma, H., Ruczyński, I., Śnieżko, S. & Gula, R. (2003) Daily patterns and duration of wolf activity in the Białowieża Forest, Poland. *Journal of Mammalogy*, **84**, 243–253.

Tokarska, M., Kawałko, A., Wójcik, J. M. & Pertoldi, C. (2009) Genetic variability in the European bison (*Bison bonasus*) population from Białowieża forest over fifty years. *Biological Journal of the Linnean Society,* **97,** 801–809.

Wójcik, J.M., Kawałko, A., Tokarska, M., Jaarola, M., Valenback, P. & Pertoldi, C. (2009) Post-bottleneck mtDNA diversity in a free-living population of European bison: implications for conservation. *Journal of Zoology*, **277**, 81–87.

Traill, L.W., Bradshaw, C.J.A. & Brook, B.W. (2007) Minimum viable population size: A meta-analysis of 30 years of published estimates. *Biological Conservation*, **139**, 159–166.

Von Arx, M., Breitenmoser-Würsten, C., Zimmermann, F. & Breitenmoser, U. (2004) *Status and conservation of the Eurasian lynx (Lynx lynx) in 2001.* KORA Bericht no 19.

Chapter 14
Exploring the Value of Wolves (*Canis lupus*) in Landscape-Scale Fenced Reserves for Ecological Restoration in the Scottish Highlands

Christopher Sandom, Joseph Bull, Susan Canney, and David W. Macdonald

Introduction

The use of fencing to enclose protected areas, and particularly to partition predators from areas where it is incompatible for them to range freely amongst agriculture or human settlement, is particularly prevalent in southern Africa. For example, Davies-Mostert et al. (2009) describe the operation of a "managed metapopulation" of African wild dogs *Lycaon pictus* comprised of 12 fenced reserves varying in size between 84 and 900 km^2. Insofar as the reintroduction of predators is an emerging priority for conservation biology (Macdonald 2009), and considering that this may often involve keeping them separate from domestic stock, the efficacy of fenced reserves is topical. Indeed, Macdonald et al. (2000) proposed that the eagerness with which western conservationists promoted the conservation of large mammals, including predators, in developing countries could usefully be matched by more radical approaches to restoration in developed countries and, in the case of the UK, that this might fruitfully involve consideration of fenced wilderness areas. Several practicalities may affect this idea, but perhaps foremost is what size fenced areas would have to be in order to deliver more or less natural communities of larger mammals for the benefits of, for example, tourism and conservation. This is an obvious question to pose in the context of frequent press articles about, for instance, the reintroduction of wolves, *Canis lupus*, to Scotland (Gilchrist 2008; Leake 2008), yet we are unaware of any systematic attempt to answer it.

C. Sandom (✉) • J. Bull • D.W. Macdonald
Wildlife Conservation Research Unit, Department of Zoology, University of Oxford, Tubney House, Abingdon Road, Tubney, Oxon OX13 5QL, UK
e-mail: chris.sandom81@gmail.com

S. Canney
Spatial Ecology and Epidemiology group, Department of Zoology, University of Oxford, Tubney House, Abingdon Road, Tubney, Oxon OX13 5QL, UK

M.J. Somers and M.W. Hayward (eds.), *Fencing for Conservation: Restriction of Evolutionary Potential or a Riposte to Threatening Processes?*,
DOI 10.1007/978-1-4614-0902-1_14,

Without mention of fenced reserves, Hetherington and Gorman (2007) have fruitfully evaluated the capacity of the Scottish countryside to sustain a population of reintroduced lynx, *Lynx lynx*, – a species successfully reintroduced in central Europe (Breitenmoser et al. 2010). Nilsen et al. (2007) modelled the likely impact on deer *Cervus elaphus* control, in terms of population regulation and economics, of reintroducing wolves to the Scottish Highlands. Furthermore, detailed simulations underpinned the reintroduction of wolves to Yellowstone National Park in the USA (Boyce 1993; Phillips et al. 2004). Turning to fencing, Davies-Mostert et al. (2009) report on the practicalities of creating and managing fenced reserves for African wild dogs in South Africa. In Britain, this constellation of possibilities has recently been brought into focus when the owner of the private Alladale Estate in the Highlands of Scotland made public (e.g. Hellen 2004) his aim to manage the estate for the purpose of restoring a more natural ecosystem. The management plan was expressed in terms of a blend of conservation and development goals, restoring and managing the Highland ecosystem, creating jobs in an economically depressed area and creating an educationally enriching recreational resource and included consideration of the feasibility of creating a fenced wilderness reserve within which large mammals, including wolves, might be maintained. Stimulated by this particular case we seek, in this chapter, to identify some of the questions that would have to be answered when evaluating the likely outcomes of creating fenced reserves for large predators.

Since this topic – the possible management of previously extirpated large carnivores such as wolves in fenced reserves in Scotland – is one open to accidental, or even wilful, misinterpretation, it seems sensible before going further to be plain about what this chapter is and what it is not. This chapter is not a call to reintroduce wolves, or other large carnivores into the wild in Scotland, and it is not even a proposal to manage them in a fenced area (itself a very different thing to a reintroduction). However, since it is the case that the topic of large predators being somehow returned to the Highlands is much discussed, and since fenced protected areas are already being used successfully in the management and conservation of carnivores – notably in southern Africa – it seems logical to discuss what steps might be involved in evaluating the pros and cons of a fenced protected area to include large native predators in the Highlands. However, even this question is bigger than we can tackle here, and we don't. Rather, we present one approach to tackling just one part of this question. The part we tackle is how the viability and functioning of populations of wolves and their prey might differ between hypothetical reserves of various sizes. We have attempted this, using methods that are transparent and explicit. Our exploration leads us to think that the wider topic is worth exploring further, but we are equally clear that even within our exploration various issues remain unresolved. For example, we do not consider genetic effects on small populations, and we do not reach a conclusion on whether the probability of the fence being breached is acceptable – we are not ignoring these questions, we simply don't yet know the answers to them.

This essay sits on an interesting continuum. A confined animal may be in a cage, a menagerie, a park, a reserve or a wilderness variously impacted by people. In some of these it may be fenced in, in others it may abut places from which it is fenced out. Along this continuum its life may be variously "natural". However, just what constitutes "natural", and just how much it is valued, is not easily defined. Furthermore, in Britain, the various designations of protected areas produce some muddling vocabulary. Landscapes are protected in National Parks and National Scenic Areas, while biodiversity is protected in Sites of Special Scientific Interest (SSSI), National Nature Reserves and Natura 2000 Sites. People gain access to nature in these areas, as well as through National Trails, Community Forests and Country Parks. Here, we do not intend to stir the hornet's nest of these designations, nor to imply any particular categorisation of the fenced protected area about which we write. Rather we ask whether such a fenced area could be helpful to conservation, to development and hence to society. If it is, then vocabulary and policy can surely be developed to deliver it.

Loss of the Native Caledonian Pine Forest of Scotland

Pollen analysis suggests that pine and birch woodland, the Caledonian Pine Forest, reached its maximum extent in the Scottish Highlands approximately 5,000–6,500 years before present (BP) when woodland may have covered 50% of the country (Smout et al. 2005). By 4,000 years BP the climate had cooled, becoming wetter and windier, and caused the Caledonian Pine Forest to retreat (Birks 1989; Smout et al. 2005). Determining the extent of reduction of woodland cover during this time is difficult but Smout et al. (2005) tentatively suggest a decline to 25% cover by 2,000 years BP. During this period, evidence suggests human settlements were restricted to coastal areas in northern Scotland and so were unlikely to have played a major role in deforestation (Kinloch et al. 1986; Birks 1989). In more recent times, anthropogenic deforestation has reduced native woodland cover. There are documented records of forest exploitation during the last 500 years at specific sites across the Highlands, including Rothiemurchus, Strathcarron, Glenorchy and The Isle of Skye (Smout et al. 2005). By 1905, only an estimated 350,000 ha (4.5% of Scotland) of woodland, native and planted, remained (Forestry Statistics 2008). These remnants were put under further strain during the First and Second World Wars when the UK was isolated from foreign timber resources. The limited domestic woodland resource was noted in 1919 and the Forestry Commission was created and charged with the task of creating a timber resource. This began a period of rapid afforestation in Scotland and woodland now covers approximately 17.2% of the country. However, only 9.9% of this total woodland cover is semi-natural and the rest is plantation, typically of non-native species such as *Picea sitchensis* (Bong.) Carrière (sitka spruce) (Forestry Statistics 2008). The remaining Caledonian Pine Forest is now a highly valued habitat designated an Annex 1 Habitat and protected under the Habitats Directive (Council Directive 92/43/EEC 1992). Thus, although climatic changes appear to be the main cause of the major decline in the native woodland, anthropogenic habitat destruction has also played a significant role, and remediating this human impact is a stimulus for the conservation and restoration of semi-natural woodland in the Highlands. Moreover, the national forestry policy has been altered to a multi-purpose approach emphasising biodiversity and recreational benefits, as well as providing a timber resource (Kanowski and Potter 1993; Warren 2002; Hobbs 2009). An effective method to meet conservation, biodiversity and recreation targets may be to re-establish a functioning Caledonian Pine Forest Ecosystem, made up of a full suite of habitats at the landscape scale using ecological restoration principles (SERI 2004), of which species reintroduction is a key component (Soulé et al. 2003; SERI 2004).

Deer Numbers and the Suppression of Natural Regeneration

Deer at high densities have been shown to suppress natural woodland regeneration through browsing and bark stripping (Palmer and Truscott 2003) and threaten remaining ancient woodland and attempts at reforestation. In Scotland, red deer, *C. elaphus*, abundance has varied considerably during the history of human occupation. During the sixteenth century they were valued quarry of royalty, and protected in deer forests (a term which doesn't necessarily mean wooded areas). Between 1750 and 1800 the Highland human population expanded rapidly and deer, by then considered a pest species, were removed in favour of upland sheep. However, by the 1870s sheep prices had collapsed. Increased interest in stalking, made fashionable by Queen Victoria who purchased Balmoral Estate in 1852, encouraged landowners to manage and develop their herds then lease their lands for sport (Yalden 1999; Warren 2002). The nineteenth and twentieth centuries, therefore, saw a considerable expansion in the deer population which by then had important socio-economic and cultural value. However, crop damage by deer became very problematic and caused conflict with crofters (a unique form of land tenure for small-scale food production in the Scottish Highlands) and farmers. Consequently, in 1959, the Red Deer Commission was established and made responsible for conservation and control of red deer (Warren 2002).

It is thought the Scottish red deer herd had reached 350,000 by the early 1990s and may have become resource-limited (Clutton-Brock et al. 2004). This led to a debate on the appropriate number of deer for the Highlands. Current population densities vary between <1 and 30/km^2 in open ground (Macdonald et al. 1998), while evidence suggests that reducing deer density to 4–8/km^2 allows woodland regeneration (Andrews et al. 2000). Although deer reductions may be achievable in some areas it requires a heavy and costly hind cull (Warren 2002; Milner-Gulland et al. 2004). Due to the difficulties of culling sufficient deer to allow woodland regeneration, the use of deer exclusion fencing has been popular. Fencing can be successful but Warren (2002) highlights eight major problems including: fencing is not a solution: it merely displaces the problem; fenced areas have reduced floral diversity; significant increases in bird mortality occur, and, in certain cases, it is counterproductive to natural regeneration as rapid development of prolific ground vegetation prevents tree establishment. These problems have led to questioning the suitability of this policy (Warren 2002). A more radical suggestion has been the reintroduction of a large carnivore, the wolf *C. lupus* to regulate and limit the deer population (Wilson 2004; Nilsen et al. 2007). It is also important to note that sheep have also played an important role in regeneration suppression and in many areas have had a greater impact than deer. Our focus on red deer here though is justified by their greater importance in the Alladale area.

The Large Carnivore Guild

During the last 2,000 years all three members of the large carnivore guild, lynx, *L. lynx*, bear, *Ursus arctos*, and wolf have been extirpated from the British Isles. Lynx bones discovered in England and Scotland have been radio-carbon dated to 1550 (±24) and 1770 (±80) years old respectively, sufficiently recent to suppose persecution and anthropogenic habitat destruction played a significant role (Hetherington et al. 2006). Bear are thought to have been lost between 1,000 and 2,000 years BP from a combination of habitat loss and persecution (Yalden 1999), whilst wolves survived in Scotland until the eighteenth century; persecution is likely to have played a significant role in their extirpation (Yalden 1999). All three are listed Annex IV species of the Habitats Directive (Council Directive 92/43/EEC 1992), which requires European Union Member States to study the desirability of reintroducing such species where they have become extinct.

Increasing consideration is also being given to species reintroduction as a tool to re-establish ecosystem function (Soulé et al. 2003), such as using the wolf to limit deer numbers (White and Garrott 2005; but see also Vucetich et al. 2005 for an alternative account) and alter foraging behaviour by altering the "landscape of fear" (Laundré et al. 2001; Manning et al. 2009; Valeix et al. 2009a, b) to allow woodland regeneration (Ripple and Beschta 2003) as has been recorded in Yellowstone National Park. However, understanding and predicting the process and effects of natural regulation of large mammals through predator–prey interactions is difficult due to the landscape scale over which they range, the consequent complexity of their environment and the disrupting influence of humans. Models created to improve understanding and predict these interactions by simulating population dynamics have become increasingly popular (Boyce et al. 2007). Mathematical modelling is an important tool in conservation when aiming to make a preliminary assessment of the likely impact or consequences of any action, especially when the cost and impracticality of doing experimental manipulations are prohibitive. It is an important caveat that the results of such explorations should be treated as no more than they are – explorations. Models are by their nature imperfect; they can provide fruitful insight, but they are always wrong to at least some degree and thus should be treated with caution. Modelling should be seen as a tool for the exploration of broad trends, and of reducing the risk of unexpected surprises when interventions are tried in reality. Models should not be considered to accurately predict population behaviour in a specific situation (Boyce et al. 2007). Nilsen et al. (2007) used such a model to consider the impact a reintroduced population of wolves might have on the Scottish Highlands and reported that their simulations suggested that a possible outcome might be a 50% reduction in the Scottish deer herd. However, many direct and indirect factors might play a role in the eventual outcome of a reintroduction, making the result difficult to predict.

Using Fences to Reduce Conflict

Reinstating the large carnivore guild, particularly wolves, to Scotland could encourage woodland regeneration through trophic cascade effects as has been witnessed by the wolf reintroduction to Yellowstone National Park (Ripple and Beschta 2004). However, reintroductions are complex and controversial (Fritts et al. 1997; Macdonald 2009). Public fears associated with the possible threats to safety and livelihood that come from the reintroduction of a predator are difficult to overcome, and consequently so is gaining political support (Wilson 2004), regardless of whether the general opinion is favourable (Nilsen et al. 2007). Fences can be used to mitigate human-wildlife conflict while also providing a conservation mechanism, facilitating research into the role and economic benefits of previously extirpated species in restoration ecology, as has been seen in the southern African wild dog "managed metapopulation". This involves reintroducing the wild dogs to numerous fenced reserves varying in size between 84 and 900 km^2, and artificially managing dispersal (achieved by translocations) between them to maintain a more or less natural social dynamic and genetic structure (Gusset et al. 2008; Davies-Mostert et al. 2009). However, fences will also cause conflict, for example by restricting public access that is protected under Scotland's Land Reform Act (2003). Here we wish to determine if this is a concept worth considering in the Scottish Highlands, and whether it could be operationally realistic from an ecological perspective. We recognise that access is a fundamental issue and that a fenced reserve would require special dispensation to be created; however, we feel that if there are strong supporting environmental, social and economic arguments then solutions could be found that would not threaten the valued and important principle behind the access legislation. We identify the following important questions which develop from the fact that the Alladale Estate, at 90 km^2, is unlikely to be large enough in itself to accommodate anything approximating a naturally functioning wolf-prey community, and hence that a first step towards a more "natural" fenced community would be to enlarge the prospective fenced area:

1. How much space is available around the Alladale Estate for a fenced reserve?
2. Is there an ecological minimum area requirement to support wolves and prey in a space-restricted environment with an 80% survival probability for 100 years?
3. Will wolves limit deer numbers within a closed and space-restricted system?
4. How to determine the best habitat and landscape to be incorporated within a fenced reserve?

Before these questions can be adequately addressed, some criteria of success of a fenced reserve must be discussed. Where on the continuum from zoo cage to unfettered wilderness are we aiming? The primary ecological purpose of reinstating a lost native, keystone, species back into an enclosed area of the Highland ecosystem is the potential to aid ecological restoration through the predation of grazers/ browsers and replacing anthropogenic culling. However, to adequately meet this goal both predator and prey must be given a fair opportunity to survive, reproduce and lead more or less natural lives. To meet these criteria, we will target the point on the scale from zoo to wilderness that offers a wolf population at least an 80% probability of surviving natural population fluctuations through stochastic variation over 100 years without the need for the introduction of new animals other than to provide new genetic material. Furthermore, the wolf population should limit the deer population to between 4 and 8 deer per km^2, in line with Scottish Natural Heritage's (SNH) deer density guidelines to allow woodland regeneration.

Approach

Alladale Estate, Sutherland, Scottish Highlands

In 2003, the 90 km^2 Alladale Estate in the Scottish Highlands was purchased by Mr Paul Lister for the express purpose of using it as a nucleus to create a fenced wilderness reserve containing lost native species. If successful, his vision would see the return of wolf, lynx, bear, wild boar, *Sus scrofa*, and European elk, *Alces alces*, to a fenced reserve as part of an ecosystem restoration project. To evaluate the proposition of releasing wolves to Scotland within a fenced reserve, we used the Alladale Estate and the surrounding Highland area as a case study.

How Much Space Is Available Around the Alladale Estate for a Fenced Reserve?

In the area around Alladale we identified regional transport infrastructure and associated buildings and classed them as unsuitable for inclusion in a fenced reserve. Using 2008 Ordnance Survey (OS) maps (1:50,000) in ArcGIS 9.2 (ESRI Inc. © 1995–2009), we marked out exclusion buffer zones around the following features: 200 m around tarmac/public roads, 500 m around railway lines, 500 m around population centres and 500 m around any building within 700 m of a tarmac/public road (Fig. 14.1).

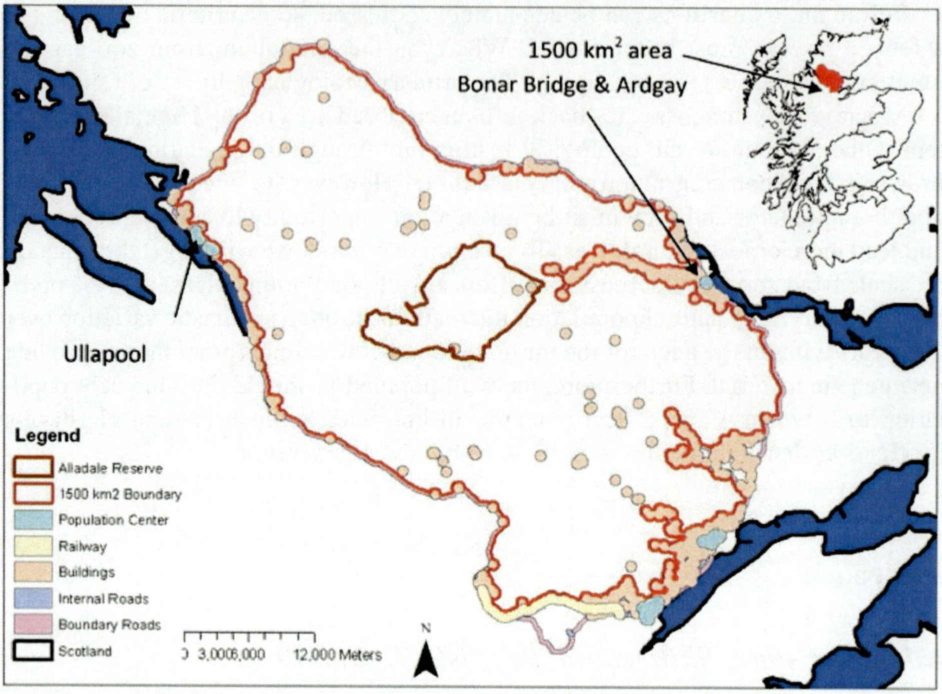

Fig. 14.1 Method of calculating the maximum area of suitable land to be included within a fenced reserve. Tarmac road infrastructure and buildings 700 m from these roads were excluded creating a maximum boundary of 1,500 km². Buildings further than 700 m from tarmac roads were deemed potentially suitable for inclusion. Map in *top right corner* depicts the location of the 1,500 km² area in Scotland

Modelling a Wolf and Deer Population Within a Closed and Space-Restricted Environment

Population Viability Analysis (PVA) is a species-specific modelling process that examines the likelihood that a population will be sustained or become extinct over a defined time period. An Individual-Based Model (IBM) is a tool, regularly used for PVA, which follows each individual within a given population, and is often tailor-made for a specific scenario. The advantage of using an IBM is the ability to investigate any specific and clearly defined management regime desired, and that it allows the impact upon the population to be understood at a relatively fine level of detail.

Such analyses have been used extensively in conservation biology; however, there is controversy concerning the best approach and the value of the results. A major criticism of the PVA approach is that it is often the occurrence of catastrophes (environmental, demographic, anthropogenic, etc.) that cause a species to go extinct, and it is difficult to evaluate the likelihood of such an event occurring (Coulson et al. 2001). As such, it is often suggested that PVA not be used to make quantitative estimates of extinction risks or minimum viable populations, but rather to compare different management regimes, or other pressures upon a species (Reed et al. 2002). Hence, whilst the results can be seen as an exploration of possible population trends and general viability given different scenarios, it should not be considered an exact predictor of extinction probability.

Given the unusual nature of the scenario of wolf release to a closed area in Scotland it is difficult to forecast what circumstances would best deliver success. PVA, with a tailor-made IBM appropriate to Alladale and surrounding area, was therefore considered an appropriate method for preliminary exploration of the viability of wolves within the hypothetical fenced reserve. The model aimed to assess the population dynamics of the wolves and the impact on prey density, in reserves of different sizes. Furthermore, it is hoped to approximate different wolf social responses when occupying a space-restricted environment. Considering social response is essential to understanding how a wolf population may function within such a fenced reserve and is an aspect that tends to be overlooked.

The model utilised was based on one developed by Nilsen et al. (2007) for wolves in the Scottish Highlands, and further by Bull et al. (2009) to explore the ecology of the closely managed and confined wild Norwegian wolf population. With the exception of Nilsson (2004), previous models of wolf population dynamics have focused on the interaction between wolves and their prey. Whilst the interaction between wolves and deer remains part of the focus of the model used here, the main emphasis of the model is on the wolf population itself. Furthermore, we consider only the demography of the population, and do not include management for genetic variation; Nilsson (2004) warns that inbreeding is a potentially serious threat to the long-term viability of the Scandinavian population and this is similarly a risk in the scenarios we consider. However, our focus is on short-term management for population viability and hence we have postponed, but not forgotten, genetic aspects (see also Liberg et al. 2005).

Modelling Predator–Prey Population Dynamics: The Reserve

The current available land area at Alladale, 90 km^2, provides a minimum potential reserve size. Our first question was what minimum area would be needed to sustain a pack of wolves. Data on deer predation by wolves in temperate climates (Mech and Boitani 2003) lead us to estimate that in temperate areas an average pack territory size is around 200 km^2. This provides a starting point for estimating the minimum sized unit for a reserve accommodating wolves in Scotland. For a larger area, a comparison was drawn with Isle Royale in Lake Superior, on the U.S./Canadian border, which covers an area of 544 km^2 and has supported an isolated wolf population for over 50 years (Peterson et al. 1998; Vucetich and Peterson 2004a). Finally, in order to consider the largest reserve size possible, in our given study area, we used the conservative estimate of the maximum available area of 1,200 km^2 that includes the Alladale Reserve, which given our assumed basic unit, represents the potential for around six wolf packs. In summary, we explore four reserve sizes in our exploration of the consequences of scale: 90, 200, 600 and 1,200 km^2.

The initial populations were varied for each size of enclosure that was modelled: 90 km^2 = an alpha (breeding) pair; 200 km^2 = 1 pack of an alpha pair and 1 male and 1 female sub-dominants; 600 km^2 = 2 packs each consisting of an alpha pair and a pair of sub-dominants; 1,200 km^2 = 3 packs each consisting of an alpha pair and a pair of sub-dominants. These are somewhat arbitrary decisions, but remember that our purpose is to explore illustrative options and to learn from them. In the two largest reserves, fewer packs than could theoretically be contained within the reserve, given our assumed average territory size of 200 km^2, were introduced to provide an opportunity for the wolves to expand naturally.

The Wolves

In the model, the natural life stages of the wolf are categorised to give four basic types: cubs, sub-dominant adults, dispersers and alpha wolves. The second and third life stages; "sub-dominant adults" and "dispersers" are considered to form collectively a pool of sub-dominant animals. The life cycle of the wolf is then divided into four basic "natural" demographic processes, which are modelled in the following order: dispersal, establishment of any new territories, reproduction and survival.

The imaginary wolves released into the virtual fenced area were treated as a completely isolated population, and there was no spatial demographic substructuring of the population as a whole. Each individual within the IBM is uniquely identified and has an age, sex, stage and pack membership associated with it. Transitions between stages are related to age and social status, and all transitions take place stochastically for each individual in random order. Each pack consists of a dominant pair and their offspring; only the dominant pair can reproduce.

The Deer

It is important to consider the deer population in this model for two reasons: firstly, in a closed ecosystem, over-exploitation of prey by predators will lead to low prey availability and increased wolf mortality; secondly, we wish to consider the impact wolves have on deer density because of the detrimental effect deer can have on woodland regeneration. To keep the model as simple as possible, the deer population is represented as the total number of deer predicted to be present in the reserve each year with no population sub-structuring. The initial starting population is based on the assumption that there is a deer density of c.11 deer per km^2 (as currently exists at time of writing on the Alladale Estate; I. MacNeill personal communication). The average deer density over the maximum potential study area (all of the Estates within the North Ross Deer Group) was 12 deer per km^2 in 2008 (Taylor 2008). These values were deemed sufficiently similar to justify the use of the Alladale deer density for analysis at each of the described enclosure sizes to facilitate the direct comparison in the population dynamics.

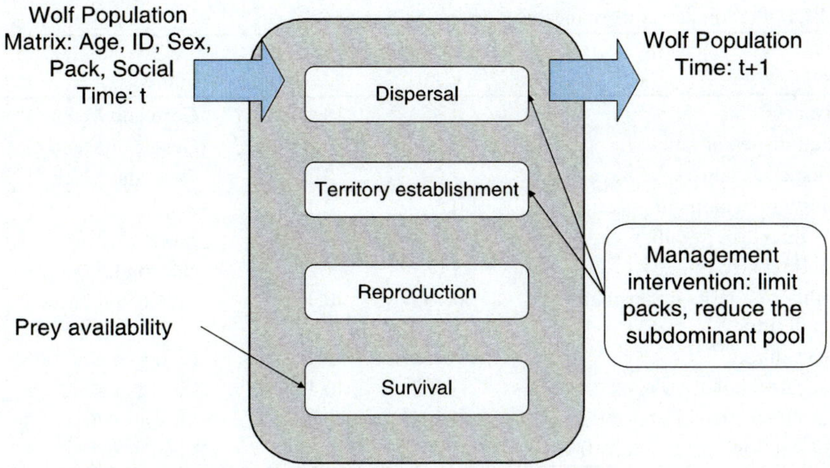

Fig. 14.2 Graphical depiction of the wolf model

The Model

The model is designed to output the average population trend over a 100-year period in all scenarios The model was coded in R (R Development Core Team 2007) and each simulation was run 50 times over 100 years in order to generate the main results; 50 was the minimum number of repetitions considered statistically necessary to keep the confidence intervals acceptably narrow, and this minimum was used because of the long time taken to run the models.

Modelling the Wolf Population

Each year, each pup from the previous year can disperse from or remain in the natal pack as a sub-adult (i.e. join the sub-dominant pool in the population), or die. The existing sub-dominant adults can disperse, die or remain in the pack. Dominant animals have a probability of survival and a reproductive rate. If a dominant animal dies, it is replaced by the oldest sub-dominant animal of the right sex in the pack. If there is no suitable animal available within the pack, the position remains vacant until it is filled by a dispersing animal or a young individual moves up into the sub-dominant adult stage. The pack cannot reproduce until a new dominant pair is established. If both dominant animals die, the rest of the pack disperses. This latter assumption was made primarily to simplify the model, but is considered a reasonable approximation with regards to the literature: Fuller et al. (2003) support this assumption, and in Brainerd et al. (2008) the loss of both breeders was noted to cause pack dissolution in 85% of cases, and in only 9% of cases did a pack reproduce the following season under such circumstances.

After dispersing, each disperser over 2 years old has a probability of establishing a territory (the second step in the yearly cycle, Fig. 14.2). If an animal is

Table 14.1 Parameters and sources used in the wolf model

Parameter	Baseline value	Standard deviation	Source
Cub dispersal rate	0.35	0.15	Gese and Mech (1991)
Sub-adult dispersal rate	0.50	0.25	Gese and Mech (1991)
Adult dispersal rate	0.90	0.20	Gese and Mech (1991)
Probability of establishing a territory	0.8	0.1	Pedersen et al. (2005)
"Small" litter size (# cubs)	2	1	Mech (1970)
"Large" litter size (# cubs)	5	1	Nilsson (2004)
Probability of a large rather than a small litter	0.73	0.15	Calculated based on Pedersen et al. (2005)
Cub survival rate	0.903	0.15	Liberg et al. (2008)
Survival rate for wolves aged 2–8 years	0.903	0.17	Liberg et al. (2008)
Survival rate for wolves aged 9 years	0.40	0.08	Mech (1970)
Survival rate for wolves aged 10 years	0.25	0.05	Mech (1970)

establishing a territory, it first looks for an existing pack with a missing animal of the correct sex, or otherwise might establish a new territory in which to wait for a mate. If, by chance, it is unsuccessful in that year it remains in the disperser pool. It is assumed that the probability of any disperser being successful in establishing a new territory is constant, and values were obtained from the literature; density dependence for the wolf population was ensured via other functions (see "survival" below).

Once the territory establishment process is complete, the third step, reproduction, takes place. Dominant pairs may reproduce, producing a large (5 pups; Nilsson 2004) or small litter (2 pups; Mech 1970) with a given probability. The fourth step is to apply the survival function, which removes some wolves from the population. Every wolf in the population has some probability of dying in any 1 year and basic survival rates are age-dependent, not related to other factors such as the life stage. This is unrealistic as, for instance, dispersing wolves, in particular, are generally likely to be at slightly greater risk of mortality than wolves resident in a territory (Pletscher et al. 1997). However, we judged it a warranted simplification of the model as the key factor governing survival rates is the availability of prey and this is considered separately in the model.

The basic ecological parameters used to characterise the wolf population were taken from various sources in the literature (Table 14.1). The values used represent best available estimates, largely from extensive observations in the wild. Furthermore, they are taken from wolf populations living in the Scandinavian population wherever possible, which is considered similar to Scotland in climate, topography and, to some degree, ecology.

Natural survival rates are modified according to prey availability, using a series of relationships, described below, previously shown to be useful in Nilsen et al. (2007), for the hypothetical case of wolves in the Highlands. The deer population is modelled, as described in the section titled "Modelling the Deer Population", and then each year the wolves remove a number of deer from the overall stock. The number removed, k (per capita kill rate), is calculated using (14.1), where a is the asymptote which the kill rate approaches, h the deer density at which the kill rate reaches half the asymptotic value, and P is the deer density ($h = 0.5$ deer per km^2; set low as kill rates are relatively constant across a wide range of deer densities), as observed in data from the Yellowstone area (Smith et al. 2004).

An element of stochasticity is then introduced via (14.2), which calculates the actual kill rate for any 1 year, where ε is a normally distributed random variable with mean zero and a standard deviation of 0.05. In every time step, the wolves remove the number of deer obtained by multiplying the realised kill rate (k') by the number of individual wolves remaining in the population

$$k = \frac{aP}{h + P}, \tag{14.1}$$

$$k' = ke^{\varepsilon}. \tag{14.2}$$

Given the remnant deer population, the impact upon wolf survival is then calculated via (14.3), in which the realised survival rate (S_i) during any one particular time step is found by multiplying the basic age-specific survival rate (s_i) by a ratio in which g is the deer:wolf ratio at which survival is half of the maximum, and P the deer:wolf ratio.

$$S_i = \frac{s_i P}{g + P}. \tag{14.3}$$

Wolf Sociological Response Scenarios to High Wolf Density

Research supports our assumption that the primary mechanism limiting the density of the wolf population is the availability of prey (Peterson et al. 1998; Mech and Boitani 2003), especially given the likely lack of pressures such as poaching or extreme weather in the Highlands. However, wolf sociology could contribute to limiting the population by limiting pack density, and therefore reproductive potential, particularly at low prey density, and increasing wolf mortality as a result of pack conflict. The likely impact of this factor is unknown. Between 1959 and 1974, the wolf population on Isle Royale appeared limited to between 17 and 31 wolves in no more than 2 packs, and between 1982 and 1986 fluctuated between 14 and 24 wolves in 3 packs (pack density = 181 km^2 per pack) (Peterson et al. 1998; Vucetich and Peterson 2004b). However, in the intervening period between 1975 and 1980 the wolf population increased steadily and fluctuated between 34 and 50 wolves, the highest wolf density ever recorded, in as many as 5 packs (pack density = 108 km^2 per pack) before crashing from 50 wolves to 14 between 1980 and 1982 (Peterson and Page 1988). To consider the importance of pack density to predator-prey interactions within a fenced reserve three scenarios were investigated separately by applying them to the model at the end of every year (Table 14.2). These scenarios represent a range of possible wolf sociological responses that limit a maximum

Table 14.2 Applied candidate management strategies that limit wolf population by removing animals when pack density exceeds set limits

Scenario	Description
Unlimited Pack Density	The wolf population is not limited by wolf sociology
Intermediate Pack Density	New packs that attempt to form out of the sub-dominant pool when insufficient space is available to allow each pack 100 km^2 are removed permanently from the population
Limited Pack Density	New packs that attempt to form out of the sub-dominant pool when insufficient space is available to allow each pack 200 km^2 are removed permanently from the population. This mechanism prevents pack formation beyond the specified pack carrying capacity and increases mortality in the sub-dominant pool

pack density of 1 pack per 200 km^2 in the Limited Pack Density scenario through to Unlimited Pack Density by sociological factors. Should the natural response be found not to be as severe as the Limited Pack Density scenario, the Limited and Intermediate Pack scenarios could also represent artificial management of the wolf population. Animals could be removed from the population if that population was deemed too dense and if wolf social responses were proving ineffective at limiting wolf density and the high wolf densities were posing a risk to the long-term survival probability of the wolf population.

The candidate scenarios remove the alpha pairs from newly established territories at the end of the year and before they begin to reproduce. The model records the number of wolves taken out of the population and the year in which this was necessary, calculating the management effort required under that specific scenario. Considering a range of scenarios allows us to explore the importance of wolf density in limiting deer density.

Modelling the Deer Population

For each year in the model the deer population was subjected to three life phases: recruitment, "natural" mortality and predation by wolves. The recruitment phase involved multiplying the total population by 1.2, with some stochastic variation, at each time step in the model. This recruitment rate of 20% has been based on data collected by Clutton-Brock et al. (1997) from the Isle of Rum population, and assumes that around 50% of the population are sexually mature females, of which approximately 40% will reproduce in any single year. Deer mortality, also estimated from Clutton-Brock et al. (1997), was set at a rate of 6% of the population with stochastic variation. Annual predation is k, the kill rate in the wolf model. The deer population goes through each of these phases annually. This general approach offers a basic representation of how the deer population might respond to the presence of wolves, but, in reality, a number of other factors would come into play for the deer; such as the fact that wolves would likely target certain age groups or that deer would not be uniformly distributed across the range. However, it is important to keep an exploratory model such as this simple as well as robust, and it was felt that these three phases give a fair approximation of the relevant fundamental life stages of the deer.

Table 14.3 Land suitability scoring scale applied to land classification data layers

Score	Explanation
−2	This land is highly unsuitable for inclusion within the fenced reserve. Inclusion would be likely to substantially increase expense, disrupt human activities and increase unrelated human disturbance inside the reserve
−1	This land is unsuitable for inclusion and may cause some increase in expense and/or disrupt human activities, and/or increase unrelated human disturbance inside the reserve
0	This land has no beneficial or detrimental implications if included within the fenced reserve
1	There is some direct or indirect ecological benefit to some of the species proposed to be reintroduced or their prey
2	A primary habitat or foraging resource for more than one of the species proposed to be reintroduced or their prey
3	Established woodland habitat
4	Ancient semi-natural woodlands

The deer population is further refined via the incorporation of a density dependence function. Empirical evidence suggests a carrying capacity of approximately 23/km^2 in a habitat such as Alladale (Gorman 2007), although greater estimates have been made. The modelled deer population is not capped at this limit; however, from this density upwards, the survival rate at each year is reduced by a factor equal to the ratio of the actual deer density to this "maximum" deer density. Again, whilst this is a relatively crude method of representing density dependence, it ensures that deer survival is increasingly limited over a given population threshold, and that this threshold is defined by empirical evidence and hence should indirectly incorporate the effects of other lesser factors.

By thus establishing the deer population as a simple factor, we hope to illustrate the importance of wolf population density on deer population dynamics, which is one of the central aims of the model. While these illustrations hopefully serve a didactic function, we realise that in reality the predator–prey relationships are likely to be considerably more complex than our depictions.

Determining Land Suitability

Within our maximum area, land suitability for a fenced reserve containing restored predator and herbivore guilds varies. Land suitability was assessed in terms of: Habitat type; Land use; Elevation and SSSI or Special Area of Conservation (SAC) or Special Protection Areas (SPA). Red deer density and distribution (as the primary prey species for wolves) is an important consideration in land suitability and does vary within the North Ross Deer Group; however, it is not included in this assessment as current distribution is heavily affected by culling efforts, internal fences and human densities, all of which would be altered by the establishment of the fenced reserve. GIS data layers were obtained for the potential predictors of land suitability and each constructed of a grid of 25 × 25 m cells, as this was our coarsest data layer. Each cell was given a score between −2 and +4 for each data layer feature contained within it according to the scale described in Table 14.3. All the data layers were overlaid and the scores summed to create a land suitability map of the combined score per cell, which resulted in a range of scores from –4 to +4. By considering the spatial distribution of these cells we can consider the geography of land suitability for a fenced reserve containing restored predator and herbivore guilds within our study area.

Assigning Land Suitability Scores

Habitat and Land Use

Eighteen broad habitat or land cover classifications of the 2000 Land Cover Classification (LCC) map of Scotland (Fuller et al. 2002) were identified within the study site and each was scored separately. Woodland cover offers shelter to the carnivore guild and shelter and foraging to their prey and received a +2 score. Comparison of the LCC map with the 2008 Ordnance Survey (OS) map reveals some discrepancy in the current extent of woodland cover, probably due to new plantations being displayed in the more recent OS map. As it is desirable to include woodland of variable age, where the OS woodlands overlapped with LCC map woodlands the cell received a +3 score, as it is assumed that these areas were well established with more ecological benefits. A further woodland data layer was available in the form of ancient semi-natural woodland sites mapped in the 1970s from the Scottish Natural History. Ancient semi-natural woodland that overlapped LCC and OS woodland received a score of +4, as it was assumed to be highly favourable habitat.

Unimproved grassland provides an important foraging habitat for resident grazers, such as deer. It is essential for maintaining a healthy deer population and may provide alternative foraging habitat to woodland thereby helping reduce browsing pressure and so received a +2 score. Open heath is another important foraging habitat for red deer and receives a +1 score.

Other land classifications found within the reserve (including: bracken, bog, open inland water, montane habitats and open bare ground) were considered to be of neither benefit nor detriment to the species being introduced and their immediate prey, and thus received a land suitability score of 0.

Buildings

Thirty-seven isolated buildings or small clusters of buildings were located within the maximum boundary of the reserve. The buildings have unknown use, but they will be owned by the estate on which they lie. It is conceivable that some of these buildings are leased to crofters; however, due to their isolated nature it is assumed that most are estate lodges, staff accommodation, highland bothies (freely available overnight accommodation for walkers) or estate management buildings. Areas within 500 m of these buildings received a –2 land suitability score.

Agriculture

Agricultural land, including arable and improved grassland, has been scored at −2 as these land uses would conflict with a fenced reserve. It is, however, likely that some of this land could be considered for inclusion if the land was purchased fairly and was not in use by tenants/crofters where inclusion within a reserve would prevent them earning their living. Establishing this would require a ground survey.

Leisure

In Scotland, hills over 914 m (3,000 ft) are called Munros. As the largest hills in the UK, they attract numerous visitors each year, including a specialist group of "Munro baggers" who aim to ascend all of the 284 Munros in Scotland. A specialist group that desires access for the specific activity of climbing Britain's tallest hills would considerably increase the level of access required for activities unrelated to the fenced reserve. This will increase the pressure on access points, increase disturbance to the reserve and inconvenience those requiring access. For this reason ground above 914 m received a −2 land suitability score. Corbetts, hills that range in height between 762 and 914 m (2,000 and 3,000 ft), received a −1 score.

Sites of Special Scientific Interest (SSSIs)

Within the maximum area 11 sites are designated as an SSSI and/or SAC or SPA. Although we are exploring the hypothetical release of native species to the area as part of an ecosystem restoration project, these introductions may pose a threat to the ecological integrity of specifically protected sites that have been designated in the absence of these species, and often to benefit species that might be impacted by the proposed releases (e.g. ground nesting birds or perhaps even Scottish wildcats, *Felis silvestris*). Each protected area therefore requires an environmental impact assessment to determine its suitability for inclusion. Without this information we have scored these areas as −1 due to the potentially detrimental ecological impacts.

Results

Is There an Ecological Minimum Area Requirement to Support Wolves and Prey in a Space-Restricted Environment with an 80% Survival Probability for 100 Years?

In all scenarios, a 90 km^2 reserve could not support a wolf population for more than a maximum of 12 years (under the Limited scenario) and on those grounds we conclude that this is an inadequate reserve size for the purpose of creating a fenced

Table 14.4 Wolf extinction probability in the range of fenced reserve sizes (200, 600, and 1,200 km²) modelled using each Pack Density scenario

		Unlimited Pack Density scenario			Intermediate Pack Density scenario			Limited Pack Density scenario		
Reserve area (km²)		200	600	1,200	200	600	1,200	200	600	1,200
No. of years after stocking	5	49	50	50	46	50	50	46	50	50
	10	35	50	50	34	50	50	27	49	50
	15	20	46	50	26	50	50	15	48	50
	20	12	43	50	14	48	50	9	48	50
	30	5	35	49	6	47	50	4	47	50
	40	3	28	46	4	39	49	0	47	50
	60	2	12	27	2	23	25	0	45	49
	80	0	6	19	0	18	13	0	45	48
	100	0	0	9	0	12	7	0	44	48

Data displayed are the number of repetitions with remaining wolves (out of a maximum of 50) in the given year from wolf release

wilderness reserve as it does not meet our success criterion of an 80% wolf survival probability for 100 years. An increase in reserve size from 200 to 600 km² significantly increased the probability of modelled wolf survival particularly under the Limited scenario. For example, in the Limited scenario no wolves survived within the 200 km² reserve after 40 years, whilst 88 and 96% of repetitions recorded surviving wolves after 100 years within the 600 and 1,200 km² reserves, respectively (Table 14.4), with both the 600 and 1,200 km² reserves meeting our wolf survival probability success criterion under the Limited scenario. Small population sizes in the 200 km² area make the population highly susceptible to extinction through stochastic variation. For example, in the 27 repetitions with an extant wolf population after 10 years the mean wolf population between 5 and 10 years was 5 wolves, (s.d. = 2.5), whilst the mean wolf population over the same period in the 600 km² reserve was 18.4 wolves (s.d. = 4.7) and in the 1,200 km² was 41.3 wolves (s.d. = 6.5).

Can Wolves Limit Deer Numbers Within a Closed and Space-Restricted System?

To assess whether wolves would limit and regulate the deer population at low densities we considered the Unlimited, Intermediate and Limited scenarios in each of the reserve sizes. In the 90 and 200 km² reserves the modelled wolves did not persist long enough to allow us to determine the probability of them limiting or regulating the deer population. To consider the impact of introducing wolves into the 600 and 1,200 km² reserves, the hypothetical deer-carrying capacities of the reserves were initially determined by running the model without a wolf population. A mean deer density of 25.8/km² (s.d. = 0.1) was recorded between years 20 and 60 for the

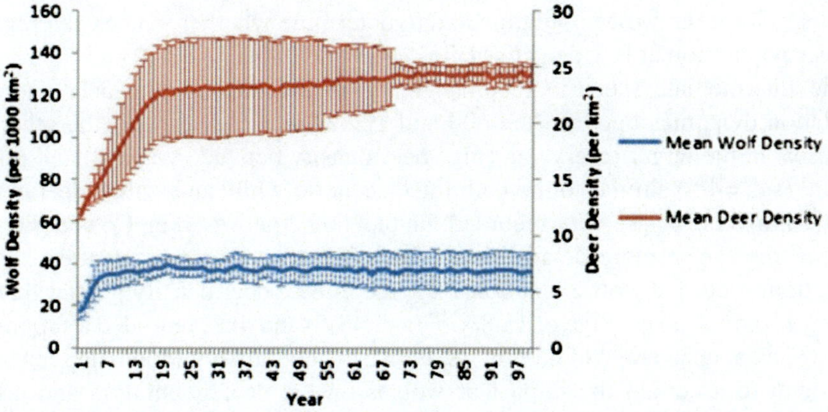

Fig. 14.3 Average deer and wolf densities under the Limited scenario in the 1,200 km² reserve. Simulations were excluded from analysis of the deer population from the point a wolf extinction was recorded, data shown are the mean and standard deviations

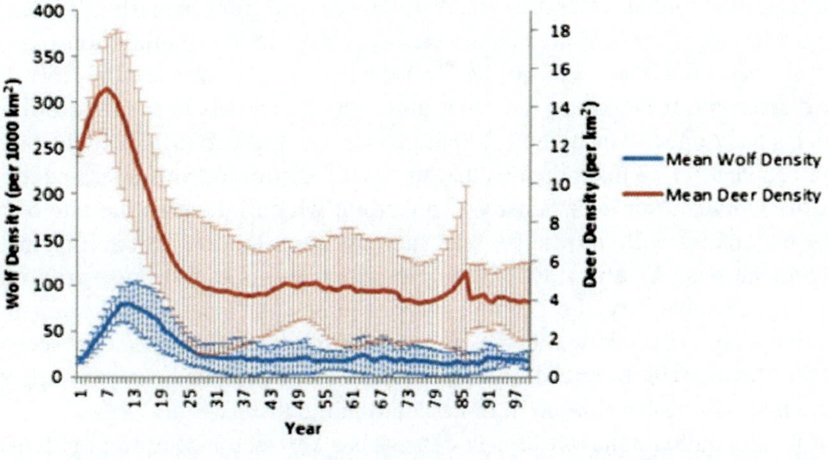

Fig. 14.4 Average deer and wolf densities under the Unlimited scenario applied in the 1,200 km² reserve. Simulations were excluded from analysis of the deer population from the point at which a wolf extinction was recorded, data shown are the mean and standard deviations

600 and 1,200 km² reserves. When wolves were introduced and the Limited scenario applied, average deer densities of 24.2/km² (s.d. = 0.8) and 23.2/km² (s.d. = 4.0) were recorded in the 600 and 1,200 km² reserves respectively over the same time period and thus did not meet our success criterion. When the wolves were subjected to the Unlimited scenario mean deer densities were limited to the target deer densities; 7.6/km² (s.d. = 3.0) and 4.6/km² (s.d. = 2.1). These results may be attributed to differences in wolf population dynamics. Using the 1,200 km² reserve as an example, the average maximum recorded wolf density in the 50 simulations in the Unlimited scenario was 104/1,000 km² (s.d. = 20.4) while under the Limited scenario it was 48/1,000 km² (s.d. = 2.0). Setting the model to limit pack density

appears to limit the wolf population and to determine whether wolves can regulate the deer population at low densities (Figs. 14.3 and 14.4).

The Intermediate scenario records greater discrepancy in the deer and wolf population dynamics between the 600 and 1,200 km^2 reserves than the other two scenarios. In the larger reserve, average deer density between years 20 and 60 was 6.6/km^2 (s.d. = 4.5) similar to the Unlimited scenario while an average deer density of 11.2/km^2 (s.d. = 6.7) was recorded in the 600 km^2 reserve. Considering the 600 km^2 reserve in more detail reveals that after 40 years of occupation 22% of simulations recorded wolf extinction, 32% recorded a deer density greater than the starting 11/km^2 at an average of 18.8/km^2 (s.d. = 3.5) and 46% recorded a suppressed deer density at an average of 6.0/km^2 (s.d. = 3.5). Under the Intermediate scenario, it is difficult to ascertain the impact of wolves on the deer population and predict whether they would regulate deer population at a high or low density. In comparison, 90% of simulations of the Intermediate scenario in the 1,200 km^2 reserve suppressed deer density below the starting density. This was recorded despite average maximum wolf density in the 600 km^2 reserve of 91 wolves per 1,000 km^2 compared to 78 recorded in the 1,200 km^2 reserve.

Wolf survival probabilities are also altered by the different scenarios (Table 14.4). Again using the 1,200 km^2 area as an example, only 18% of simulations recorded wolf existence after 100 years in the Unlimited scenario compared to 96% in the Limited scenario. It is also of note that under the Intermediate scenario only 14% recorded surviving wolves in the 1,200 km^2 reserve compared to 24% in the 600 km^2 reserve (Table 14.4). This indicates that under the Unlimited and Intermediate scenarios the wolves limit deer density to an extent which increases the rate of prey-density dependent wolf mortality, thus increasing extinction probability in these small populations. As a result the scenario which meets our success criterion for wolf survival probability, the Limited scenario, does not meet the subsequent deer density success criterion while the opposite applies to the other two scenarios. However, this may be as a result of our simplistic one predator one prey approach, an issue that will be considered in more detail in the discussion.

Our results indicate that wolf pack density is a key factor determining their ability to regulate deer density. We performed a sensitivity analysis on the probability of packs successfully forming by halving the pack formation success rate to 0.4 and ran the model using the Unlimited scenario in the 1,200 km^2 reserve. Deer density between 20 and 60 years was 23.8/km^2 (s.d. = 0.9), average wolf abundance in the first 20 years was 36/1,000 km^2 (s.d. = 14) and 86% of simulations recorded surviving wolves after 100 years. These results reflect the results in the Limited scenario and we can conclude that pack formation probability is a critical factor in determining the impact wolves may have on deer abundance and wolf survival probability.

Determining Preferential Land Suitability

After visual examination of the map created by the land suitability scoring, the study site was further subdivided into five distinct zones (Fig. 14.5). These zones

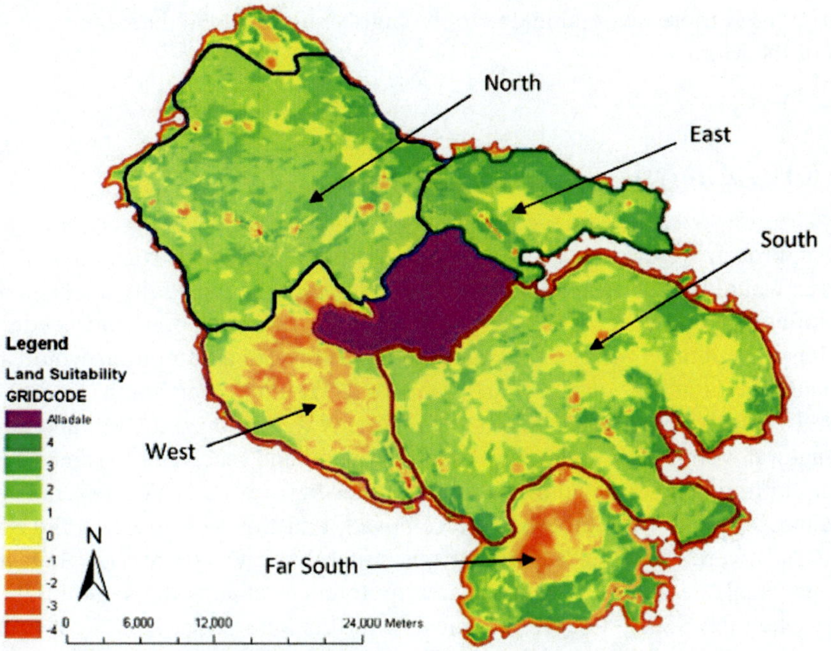

Fig. 14.5 Land suitability map of the maximum area potentially available to be fenced. The area has been divided into five categories used to help determine the most preferential areas to fence

Table 14.5 Table displaying the total area and perimeter length of each subdivided area as well as the proportion of suitable, no cost and unsuitable land within each area

	North	South	East	West	Far South
Area (km^2)	384.41	472.25	145.66	173.45	132.45
Suitable (%)	85.48	67.62	83.45	25.82	49.44
No cost or benefit (%)	13	31.19	15.63	44.44	24.61
Unsuitable (%)	1.6	1.2	0.97	29.91	25.97
Rank preference	1	2	3	5	4
Perimeter (km)	91.4	126	52	72.7	48.8

Rank preference was calculated by subtracting the Unsuitable area from Suitable area for each zone with the highest ("best") score ranked 1 and the lowest ("worst") scored ranked 5

were selected as general directions of possible expansion from the current Alladale boundary while grouping areas of generally suitable or unsuitable land. Under our assumptions, "North" and "East" offer the most suitable zones (Fig. 14.5; Table 14.5). Combining the existing Alladale Estate with the "East" zone, or a combination of parts of the "East" and "North" zones, would create the most suitable area for an enclosure of 200 km^2. Combining Alladale, "East" and "North" provides the most suitable 600 km^2 area. To create the 1,200 km^2 fenced area the "South" would need to be added along with either the "West" or "Far South" zones. These latter zones are both, under our assumptions, less suitable for inclusion due to the presence of Munros, Corbetts, protected areas and less favourable habitat. Of the two, the "Far

South" zone is more favourable due to the highly suitable land that is present to the south of the Munro.

Sensitivity Analysis

If our approach were to be developed further, then more sophisticated sensitivity analyses would be necessary. To illustrate the principle, a sensitivity analysis of the most influential and uncertain land suitability scores, woodland, unimproved grassland, leisure and SSSI, was carried out to test whether our assumptions would affect the resulting suitability of the "East", "North", "South", "Far South" and "West" zones. To test the importance of woodland scores we removed the OS and ancient semi-natural woodland data layers, so that all woodland received the same score of +2, for unimproved grassland the score was reduced from +2 to +1, the scores of −2 for Munros and −1 for Corbetts were removed, and the −1 scores for the SSSI's (SPA/SAC) were removed. Repeating our suitability analysis reveals that, under each new analysis, the rank order of area preference remains the same in all tests except when the SSSI −1 score is removed. In this latter case, the "West" zone is promoted above the "Far South" zone. The "North", "East" and "South" suitability and unsuitability scores vary by no more than 4% between each test. There is greater variation in the "Far South" and "West" zones, with suitability increased in both zones, when the −1 score for SSSI's are removed. Two large SPAs, which are protected for their importance to ground nesting birds and are situated within the zones "West" and "Far South", cause this alteration in zone suitability. It is encouraging to note that, should an environmental impact assessment find that introduced species pose low risk to the ecological integrity of these protected sites, then these areas may pose less suitability issues than the results initially indicated. This analysis suggests that our conclusions are fairly robust, and that the "North" and "East" zones offer the greatest suitability. Other suitability scores were not investigated, as they would involve unrealistic parameter values, and were therefore not deemed a priority for this analysis.

Discussion

To remind the reader, our purpose here has not been to advocate the establishment of a fenced protected area, or the inclusion of wolves within it. Rather, we have sought to demonstrate that the formulation of any such proposal should be based on identifying the key issues and then exploring each systematically, transparently and explicitly. We have shown that this can be done, as illustrated by the one question we have tackled, which concerns the effect of scale on the possible dynamics of wolves and their prey in a hypothetical fenced reserve. Others may demonstrate that there are better ways of tackling even this question, but we have illustrated a starting point.

The Importance of Scale

In this exploration of the importance of scale for fenced reserves our results suggest that by increasing reserve size from 200 to 600 km^2 the probability of a wolf population surviving natural stochastic variation is significantly improved, with the 600 km^2 reserve meeting our 100 year wolf survival success criterion, under the Limited Pack Density scenario. The existence of a wild wolf population on the 544 km^2 Isle Royale for the last 50 years indicates that it is possible for wolves to persist in such an area (Vucetich and Peterson 2004a). These values don't provide a definitive minimum required area but offer some insight into the range of scales worth further consideration. The range of scales explored here, and in the interesting context of an approach to rewilding, researching the role of the wolf to the ecological restoration of Scotland and a tourism boost, could be seen as part of the progression along a continuum running from a level of management intervention that might more closely approximate a safari park to a more ecologically functional reserve.

Predator–Prey Population Dynamics in a Space-Restricted Environment

One of several important reasons for considering the introduction of the wolf to a fenced reserve in Scotland was the potential to regulate deer numbers, or alter their behaviour, to allow forest regeneration. Wolf and deer population dynamics explored in the three scenarios applied to our model indicate that limiting pack density, and thus maximum wolf population, is influential in determining whether wolves could regulate the deer population at a low density. In the Unlimited scenario, where no sociological factors operated in the model of the wolves' behaviour, the wolf population was limited only by prey density. In our model, using this scenario in the 1,200 km^2 reserve, mean maximum wolf density reached 105 wolves per 1,000 km^2, which is greater than the maximum wolf density of approximately 92 wolves per 1,000 km^2 that has been recorded in the wild, on Isle Royale (Peterson and Page 1988). These wolf densities introduced a top-down process that limited the deer population. However, this also reduced deer density to an extent that greatly increased the probability of wolf extinction in these small populations in our modelled single predator-prey relationship. When pack density was limited at one territory per 200 km^2, the Limited scenario, the wolf population was limited to a density that could not limit deer numbers. However, in doing so wolf extinction probability was significantly reduced in comparison with the Unlimited scenario. The Intermediate scenario recorded an average maximum density of 78 wolves per 1,000 km^2, less than the maximum recorded in the wild, and in 90% of simulations the deer population had been reduced below the starting deer density 40 years after wolves had been released. Variation in population dynamics recorded between the 600 and 1,200 km^2 reserves using the Intermediate scenario indicates the difficulties in predicting these relationships and also appears to indicate a greater probability that wolves will regulate the deer population at a low density in a larger reserve. It is likely that in reality a combination of top-down and bottom-up processes would fluctuate in their importance for governing the predator–prey relationship (Vucetich and Peterson 2004b). In the period after the reintroduction of the wolf to Yellowstone National Park (1995–2004) the elk *C. elaphus* herd decreased in size by around 50% (Vucetich et al. 2005; White and Garrott 2005). However, the cause of this decline is debated, with White and Garrott (2005) emphasising the importance of the wolves while Vucetich et al. (2005) emphasise the importance of severe climate and hunting by humans during this period.

An element of wolf predation that has not been considered in our model but merits further consideration is the availability of other prey species within the reserve. Although red deer are likely to be the wolves' primary prey they may also utilise roe deer, *Capreolus capreolus*, wild boar and European elk, amongst others (Jędrzejewski et al. 2002). Under the Unlimited scenario the probability of wolf extinction was increased as a result of reduced deer density. Other prey sources could provide the opportunity for wolves to switch prey species at low red deer densities in a type 3 functional response, offering an increased probability of wolf survival and perhaps allowing the red deer population to recover, leading to a more stable population dynamic with greater potential of meeting both of our success criteria. To further this analysis, a multiple predator–prey relationship needs to be considered along with non-lethal control mechanisms such as the "landscape of fear;" which affects deer browsing behaviour and consequently vegetation structure, composition and distribution (Manning et al. 2009) that is not achieved when the deer are culled by humans because of the closed seasons. Another important ecological consideration is the potential for wolves to impact smaller carnivore populations such as the red fox *Vulpes vulpes*, and Scottish wildcat through intra-guild hostility (Macdonald and Sillero-Zubiri 2004). However, the complex and localised nature of these interactions makes it difficult to predict without research in the Scottish context. In considering the feasibility of wolves controlling the deer population through both lethal and non-lethal means in Scotland, Manning et al. (2009) have highlighted the national and international scientific value and opportunity of creating a fenced or island reserve in the Scottish Highlands, as we have described here, as a controlled restoration ecology experiment. The aim would be to gather information regarding the role of wolves in ecosystem restoration, including key trophic levels (plant–herbivore–predator), to inform any future decision regarding wolf reintroduction and to provide a model for other restoration ecology projects in Scotland and internationally.

The Role of Management Interventions Within Small Isolated Populations

Management could intervene to limit the wolf population if it transpired that wolf sociology didn't limit wolf density sufficiently to prevent volatile population dynamics.

Such intervention might, for example, be helpful if wolves turned out to have a positive effect on woodland regeneration; for example, it might transpire that wolves influenced deer foraging behaviour by altering the "landscape of fear" (Valeix et al. 2009b), without limiting deer numbers to the 4–8/km^2 currently thought necessary for habitat regeneration (Andrews et al. 2000). Management of the wolf population might include supplementing a depressed wolf population or removing individuals at high densities. Additionally, genetics were not considered within our model and, as has been seen in the Isle Royale population, genetic deterioration can have severe morphological consequences (Räikkönen et al. 2009). Intervention might be necessary to prevent inbreeding depression through introducing new genetic material. This could involve replacing a pack or manipulating the hierarchy, bearing in mind the evidence that wolves avoid inbreeding when possible (Mech and Boitani 2003).

Thinking about such interventions is already quite advanced in the case of the *African wild dog* managed metapopulation, which has also drawn attention to the resulting difficult ethical questions (Davies-Mostert et al. 2009). For example, should wild wolves be caught on a regular basis to supplement a heavily managed and enclosed population? Should captive-bred animals be released into a semi-wild situation where they must fend for themselves? What should be done with the surplus wolves? The wild dog managed metapopulation involves over 12 separate reserves, offering the opportunity to move individuals or packs between locations (Davies-Mostert et al. 2009). Managing numerous small populations in this way may alleviate some ethical issues as well as increasing the effective size of the population and increasing the genetic pool. However, this solution may not be appropriate in a space-restricted country such as Scotland, unless it is possible to achieve co-operation between a number of large reserves in several different countries. Interventions would doubtless not be confined to wolves, but would extend to their prey if the balance of predator and prey required adaptive management, which might be fine-tuning, but could also occasionally be radical.

Difficulties in Creating a Fenced Reserve

Our investigation of the human, physical, biological and geographical suitability of the area that surrounds Alladale for a fenced reserve are declaredly simplistic. On the other hand, and in comparison to the copious hot air often exhaled on this topic, we have sought to be explicit about the types of questions that need to be addressed, and to illustrate a transparent approach to tackling them. Although we have been clear that at this stage we are not proposing a fenced wilderness reserve, but rather that we are proposing that one should be thought about seriously, it seems plausible that the rewards for biodiversity, ecological learning, local development and economics – as increasingly developed in southern Africa – are worth serious consideration. We conservatively estimate that there is enough space for a 1,200 km^2 reserve. Importantly, there would be relatively low disturbance to the human population (although considerable to those involved) as only 37 individual buildings or small clusters of buildings exist within this area. If the notion of a fenced wilderness area in the Highlands of Scotland is thought to merit further consideration, then to nudge it towards the status of a proposal would require a level-headed and methodical approach to diverse issues, ranging from the legalities of ownership rights, access and veterinary care, economics and local development issues, to the practicalities, feasibility and finances of fencing, and contingency plans for breaches.

The Animal Protection Act (1911, 1912) UK prevents predators being fed live vertebrate prey in captivity and thus prevents predator and prey occupying the same enclosure. At what point does an enclosure become a reserve? We have described an approach to the scale-dependent question of when the interactions of predators and prey might be considered "natural" or "in the wild". The parallel between the reality of Isle Royale, the interventions of the wild dog metapopulation and the virtual world we have explored in our models, combine to show that exactly the same area could function as a wilderness with or without a fence. To cope with such an area being surrounded by a fence rather than a sea might require altering the law, adapting regulation and tinkering with vocabulary. That in itself should not be a problem insofar as the law is there to serve the desires of society, and society might have aesthetic, cultural and financial grounds for wishing such a fenced protected area to exist.

Similarly, under the current status quo creating a fenced reserve would be contentious insofar as it diminished the public's right of access under Scotland's Land Reform Act (2003). As we have several times repeated, this essay is not a proposal to create a fenced reserve, but rather an attempt to expose some of the issues that would have to be addressed were such a proposal to take shape. In this context we obviously realise that creating a barrier restricts legally protected access rights, and we believe these rights are important. Nonetheless, it is clear that biodiversity conservation, concerns for the wider environment and for human development and well-being within it, will all require radical thinking and challenging action in the coming years. If that radical thinking takes society in a direction that requires a new purpose from the law, then the law can be made fit for that purpose. In particular, it does not strike us as silly to imagine that the sorts of people who rejoice in rights of access to Scotland's marvellous countryside might also be amongst those exhilarated by the prospect of engaging with some variant of the type of biodiversity reserve we have been discussing. The relatively new access legislation has been the result of considerable effort, particularly on the part of the rambling community. This effort has rewarded the public with relatively unrestricted access to enjoy the dramatic Scottish landscape. It is important legislation, and we think the principle behind it should be treasured. That is not to say that it should unbendingly trump or curtail other innovative options for fostering a biodiverse, naturally functioning countryside that provides society with even more diverse and exciting experiences of nature. Solutions to integrate such aspirations would doubtless require a lot of discussion, and the development of ingenious mechanisms to facilitate multiple uses and diverse outcomes. It is not our purpose here to dictate the terms of the debate, or to prejudge its outcome but simply to suggest that it is one worth having.

Gusset et al. (2008) indicated that the electrified fences used to contain wild dogs were not impenetrable, with at least eight reserves experiencing a breach between

1998 and 2006; an observation supported by Hayward and Kerley (2009) in a broader study of the use of fences for conservation. The issue of fence security is key. Escapes would pose an immediate threat to sheep farming interests, public peace of mind and the safety of companion animals and perhaps even people. Legally, a native predator breaching the fence could constitute a de facto reintroduction. Scotland's harsh climate makes it obvious that sooner or later an escape would be likely. Heavy rain erodes fencelines and snow drifts can provide a convenient ramp to escape. These risks might be minimised by a combination of ingenious technology, thoughtful experience, and money, but it would be folly to claim the risk would be zero. Better to consider seriously a plan for identifying and quantifying risks and their consequences, and how they might be managed and insured against, for instance careful fence line selection along with secondary fencing can be used to minimise and manage the build up of snow drifts away from the main fence, shock collars could be used on predators to prevent them getting too close to the fence while GPS technology can be used to monitor predator location in and outside the reserve should capture and recovery be required, and to set these negatives against a similarly down-to-earth analysis of the benefits of creating such a protected area. In first raising the question of fenced areas as part of a radical strategy for nature conservation in Britain, Macdonald et al. (2000) pointed out the unattractive double standards of developed countries urging those in the less developed world to welcome large predators into their own back yards. Fenced reserves may offer one relatively comfortable step towards evening out this imbalance, and there is abundant experience to be drawn on in South Africa and Namibia.

Lessons from southern Africa reveal that the 800 km fence of the Kruger National Park cost US$31,250/km when it was constructed in the 1960s and 1970s (Hayward and Kerley 2009). The cost of fencing varies considerably according to purpose. Hayward and Kerley (2009) report variation between US$170 and $100,000/km although the latter was used to exclude small mammals. This represents only the initial costs, and a long-term commitment to maintenance would ensue. The costs would doubtless be immense, but before dismissing them it would be sensible to consider the potential profits on the balance sheet of creating a wilderness reserve that would be unique in the Northern Hemisphere, with the potential to generate employment and revenue in a region that is currently economically depressed (see Lindsey et al. 2005). The necessary cost–benefit analysis would hinge crucially on the conservation, educational, economic and ecological research and social benefits achieved. Although our model suggests that larger reserves are better suited to meeting our 100 year success criteria the extended time frame means a staged approach could be considered. The initial area fenced may be smaller than ideal for long-term success but could then be expanded over time. This would allow lessons to be learnt while managing a smaller reserve with fewer animals, less fencing – and so reducing start up and maintenance costs – while causing less disturbance to human activity in the area. By dealing in smaller, more manageable areas, each step towards a radical proposal could be more achievable (Manning et al. 2006).

Hayward and Kerley (2009) highlighted a number of threats associated with use of fences for conservation, including disruption to migratory patterns, exceeding natural carrying capacity, sub-dividing populations, reducing genetic pool, mortality along fencelines and restriction of animals to escape fire. These, together with the issues associated with access mentioned above (Land Reform (Scotland) Act 2003) and other landuse conflicts such as sheep farming and other livelihoods that maybe threatened by such an enterprise, would obviously require thoughtful evaluation. Our purpose here has been to demonstrate just one approach to one of these many issues – in this case the impact of scale on the viability and "naturalness" of a fenced wolf population. Each of the other relevant issues could be scrutinised with similar transparency and should be, if this notion is to take shape as a serious proposal. Nonetheless, we note that the Scottish Highlands is already a fenced landscape, primarily for agricultural and forestry purposes and the associated problems have been considered (Warren 2002). The problems associated with fencing are typically related to the fence itself, and some may thus diminish with the more favourable area-to-perimeter ratio typical of larger areas.

Conclusion

We undertook this exercise because we were impressed that discussion of wolves being returned to Scotland generated fiery debate, but rather little of the plodding evaluation that is often essential to reaching an informed decision. Ours has been a scoping study, and has focused on just one of the many aspects that would be necessary for a comprehensive review of this topic. Our purpose has been neither to promote a proposal nor to advocate any particular outcome, but rather to illustrate one way of teasing out possible answers to relevant questions. This will have been useful if it stimulates others to tackle other relevant questions in similarly transparent and impartial ways, with the goal of evaluating whether the fenced reserve concept is worth further consideration in Scotland to facilitate species reinstatement, ecosystem restoration and to restore ecosystem function. Although we do not, nor have we attempted to, offer definitive answers to our key questions, our results do offer some insight into the ecological feasibility of creating a landscape-scale fenced reserve as a tool for ecological restoration in Scotland. This has been a didactic exploration, intended to illustrate the questions that need to be tackled if this issue, the subject of much superficial rhetoric in the press, is to be treated seriously. Furthermore, we argue that it merits serious treatment. The approaches we have outlined are declaredly simplistic, but they are intended to lay foundations that could fruitfully be refined. A range of fenced reserve scales has been identified that could potentially allow a wolf population to function, although artificial management is likely to be required. The required size is hypothetically achievable within the area around the Alladale Estate. It is unclear whether wolves would limit deer abundance although the potential is there if wolf population is primarily regulated by prey density, especially in larger reserves. Conservation without the use of fences is desirable and is surely a general long-term goal. However, for the foreseeable future the reintroduction of large predators to Scotland, with its potential benefits to biodiversity conservation and economic development, seems most likely in a fenced area. We suggest that the idea of a large, fenced wilderness reserve in the Scottish Highlands merits serious consideration.

Acknowledgements We would like to thank Mr. Paul Lister and all the Alladale staff for their continued support. We are also grateful to Adrian Manning, Hans Kruuk, Steve Rushton, Mark Boyce, Mike Daniels, Mark Stanley-Price and Joelene Hughes for their invaluable contributions in reviewing earlier manuscripts. We would also like to thank E.J. Milner-Gulland for her insightful comments as well as the Imperial College London Conservation Scientists who also provided useful feedback.

References

Andrews, J., Miller, G.R. & Armstrong, H.M. (2000). The effects of mammalian herbivores on natural regeneration of upland, native woodland. Information and Advisory Note 115. Scottish Natural Heritage, Battleby.

Birks, H.J.B. (1989) Holocene isochrene maps and patterns of tree-spreading in the British-Isles. *Journal of Biogeography,* **16**, 503–540.

Boyce, M. (1993) Predicting the consequences of the wolf recovery to ungulates in Yellowstone National Park. *Ecological issues on reintroducing wolves into Yellowstone National Park* (ed. R.S. Cook), pp. 234–269. Scientific Monograph NPS/NRYELL/NRSM-93/22. USDI National Park Service, Denver.

Boyce, M.S., Rushton, S.P. & Lynam, T. (2007) Does modelling have a role in conservation? *Key Topics in Conservation Biology* (eds. D.W. Macdonald & K. Service), pp. 134–144. Blackwell Publishing, Oxford.

Brainerd, S.M., Andren, H., Bangs, E.E., Bradley, E.H., Fontaine, J.A., Hall, W., Iliopoulos, Y., Jimenez, M.D., Jozwiak, E.A., Liberg, O., Mack, C.M., Meier, T.J., Niemeyer, C.C., Pedersen, H.C., Sand, H., Schultz, R.N., Smith, D.W., Wabakken, P. & Wydeven, A.P. (2008) The effects of breeder loss on wolves. *Journal of Wildlife Management,* **72**, 89–98.

Breitenmoser, U., Ryser, A., Molinari-Jobin, A., Zimmermann, F., Haller, H., Molinari, P. & Breitenmoser-Würsten, C. (2010) The changing impact of predation as a source of conflict between hunters and reintroduced lynx in Switzerland. *The Biology and Conservation of Wild Felids* (eds. D.W. Macdonald & A. Loveridge), pp. 493–506. Oxford University Press, Oxford.

Bull, J., Nilsen, E.B., Mysterud, A. & Milner-Gulland, E.J. (2009) Survival on the border: A population model to evaluate management options for Norway's wolves *Canis lupus. Wildlife Biology*, **15**, 412–424.

Clutton-Brock, T., Illius, A., Wilson, K., Grenfell, B., MacColl, A. & Albon, S. (1997) Stability and Instability in Ungulate Populations: An Empirical Analysis. *The American Naturalist,* **149**, 195–219.

Clutton-Brock, T. H., Coulson, T. & Milner, J. M. (2004) Red deer stocks in the Highlands of Scotland. *Nature,* **429**, 261–262.

Coulson, T., Mace, G.M., Hudson, E. & Possingham, H. (2001) The use and abuse of population viability analysis, *Trends in Ecology and Evolution*, **16**, 219–221.

Council Directive 92/43/EEC (1992) of 21 May 1992 on the conservation of natural habitats and of wild fauna and flora. http://europa.eu.int/comm/environment/nature/habdir.htm.

Davies-Mostert, H., Mills, M.G.L. & Macdonald, D.W. (2009) A critical assessment of South Africa's managed metapopulation recovery strategy for African wild dogs and its value as a template for large carnivore conservation elsewhere. *Reintroduction of Top-Order Predators* (eds. M.W. Hayward & M.J. Somers), pp. 10–42. Wiley-Blackwell Publishing, Oxford.

Forestry Statistics (2008) A compendium of statistics about woodland, forestry and primary wood processing in the United Kingdom. Forestry Commission, National Statistics.

Fritts, S.H., Bangs, E.E., Fontaine, J.A., Johnson, M.R., Phillips, M.K., Koch, E.D. & Gunson, J. R. (1997) Planning and implementing a reintroduction of wolves to Yellowstone National Park and central Idaho. *Restoration Ecology,* **5**, 7–27.

Fuller, R.M., Smith, G.M., Sanderson, J.M., Hill, R.A., Thomson, A.G., Cox, R., Brown, N.J., Clarke, R.T., Rothery, P. & Gerard, F.F. (2002) Countryside Survey 2000 Module 7. Landcover Cover Map 2000: Final Report. NERC/Centre for Ecology and Hydrology.

Fuller, T.K., Mech, L. D. & Cochrane, F. D. (2003) *Wolves behaviour, ecology, and conservation.* University of Chicago Press, Chicago.

Gese, E. M. & Mech, L.D. (1991) Dispersal of wolves (*Canis lupis*) in Northeastern Minnesota, 1969–1989. *Canadian Journal of Zoology-Revue Canadienne De Zoologie,* **69**, 2946–2955.

Gilchrist, J. (2008) Should we keep the wolf from the door. *The Scotsman*. http://thescotsman.scotsman.com/features?articleid=3752204.

Gorman, M. (2007) Restoring ecological balance to the British mammal fauna. *Mammal Review,* **37**, 316–325.

Gusset, M., Ryan, S.J., Hofmeyr, M., van Dyk, G., Davies-Mostert, H. T., Graf, J. A., Owen, C., Szykman, M., Macdonald, D.W., Monfort, S.L., Wildt, D.E., Maddock, A.H., Mills, M.G.L., Slotow, R. & Somers, M.J. (2008) Efforts going to the dogs? Evaluating attempts to re-introduce endangered wild dogs in South Africa. *Journal of Applied Ecology,* **45**, 100–108.

Hayward, M.W. & Kerley, G.I.H. (2009) Fencing for conservation: Restriction of evolutionary potential or a riposte to threatening processes? *Biological Conservation,* **142**, 1–13.

Hellen, N. (2004) Laird plans to unleash wolves on the Highlands. *The Times Online.* http://www.timesonline.co.uk/tol/news/uk/article393615.ece.

Hetherington, D.A. & Gorman, M.L. (2007) Using prey densities to estimate the potential size of reintroduced populations of Eurasian lynx. *Biological Conservation,* **137**, 37–44.

Hetherington, D.A., Lord, T.C. & Jacobi, R.M. (2006) New evidence for the occurrence of Eurasian lynx (*Lynx lynx*) in medieval Britain. *Journal of Quaternary Science,* **21**, 3–8.

Hobbs, R. (2009) Woodland restoration in Scotland: Ecology, history, culture, economics, politics and change. *Journal of Environmental Management,* **90**, 2857–2865.

Jędrzejewski, W., Schmidt, K., Theuerkauf, J., Jędrzejewska, B., Selva, N., Zub, K. & Szymura, L. (2002) Kill rates and predation by wolves on ungulate populations in Biaowiezia Primeval Forest (Poland). *Ecology,* **83**, 1341–1356.

Kanowski, P. & Potter, S. (1993) Making British Forest Policy Work. *Forestry,* **66**, 233–247.

Kinloch, B., Westfall, R. & Forrest, G. (1986) Caledonian Scots Pine: Origins and Genetic Structure. *New Phytologist,* **104**, 703–729.

Laundré, J.W., Hernández, L. & Altendorf, K.B. (2001) Wolves, elk, bison: re-establishing the landscape of fear in Yellowstone National Park, USA. *Canadian Journal of Zoology-Revue Canadienne De Zoologie,* **79**, 1401–1409.

Leake, J. (2008) If you go down to the woods…Support is growing for 'rewilding', which could see bears, wolves and elk roaming Britain. *The Times Online*. http://www.timesonline.co.uk/tol/news/environment/article5404100.ece.

Liberg, O., Andren, H., Pedersen, H.C., Sand, H., Sejberg, D., Wabakken, P., Akesson, M. & Bensch, S. (2005) Severe inbreeding depression in a wild wolf (*Canis lupus*) population. *Biology Letters,* **1**, 17–20.

Liberg, O., Sand, H., Pedersen, H.C. & Wabakken, P. (2008) Dodlighet och illegal jakt i den skandinaviska vargstammen. Rapport från Grimsö forskningsstation och Viltskadecenter. 42 Pp.

Lindsey, P.A., Alexander, R.R., du Toit, J.T. & Mills, M.G.L. (2005) The potential contribution of ecotourism to African wild dog Lycaon pictus conservation in South Africa. *Biological Conservation,* **123**, 339–348.

Macdonald, D.W. (2009) Lessons learnt and plans laid: seven awkward questions for the future of reintroductions. *Reintroduction of Top-Order Predators* (eds. M.W. Hayward & M.J. Somers), pp. 411–448. Wiley-Blackwell, Oxford.

Macdonald, D.W., Mace, G.M. & Rushton, S.P. (1998) *Proposals for the future monitoring of British mammals.* Department of the Environment, Transport and the Regions, London.

Macdonald, D.W., Mace, G.M. & Rushton, S.P. (2000) British mammals: is there a radical future? *Priorities for the Conservation of Mammalian Diversity: Has the Panda had its day?* (eds. A. Entwhistle & N. Dunstone), pp. 175–206. Cambridge University Press, Cambridge.

Macdonald, D.W. & Sillero-Zubiri, C. (eds) (2004) *Biology and Conservation of Wild Canids*. Oxford University Press, Oxford.

Manning, A.D., Gordon, I. J. & Ripple, W.J. (2009) Restoring landscapes of fear with wolves in the Scottish Highlands. *Biological Conservation,* **142**, 2314–2321.

Manning, A.D., Lindenmayer, D.B. & Fischer, J. (2006) Stretch goals and backcasting: Approaches for overcoming barriers to large-scale ecological restoration. *Restoration Ecology,* **14**, 487–492.

Mech, L.D. (1970) *The Wolf*. Natural History Press, New York.

Mech, L.D. & Boitani, L. (eds) (2003) *Wolves: Behaviour, Ecology and Conservation*. University of Chicago Press, Chicago.

Milner-Gulland, E., Coulson, T. & Clutton-Brock, T.H. (2004) Sex differences and data quality as determinants of income from hunting red deer *Cervus elaphus*. *Wildlife Biology,* **10**, 187–201.

Nilsen, E.B., Milner-Gulland, E.J., Schofield, L., Mysterud, A., Stenseth, N. C. & Coulson, T. (2007) Wolf reintroduction to Scotland: public attitudes and consequences for red deer management. *Proceedings of the Royal Society B: Biological Sciences,* **274**, 995–1003.

Nilsson, T. (2004) Integrating effects of hunting policy, catastrophic events, and inbreeding depression, in PVA simulation: the Scandinavian wolf population as an example. *Biological Conservation,* **115**, 227–239.

Palmer, S. & Truscott, A. (2003) Browsing by deer on naturally regenerating Scots pine (Pinus sylvestris L.) and its effects on sapling growth. *Forest Ecology and Management,* **182**, 31–47.

Pedersen, H.C., Wabakken, P., Arnemo, J. M., Brainerd, S.M., Broseth, H., Gundersen, H., Hjeljord, O., Liberg, O., Sand, H., Solberg, E.J., Storaas, T., Stromseth, T.H., Wam, H. & Zimmermann, B. (2005) *Det Skandinaviske ulveprosjektet - Skandulv: Oversikt over gjennomforte aktiviteter i 2000 – 2004*. Norwegian Institute of Nature Research **117**, Trondheim.

Peterson, R.O. & Page, R.E. (1988) The rise and fall of Isle Royale wolves 1975–1986. *Journal of Mammalogy,* **69**, 89–99.

Peterson, R.O., Thomas, N.J., Thurber, J.M., Vucetich, J.A. & Waite, T.A. (1998) Population limitation and the wolves of Isle Royale. *Journal of Mammalogy,* **79**, 828–841.

Phillips, M.K., Bangs, E.E, Mech, L.D, Kelly, B.T. & Fazio, B.B. (2004) Grey Wolves-Yellowstone. Extermination and recovery of red wolf and grey wolf in the contiguous United States. *The biology and conservation of wild canids* (eds. D.W. Macdonald & C. Sillero-Zubiri), pp. 297–309. Oxford University Press, Oxford.

Pletscher, D.H., Ream, R.R., Boyd, D.K., Fairchild, M.W. & Kunkel, K.E. (1997) Population dynamics of a recolonizing wolf population. *Journal of Wildlife Management,* **61**, 459–465.

R Development Core Team (2007) R: A language and environment for statistical computing. R Foundation for Statistical Computing, Vienna, Austria. http://www.R-project.org.

Räikkönen, J., Vucetich, J.A., Peterson, R.O. & Nelson, M.P. (2009) Congenital bone deformities and the inbred wolves (*Canis lupus*) of Isle Royale. *Biological Conservation,* **142**, 1025–1031.

Reed, J.M., Mills, L.S., Dunning, J.B., Menges, E.S., McKelvey, K.S., Frye, R., Beissinger, S. R., Anstett, M. C. & Miller, P. (2002) Emerging issues in population viability analysis. *Conservation Biology,* **16**, 7–19.

Ripple, W.J. & Beschta, R.L. (2003) Wolf reintroduction, predation risk, and cottonwood recovery in Yellowstone National Park. *Forest Ecology and Management,* **184**, 299–313.

Ripple, W.J. & Beschta, R.L. (2004) Wolves and the Ecology of Fear: Can Predation Risk Structure Ecosystems? *Bioscience,* **54**, 755–766.

SERI (Society for Ecological Restoration International). (2004) *The SER international primer on ecological restoration*. SERI, Washington, D.C.

Smith, D.W., Drummer, T.D., Murphy, K.M., Guernsey, D.S. & Evans, S.B. (2004) Winter prey selection and estimation of wolf kill rates in Yellowstone National Park, 1995–2000. *Journal of Wildlife Management,* **68**, 153–166.

Smout, T., MacDonald, A. & Watson, F. (2005) *A History of the Native Woodlands of Scotland, 1500–1920*. Edinburgh University Press Ltd., Edinburgh.

Soulé, M.E., Estes, J.A., Berger, J. & Del Rio, C.M. (2003) Ecological effectiveness: Conservation goals for interactive species. *Conservation Biology,* **17**, 1238–1250.

Taylor, G. (2008) Deer Count Report for the North Ross Deer Management Group. Deer Commission for Scotland, Scotland.

Valeix, M., Fritz, H., Loveridge, A.J., Davidson, Z., Hunt, J.E., Murindagomo, F. & Macdonald, D.W. (2009a) Does the risk of encountering lions influence African herbivore behaviour at waterholes? *Behavioral Ecology and Sociobiology,* **63**, 1483–1494.

Valeix, M., Loveridge, A.J., Chamaille-Jammes, S., Davidson, Z., Murindagomo, F., Fritz, H. & Macdonald, D.W. (2009b) Behavioral adjustments of African herbivores to predation risk by lions: Spatiotemporal variations influence habitat use. *Ecology,* **90**, 23–30.

Vucetich, J.A. & Peterson, R.O. (2004a) Grey Wolves – Isle Royale. *The biology and conservation of wild canids* (eds. D.W. Macdonald & C. Sillero-Zubiri), pp. 285–296. Oxford University Press, Oxford.

Vucetich, J.A. & Peterson, R.O. (2004b) The influence of top-down, bottom-up and abiotic factors on the moose (Alces alces) population of Isle Royale. *Proceedings of the Royal Society of London. Series B: Biological Sciences,* **271**, 183–189.

Vucetich, J.A., Smith, D.W. & Stahler, D.R. (2005) Influence of harvest, climate and wolf predation on Yellowstone elk, 1961–2004. *Oikos,* **111**, 259–270.

Warren, C. (2002) *Managing Scotland's Environment*. Edinburgh University Press, Edinburgh.

White, P.J. & Garrott, R.A. (2005) Yellowstone's ungulates after wolves - expectations, realizations, and predictions. *Biological Conservation,* **125**, 141–152.

Wilson, C.J. (2004) Could we live with reintroduced large carnivores in the UK? *Mammal Review,* **34**, 211–232.

Yalden, D. (1999) *The History of British Mammals*. T. & A.D. Poyser, London.

Chapter 15
The Influence of Land Use and Fences on Habitat Effectiveness, Movements and Distribution of Pronghorn in the Grasslands of North America

C. Cormack Gates, Paul Jones, Michael Suitor, Andrew Jakes, Mark S. Boyce, Kyran Kunkel, and Kevin Wilson

Introduction

The pronghorn *Antilocapra americana* is an open-country grassland and shrub-steppe obligate and the sole surviving member of a taxonomic family unique to North America. Adapted to outrun American cheetahs, *Miracinonyx trumani*, an extinct predator that once roamed the North American plains (Byers 1997), the pronghorn can reach a top speed of nearly 100 km/h, making it the fastest land mammal on the continent. With its exceptionally large eyes set far back on the skull it can detect movements up to 5 km away, and with a burst of speed it can quickly deter any modern predator from giving chase. Despite these and other adaptations to the prairie landscape, the pronghorn is ill equipped to deal with the agro-industrial and social transformation of the mixed-grasslands of the Northern Great Plains, which began at the turn of the nineteenth century (O'Gara and McCabe 2004).

Today, the landscape mosaic of native prairie and cultivated fields comprising the northern mixed-grasslands largely reflects the early agricultural history of the region. More broadly, the prairies continue to be dramatically altered by the cumulative effects of cultivation, irrigation, roads, petroleum and natural gas development,

C.C. Gates (✉) • M. Suitor • A. Jakes
Faculaty of Environmental Design, University of Calgary, Calgary, AB, Canada
e-mail: ccgates@nucleus.com

P. Jones
Alberta Conservation Association, Lethbridge, AB, Canada

M.S. Boyce
Department of Biological Sciences, University of Alberta, Edmonton, AB, Canada

K. Kunkel
World Wildlife Fund, Northern Great Plains Office, Bozeman, MT, USA

K. Wilson
Venture North EcoRec Enterprises Ltd, Edmonton, Canada

M.J. Somers and M.W. Hayward (eds.), *Fencing for Conservation: Restriction of Evolutionary Potential or a Riposte to Threatening Processes?*,
DOI 10.1007/978-1-4614-0902-1_15,

mining, water development, urban expansion and exurban development, electrical transmission lines, fences and other developments (Barrett and Vriend 1980; Czech et al. 2000; Forrest et al. 2004). In Saskatchewan and Alberta, only ~20 and ~40%, respectively, of native mixed-grasslands remain untilled (Gauthier and Wiken 2003). As development progressed following settlement for farming and the human population grew, roads associated with industry and rural access improvement increased. By the mid-1990s more than 90,000 km of highways, roads and natural gas well-site access trails existed in the Grassland Natural Region of Alberta (Alberta Environmental Protection 1997).

Although pronghorn can persist in low numbers in cultivated areas, they are most common in large open native rangelands where they are able to satisfy life history requirements, including migrating in response to landscape and regional-scale variations of forage availability in winter. Human activities from the period of agricultural settlement during the early years of the twentieth century to contemporary infrastructure and industrial uses of the landscape have had a profound effect on the distribution and abundance of pronghorn. Here, we review the historical depletion and partial recovery of pronghorn and the contemporary influence of fences, roads and other infrastructure and human activities on the spatial ecology of pronghorn. We comment on the importance of land-use planning and measures that can mitigate the negative impacts of land use, fences and roads on ecological cohesion of landscapes to sustain the pronghorn as a common species.

Pronghorn Ecology and Habitat Requirements

Remaining mixed-grasslands of south-western Saskatchewan, north-eastern Montana and south-eastern Alberta provide habitat for pronghorn and other native prairie obligates (Forrest et al. 2004). The landscape between the Missouri River in Montana and the Parklands in Alberta and Saskatchewan contains several large, relatively intact grassland landscapes (Forrest et al. 2004). The dominant sagebrush species in this area, particularly north of the Milk River in Montana, is silver sagebrush *Artemisia cana* (Jones et al. 2005). This palatable shrub becomes critical winter forage for pronghorn, particularly during harsh winters (Martinka 1967). Winter ranges have been identified for pronghorn in Alberta and Montana. These areas provide shrub cover and other winter forage (evergreen forbs) important for pronghorn in winter (Dirschl 1963; Martinka 1967; Mitchell and Smoliak 1971; Barrett 1980, 1982).

Pronghorn aggregate in restricted areas during severe winter conditions more than in normal winters (Barrett 1982). They have a low chest height and a small hoof with the heaviest foot loading of any North American ungulate (Guthrie 1990). Hence, they are poorly adapted to cope with moving through or foraging in snow (Telfer and Kelsall 1984). Severe winter storms and prolonged heavy snowfall with reduced access to browse sporadically causes heavy mortality in pronghorn

(Compton 1970; McKenzie 1970; West 1970; Wishart 1970; Oakley 1973). When snow is present, pronghorn seek areas that have blown free or areas with low snow cover through which forage protrudes (Dirschl 1963; Martinka 1967; Mitchell and Smoliak 1971; Bruns 1977). After heavy snowfall, pronghorn have been observed to undertake long-distance movements in search of areas with better forage availability (Creek 1967; Yoakum 1978; Guenzel 1986; Raper et al. 1989; Sawyer and Lindzey 2000). Such movements may be a facultative response (Baker 1978; a response to unpredictable spatial resource availability) to a severe storm (Martinka 1967) or an annually repeated calculated movement through an invariant corridor from an area where snow accumulates predictably every year (e.g. the Jackson Hole/Green River herd in Wyoming; Berger 2004; Berger et al. 2006). In a study in Oregon, Dalton (2009) concurred with White et al. (2007; Yellowstone ecosystem) that pronghorn do not return to the same winter range consistently because they only migrate as far as necessary to find suitable environmental conditions. Regardless of where they overwinter, in large landscapes female pronghorn exhibit high fidelity to fawning areas within their summer range, returning each spring to the general area where they successfully fawned in previous years (Autenrieth and Fichter 1975; Mitchell 1980; Byers 1997). In an example from Idaho, all 22 collared female pronghorn followed for 2 years returned to the same areas they had occupied the previous summer (Hoskinson and Tester 1980).

Maintaining large tracts of native prairie, traditional and critical seasonal ranges and corridors to sustain long-distance movements of this species are important targets for conserving the ecological cohesion of landscapes (*sensu* Opdam et al. 2003). Pronghorn select large, open landscapes (O'Gara and Yoakum 2004). The largest populations of pronghorn in the mixed-grasslands are associated with expansive tracts of native prairie (Martinka 1967; Mitchell 1980; Sheriff 2006). Based on 23 years of survey records from Alberta, Sheriff (2006) found a significant log-linear relationship between the density of pronghorn in provincial management units and proportion of the landscape in native cover (Fig. 15.1). This landscape cover variable alone explained ~56% of variation in pronghorn densities. The lowest densities were in management units in south central Alberta that have been extensively converted (>75%) for dryland crop production. Although pronghorn exhibited lower densities in cultivated landscapes, they produced more offspring (fawns at heel in July) than in native landscapes (Sheriff 2006). Under the regulated management system of the province (Alberta Fish and Wildlife 1990), doe hunting is either not allowed or is set proportionately very low in croplands, and does not explain the low density of pronghorn in these areas. In Alberta, cropland-dominated landscapes appear to be population sinks that may provide surplus forage for the few animals residing there, supporting high productivity, but compensated by high mortality.

During a 3-year period between 2003 and 2007, the Alberta Conservation Association recovered GPS receivers from 65 pronghorn collared in Alberta with the objective of tracking movements and studying habitat selection. Results of the study confirmed a strong association of pronghorn with native prairie and provided insights into movement patterns. Among pronghorn residing in mixed or predominantly

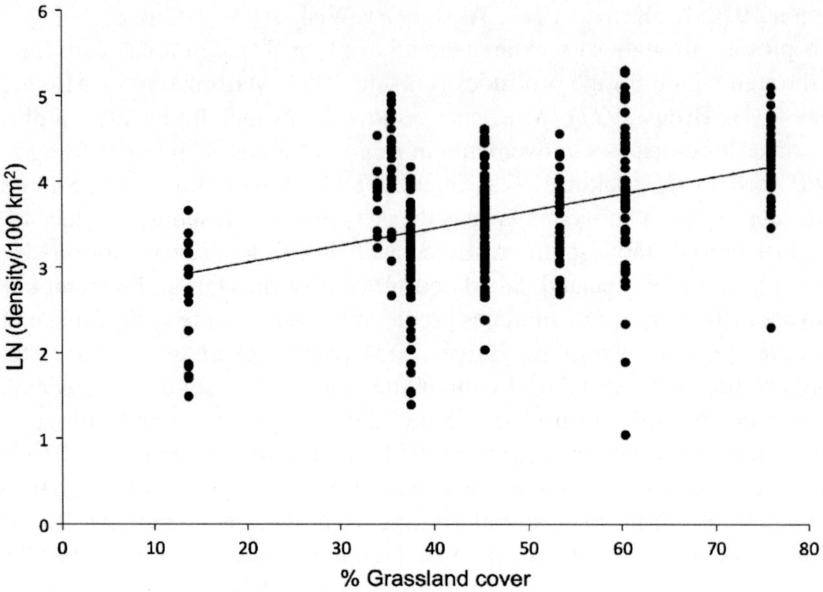

Fig. 15.1 Relationship between pronghorn density and percent native grassland cover in pronghorn management units in Alberta for the period 1985–2003. Source: Sheriff (2006)

native landscapes, about one-third undertook long-distance movements between winter ranges and fawning locations in summer range. They moved an average of 204 ± 112 km (n = 17). Pathways followed during spring migration were strongly directional (mainly south to north) and linear, deviating only where a major highway and railway corridor, fences, urban areas or steep terrain deflected them (Fig. 15.2). One exceptional movement originated from a winter-range capture location near the Montana border and followed a 445-km pathway northwards into west-central Saskatchewan. This is a record movement for the species; previously Sawyer et al. (2005) recorded a seasonal movement of 258 km between winter and summer range in Wyoming.

Terrain can have an important influence on pronghorn movements. In south-eastern Oregon, a river and associated road profoundly influenced pronghorn movements (Dalton 2009). Pathways followed by migrating pronghorn in Alberta were influenced by terrain features, particularly by steep banks of the deeply incised V-shaped valleys of the Saskatchewan and Red Deer Rivers. River crossing sites are limited in number and are characterized by shallow grades or flats on either side of the river, and were connected to native grassland habitat in the adjacent uplands. Pathways of 4-h point locations of radio-collared pronghorn that crossed the river appeared to be aligned with these likely crossing sites. Among anecdotal information obtained by Alberta researchers were several reports of pronghorn using small low-traffic bridges to move across rivers, indicating pronghorn can learn to use some human structures. Sawyer and Rudd (2005) reviewed anecdotal published accounts to formulate recommendations for designing road-crossing structures for pronghorn.

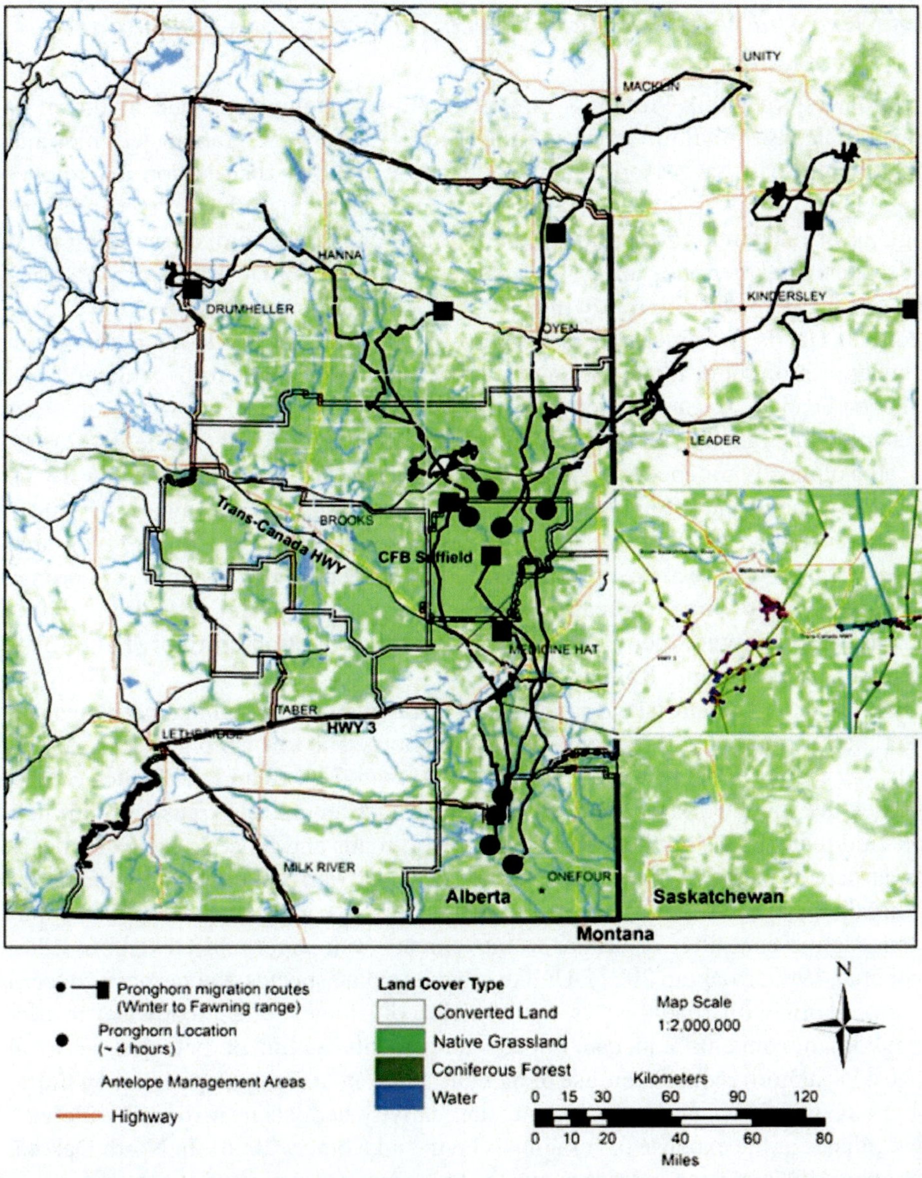

Fig. 15.2 Examples of long-distance movement pathways of pronghorn in spring in Alberta based on 4-h GPS collar location fixes. Note the strong northward directionality of movements, the funnelling effect of land use and the impeding effect of the Trans-Canada transportation corridor near Medicine Hat. Data source: Alberta Conservation Association

Land Use and Infrastructure Effects on Pronghorn Abundance

In rural western North America, effective land-use planning is one of the most important needs for wildlife conservation (Brown et al. 2005). Human activities and infrastructure can have a profound influence on wildlife distribution and movements. Urban sprawl and exurban settlement are expanding urban edges and leapfrogging of residential development and commercial zones into surrounding rural landscapes. Together, they are leading causes of biodiversity loss in North America (Hansen and Brown 2005; Baldwin et al. 2007). In Alberta, expansion of the City of Medicine Hat is incrementally infringing on the primary north-south migration pathway of pronghorn identified in our current research. The City is situated at an L-shaped bend of the Saskatchewan River that acts as a funnel for pronghorn moving south in the fall and winter from Canadian Forces Base (CFB) Suffield, a 2,600-km^2 area of mostly unfenced native prairie located north of Medicine Hat (Fig. 15.3). Mass movements of pronghorn into the City are common during harsh winter weather events as the animals strive to navigate southwards to more favourable ranges (Fig. 15.4). The area surrounding the City is becoming less porous to pronghorn movements owing to increasing commercial, residential and agricultural development combined with the main cross-Canada transportation corridor (Highway 1 and the Canadian-Pacific Railway).

Even in extensive grassland landscapes, infrastructure and human activities can affect the distribution and movements of pronghorn. The township grid of surveyed lands, a legacy of federal western settlement Acts in Canada and the United States, provides the framework for land development, land tenure, the transportation network and fence lines. As their density increases, roads and fences have had a negative effect on populations of several species and various ecological processes (Trombulak and Frissell 2000). Roadways are a major concern for pronghorn populations, acting as a filter or barrier to movements (van Riper and Ockenfels 1998; Ticer et al. 1999; Yoakum 2004). Unlike other ungulate species, the primary concern for pronghorn with roadways is not vehicle collisions; rather, roadways restrict pronghorn movements and can impede movements within or between seasonal ranges. Pronghorn reduce their use of habitats adjacent to roads, depending on traffic volume (Gavin 2006; Kolar 2009), and alter activity budgets near roads by increasing vigilance at the expense of foraging (Gavin and Komers 2006). In North Dakota, pronghorn selected open areas in non-rugged landscapes during winter and summer, and avoided primary roads during summer and secondary roads during both seasons (Kolar 2009). Similarly, in Wyoming pronghorn consistently selected home ranges with lower density of highway rights-of-way (ROW) than available (Sheldon 2005). Depending on traffic volume and width of an ROW, roads may be barriers to movements or deflect or deter passage across them (Forman and Alexander 1998). Ockenfels et al. (1997) radio-tracked 37 pronghorn females and found that fenced roads and railways were barriers to movement and influenced the shapes of home ranges. In Alberta, we observed that the transportation corridor near Medicine Hat (Fig. 15.2) presents a significant impediment to migrating pronghorn; radio-tracked

Fig. 15.3 Generalized movement corridors of GPS radio-collared pronghorn captured on winter range and winter movement pathways previously identified by Mitchell (1980) in Alberta, Canada. CFB Suffield is the Canadian Forces Base Suffield, a military training facility. Source: Alberta Conservation Association

Fig. 15.4 Pronghorn attempting to move through a suburban area on the outskirts of Medicine Hat, Alberta following a severe winter storm in November 2003. Photo: D. Eslinger

individuals (4-h GPS locations) were recorded moving back and forth along the transportation corridor for up to 10 days before finally crossing it. In southern Alberta, herds of pronghorn have been killed by trains during periods of deep snow cover. Pronghorn appear to be attracted to snow free railway tracks from which they cannot readily escape owing to fences and high snow banks.

Manufacturing of barbed-wire fences to contain domestic livestock on western rangelands was an emerging industry in 1870. In 1874, 4.5 tonnes of barbed wire was manufactured in the United States (MacCallum 1957; cited by Leftwich and Simpson 1978). By 1880, 36,000 tonnes were produced, increasing to 210,600 tonnes in 1945. Initially most fences were "drift fences", used to guide movements of livestock on extensive rangelands. Within a decade of barbed-wire fences first being erected on western rangelands, Canton (1877:48) reported that four-foot (1.2 m) high fences restricted pronghorn movements. Cross-fencing and fencing of smaller pastures became increasingly commonplace, further restricting the movement of pronghorn (Martinka 1967; Spillett et al. 1967; Hailey 1979). Although most fences are installed to control livestock, they also are constructed to control animal access to roads, highways and railroads, to protect agricultural crops, or to limit access of wildlife to private property or special land-use areas (e.g. mining operations, military installations).

Several studies have found that some fence designs, their location and density can affect pronghorn movements and distribution (Hailey and DeArment 1972; O'Gara and Yoakum 1992; Scott 1992; Sheldon 2005). Fences designed to confine

Fig. 15.5 Pronghorn struggling to move through wire fences (buck on *left*, doe on *right*). Photographed using automated cameras at Canadian Forces Base Suffield, Alberta. Source: M. Suitor, Alberta Sustainable Resource Development

traditional livestock (cattle, sheep or horses) or novel livestock (bison [*Bison bison*], elk [*Cervus elaphus*] or deer [*Odocoileus* spp.]) without regard for the ecological needs of free-ranging wild large mammals can act as partial or impermeable barriers to wildlife movements (Kie et al. 1994; Demarais et al. 2002; Yoakum 2004; Autenrieth et al. 2006; Gadd 2012). Fences also can be purposefully constructed to impede wildlife movements to protect crops (Palmer et al. 1985; Fritzell 1998) or reduce wildlife-vehicle collisions (Clevenger et al. 2001).

The effects of fences on the ecology of wildlife are expressed at different scales and levels of ecological organization, including partial to complete obstruction of daily movements, reduced access to seasonal habitat, food, cover and water, blockage or diversion of seasonal migration, increased energy demands, separation of juveniles from does, entanglement or impact injuries. Increased mortality from obstructive fences also can occur if traditional travel corridors are blocked and animals are forced into situations where vulnerability is high, such as along highways or railroad beds (Reed et al. 1974; Harrington and Conover 2006). Predators may push prey into fences to increase their chances of making a kill (Holzenbein and Marchinton 1992).

Although pronghorn have the physical ability to jump fences up to 2.5 m high (Spillett et al. 1967), they do so infrequently (Rouse 1954; Harrington and Conover 2006). Typically they crawl through or under barbed-wire fences. They have been observed to experience difficulty in crossing fences (Fig. 15.5), and may suffer injuries or become entangled in strand wire and die (Spillett 1965; Bear 1969; Oakley 1973; Kie et al. 1994; Sheldon 2005; Harrington and Conover 2006). Pronghorn are unable to pass under woven-wire fences and barbed-wire fences with low bottom wires (Yoakum 2004). On western rangelands, three or four strand barbed-wire fences with ~25 cm high bottom wires are common and represent a significant barrier to pronghorn movements. In a current study, we observed a large herd of pronghorn driven by adverse snow conditions temporarily stopped by such a fence as they

Fig. 15.6 Pronghorn exhibiting a mid-line dorsal wound likely caused by a low-strung barbed wire. Photo: Paul Jones, Alberta Conservation Association

attempted to move south of the Bodoin Wildlife Refuge in north-eastern Montana. In all seasons, we have observed pronghorns searching fence lines for feasible crossing opportunities. A well-documented catastrophe occurred in Wyoming in 1983 when several 100 migrating pronghorn died during harsh snow conditions as a result of being unable to cross a newly erected fence (Johnson 1988, cited by Cherney and Clark 2009). Injuries from barbed wire are common. Pronghorn captured in Alberta and photographed with automated cameras commonly had linear tissue injuries down the mid-line of their backs, which we inferred were caused by crawling under barb wire (Fig. 15.6).

In an analysis of minor highways and fences in northern Arizona, Bright and van Riper (2000) found fenced ROWs were greater barriers to movement than highways alone. Pronghorn will travel long distances parallel to fenced roads then cross the road where no fence is present (Bear 1969; Riddle and Oakley 1973; Sheldon 2005). In Wyoming, pronghorn selected habitats with the lowest fence densities (Sheldon 2005). Home ranges were bounded by fences in Arizona (Ockenfels et al. 1997; Ticer et al. 1999) and Wyoming (Sheldon 2005). In Alberta, we observed a distinct boundary effect on the distribution of pronghorn using CFB Suffield along the western and southern borders of the military base. This fence-line effect may have reflected the combined influence of boundary roads, fences and land use that reduced habitat quality adjacent to the military base. In Wyoming, pronghorn selected migration routes where they encountered fewer fences than random (Sheldon 2005). Similarly, in a current study in northern Montana and adjacent Saskatchewan, led by one of the authors of this chapter, movement data for radio-tracked animals indicates migrating pronghorn elect to move through areas with either no fences or highly permeable fences.

A combination of heavy traffic volume and fences can make fenced roads effective barriers or severe filters to pronghorn movements (Büechner 1950; Ockenfels et al. 1994). Fences have been shown to limit daily movements of pronghorn in winter in south-eastern Alberta and northern Montana, and accumulated snow can interfere with the ability of pronghorn to crawl under fences (Bruns 1977). Riddle and Oakley (1973) found the combined impacts of severe winter conditions and fences obstructed movements and led to increased mortality. Fences can reduce or prevent the use of previously used movement routes across highways (Büechner 1950; Ward et al. 1976; Guenzel 1986). Under adverse winter conditions in the absence of such obstructions we observed large herds of prairie pronghorn rapidly moving long distances, nose-to-tail, pushing towards distant ranges where conditions may be more favourable.

Land-Use Planning and Mitigation

As human populations continue to grow and economic activity shifts from east to west in North America, transportation infrastructure will inevitably become more elaborate. Road upgrading will include wider ROWs and top-widths, multiple-lanes, divided-lanes and fences. Consequently, the ability of pronghorn to move across transportation corridors is expected to decrease or perhaps be eliminated in some areas. To maintain connectivity within and between seasonal ranges it is increasingly important to design crossing structures that feasibly allow pronghorn to safely cross roadways (Sawyer and Rudd 2005). Placement of such structures in key linkage zones like the Trans-Canada transportation corridor near Medicine Hat and Montana Highway two area near Malta, Montana will require improving the awareness and cooperation of transportation, land management and wildlife agencies.

On a broad scale, conserving large tracts of native prairie by preventing conversion to croplands (intensive farming; e.g. Fargione et al. 2009) or large intensive industrial development (sub-surface minerals and wind energy; e.g. Sawyer et al. 2005; Pruett et al. 2009) and changing conventional livestock fences to wildlife-friendly fences (Autenrieth et al. 2006) represent important opportunities for advancing pronghorn habitat conservation and enhancement. For example, Gross et al. (1983) and Mapston (1970) tested gate structures that restrict domestic sheep but allow passage by pronghorn. They found specifically designed horizontal grills placed in fence corners effectively restricted sheep but pronghorn would leap over the grill (Fig. 15.7). An obvious solution to making impassable fences permeable to pronghorn is to leave gates open during periods when livestock are not in a pasture.

Pronghorn will readily cross under well-designed fences where there is sufficient space between the bottom wire and the ground (Gregg 1955: 57 cm; Cole 1956: 43 cm; Autenrieth et al. 2006: 41–46 cm). Most western wildlife agencies have published guidelines for wildlife-friendly fences (Arizona, Alberta, Colorado, Montana,

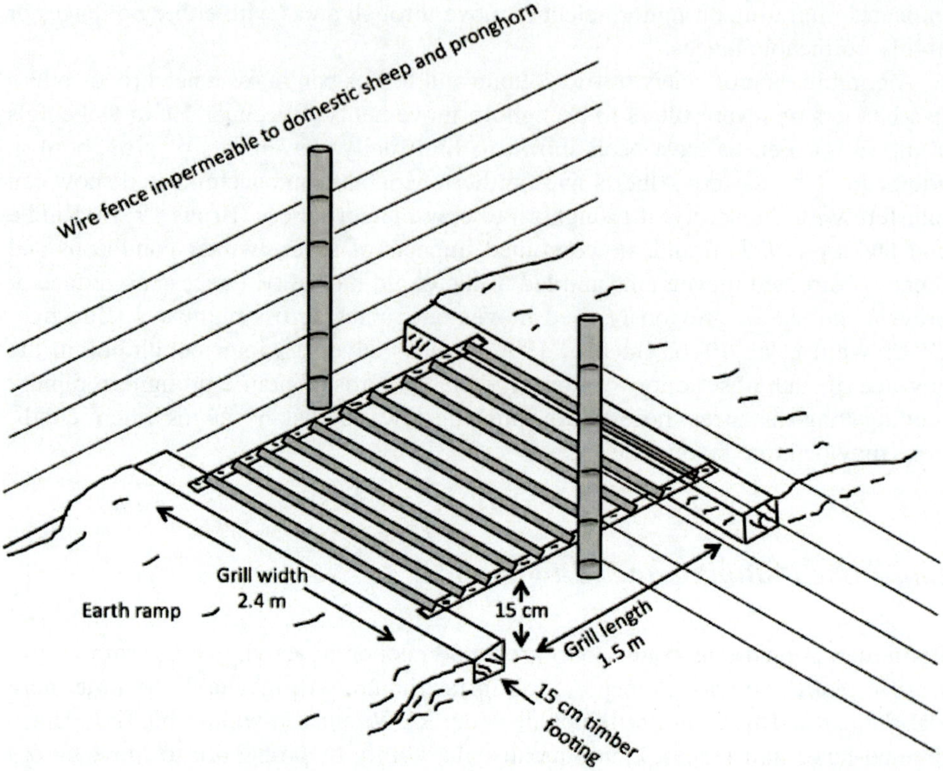

Fig. 15.7 Design and placement of a welded tubular metal grill that allows passage of pronghorn in an otherwise impermeable fence line. Adapted from Gross et al. (1983) and Mapston (1970)

North Dakota, Washington, Wyoming). As an example, the MultiSAR Program in Alberta (http://www.multisar.ca/documents/fencingBMP.pdf) provided a design for a wildlife-friendly fence that considered passage by pronghorn and deer species (Fig. 15.8).

Cooperative actions by agencies and NGOs can play an important role in improving landscape permeability for pronghorn. In Wyoming, backed by a court decision, the Wyoming Wildlife Federation successfully forced the removal of a woven-wire fence from a rancher's property after pronghorn were blocked from their winter habitat on public land and hundreds died during the tough winter of 1983–1984. Citizen groups may volunteer to modify fence lines to accommodate pronghorn movements in local areas. In the late 1980s, the California Department of Fish & Game reintroduced pronghorn into the Carrizo Plain. A joint effort between the Sierra Club, ForestWatch, Desert Survivors and the California Department of Fish & Game was organized to remove or modify fences, with the goal of restoring free pronghorn movements. Together the non-profit organizations and state and federal agencies worked to remove or modify more than 240 km of fencing. In Charles M. Russell National Wildlife Refuge in Montana fences are being removed from areas where containment of livestock is no longer necessary. In eastern Montana, the US

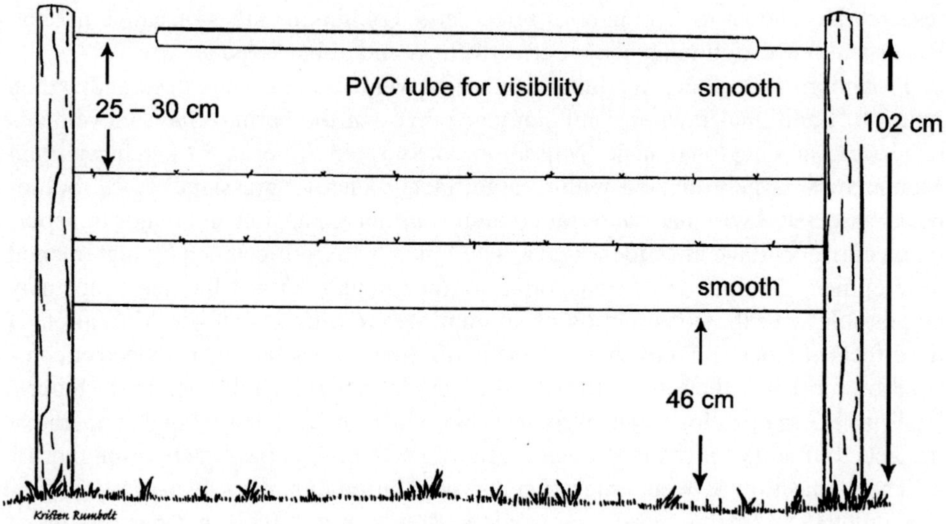

Fig. 15.8 Design of a wire fence for containing cattle yet allowing free passage of pronghorn under the bottom wire. Note use of smooth top and bottom wires to prevent injuries. Split PVC tubing can be used to increase visibility of the top wire in high frequency crossing areas used by wild ungulates. Source: Alberta Conservation Association

Bureau of Land Management is converting fences on land it manages to make them friendlier to wildlife movements.

In 2009, the Alberta Fish and Game Association, with assistance from the Alberta Conservation Association and the Alberta Fish and Wildlife Division, undertook a project to replace the bottom barbed wire with raised double-strand barbless wire along the 37 km northern boundary fence line of CFB Suffield. Officials at CFB Suffield were encouraged by the pilot project and continued the effort. To date 153 km of boundary fence has been modified. Sparked by success of the CFB Suffield projects, other initiatives have moved forward in Alberta. For example, fences are being retrofitted at Antelope Creek Ranch, a model ranch in the mixed-grasslands of Alberta.

Conclusions

Grasslands of the Northern Great Plains were substantially altered for crop production early in the twentieth century. Remaining native range is used predominantly for livestock grazing, although the energy sector has increased its footprint incrementally during the past 2 decades. Fences were first used to guide livestock movements but as pasture size decreased and the road network was elaborated, the density of fences has increased. Conventional fencing designed to contain cattle (three or four strands of barbed wire) with the bottom wire set at about 25 cm above grade is relatively impermeable to pronghorn. Woven-wire fencing is impermeable. Recent research has shown that pronghorn select large contiguous areas of native habitat. Roads and fences diminish habitat effectiveness and effective patch size.

Pronghorn may undertake long-distance movements to escape deep snow, then return to traditional fawning and summer ranges in the spring. Such movements may occur at a regional scale. Migration corridors in Alberta, Saskatchewan and Montana are largely located within large tracts of native grassland. Local bottlenecks associated with natural terrain features, urban expansion and major transportation corridors have been identified as key linkage areas threatened by incremental development. Mitigation of transportation impediments in key linkage zones may be possible with the construction of appropriate crossing structures. Mitigation of the effects of fences on movements and distribution is possible and is indeed encouraged by most jurisdictions with published guidelines for wildlife-friendly fences. Non-profit organizations can play an important role by undertaking cooperative projects to modify fences in strategic locations to facilitate passage by pronghorn.

The connection between migration and population dynamics of partially migratory ungulates remains poorly understood (Bolger et al. 2007), but obstruction of migratory pathways can be expected to result in negative impacts. Although the pronghorn remains a relatively common species in many areas of western North America, without appropriate land-use planning and management, cumulative anthropogenic changes will continue to erode habitat, and alter the suitability of movement corridors and key linkage zones supporting movements within and between seasonal ranges.

In the context of eco-regional planning, the pronghorn requires conservation of large areas of native habitat connected by corridors that allow long-distance movements to various winter ranges during harsh conditions. Plans that focus on maintaining the ecological coherence of landscapes (Opdam et al. 2003) for common species like the pronghorn should also benefit conservation of other native grassland species.

References

Alberta Environmental Protection. (1997) *The Grassland Natural Region of Alberta*. Alberta Environmental Protection, Natural Resources Service. Edmonton.

Alberta Fish and Wildlife. (1990) Management plan for pronghorn antelope in Alberta. *Wildlife Management Planning Series No. 3*. Alberta Fish & Wildlife Division, Edmonton.

Autenrieth, R.E. & Fichter, E. (1975) On the behavior and socialization of pronghorn fawns. *Wildlife Monographs*, **42**, 1–111.

Autenrieth, R.E., Brown, D.E., Cancino, J., Lee, R.M., Ockenfels, R.A., O'Gara, B.W., Pojar, T.M. & Yoakum, J.D. (2006) Pronghorn management guides. 4th edition. *Pronghorn Workshop and North Dakota Game and Fish Department*, Bismarck.

Baker, R.R. (1978) *The evolutionary ecology of animal migration*. Holmes and Meier publishers, Inc., New York.

Baldwin, R.F., Ray, J.C., Trombulak S.C. & Woolmer, G. (2007) Relationship between spatial distribution of urban sprawl and species imperilment: response to Brown and Laband. *Conservation Biology*, **21**, 546–548.

Barrett, M.W. (1980) Seasonal habitat associations of pronghorns in Alberta. *Proceedings of the Biennial Pronghorn Antelope Workshop*, **9**,174–195.

Barrett, M.W. (1982) Distribution, behavior, and mortality of pronghorns during a severe winter in Alberta. *Journal of Wildlife Management*, **46**, 991–1002.

Barrett, M.W. & Vriend, H.G. (1980) Management implications of the seasonal distributions of pronghorns and land use practices in Alberta. *Proceedings of the Biennial Pronghorn Antelope Workshop*, **9**, 196–214.

Bear, G.D. (1969) Antelope and net wire fences. *Proceedings Western Association State Game and Fish Commissioners*, **49**, 265–271.

Berger, J. (2004) The last mile: how to sustain long-distance migration in mammals. *Conservation Biology*, **18**, 320–331.

Berger, J., Cain, S. L. & Berger, K.M. (2006) Connecting the dots: An invariant migration corridor links the Holocene to the present. *Biology Letters,* **2**, 528–531.

Bolger, D.T., Newmark, W.D., Morrisson T.A. & Doak D.F. (2007) The need for integrative approaches to understand and conserve migratory ungulates. *Ecology Letters*, **10**, 1–15.

Bright J.L. & van Riper, C. (2000) *Pronghorn home ranges, habitat selection and distribution around water sources in northern Arizona.* US Geological Survey, Forest and Rangeland Ecosystem Science Center, Colorado Plateau Research Station. Technical Report. USGSFRESC/COPL/2000, Denver.

Brown, D.G., Johnson, K.M., Loveland, T.R. & Theobald, D.M. (2005) Rural land-use trends in the coterminous United States, 1950–2000. *Ecological Applications* **15**, 1851–1864.

Bruns, E.H. (1977) Winter behavior of pronghorns in relation to habitat. *Journal of Wildlife Management*, **41**, 560–571.

Büechner, H.K. (1950) Life history, ecology and range of the pronghorn antelope in Texas. *American Midland Naturalist*, **43**, 257–354.

Byers, J. (1997) *American Pronghorn: Social Adaptations and the Ghosts of Predators Past.* Chicago University Press, Chicago.

Canton, J.D. (1877) *The antelope and deer of America*. Hurd and Houghton, New York.

Cherney D.N. & Clark S.G. (2009) The American West's longest large mammal migration: clarifying and securing the common interest. *Policy Sciences*, **42**, 95–111.

Clevenger, A.P., Chruscz, B. & Gunson, K.E. (2001) Highway mitigation fencing reduces wildlife-vehicle collisions. *Wildlife Society Bulletin*, **29**, 646–653.

Cole, G.F. (1956) The pronghorn antelope, its range use and food habits in central Montana with special reference to alfalfa. Montana Department of Fish and Game, *Technical Bulletin* **516**, 1–62.

Compton, H.O. (1970) Southeastern Montana antelope population trends in relation to severe winters. *Proceedings Antelope States Workshop*, **4**, 50–54.

Creek, D.E. (1967) *A Summary of Pronghorn Antelope Investigations*. Wyoming Game and Fish Commission, Cheyenne.

Czech, B., Krausman, P.R. & Devers, P.K. (2000) Economic associations among causes of species endangerment in the United States. *BioScience* **50**, 593–601.

Dalton, K.A. (2009) *Pronghorn: Migration Triggers and Resource Selection in Southeastern Oregon*. M.Sc. Thesis, Washington State University, Pullman.

Demarais, S., DeYoung, R.W., Lyon, L.J., Williams, E.S., Williamson, S.J. & Wolfe, G.J. (2002) *Biological and social issues related to confinement of wild ungulates*. Technical Review, 02–3. The Wildlife Society, Bethesda.

Dirschl, H.J. (1963) Food habits of the pronghorn in Saskatchewan. *Journal of Wildlife Management*, **27**, 81–93.

Fargione, J.E., Cooper, T.R., Flaspohler, D.J., Hill, J., Lehman, C., McCoy, T., McLeod, S., Nelson, E.J., Oberhauser, K.S. & Tilman, D. (2009) Bioenergy and wildlife: threats and opportunities for grassland conservation. *BioScience*, **59**, 767–777.

Forrest, S.C., Strand, H., Haskins, W.H., Freese, C., Proctor, J. & Dinerstein, E. (2004) *Ocean of grass: A conservation assessment for the northern Great Plains*. Northern Plains Conservation Network and Northern Great Plains Ecoregion, World Wildlife Fund, US. Bozeman.

Forman, R.T. & Alexander, L.E. (1998) Roads and their major ecological effects. *Annual Review of Ecology and Systematics*, **29**, 207–231.

Fritzell, Jr., P.A. (1998) *A survey of Michigan Agricultural Producers' Attitudes, Perceptions, and Behavior Regarding Deer Crop Depredation to Fruit, Vegetables, and Field Crops*. Ph.D. Thesis, Michigan State University, East Lansing.

Gadd, M.E. (2012) Barriers, the beef industry and unnatural selection: a review of the impacts of veterinary fencing on mammals in southern Africa. In: M.J. Somers and M.W. Hayward (eds.) Fencing for Conservation. Springer, New York.

Gauthier, D. & Wiken, E. (2003) Monitoring the status of grassland habitats, Prairie Ecozone, Canada. *Environmental Monitoring and Assessment*, **88**, 343–364.

Gavin, S.D. (2006) *The effects of Roads on Pronghorn Antelope (Antilocapra americana) behaviour and habitat use*. M.Sc. Thesis, University of Calgary, Calgary.

Gavin, S.D. & Komers, P.E. (2006) Do pronghorn (*Antilocapra americana*) perceive roads as a predator risk. *Canadian Journal of Zoology*, **84**, 1775–1780.

Gregg, H.A. (1955) *Summer Habits of Wyoming Antelope*. Ph.D. thesis, Cornell University, Ithaca.

Gross, B., Holechek, J., Halford, D. & Pieper, R. (1983) Effectiveness of antelope pass structures in restriction of livestock. *Journal of Range Management,* **36**, 22–24.

Guenzel, R.J. (1986) *Pronghorn ecology in southcentral Wyoming*. M.S. Thesis, University of Wyoming, Laramie.

Guthrie, R.D. (1990) *Frozen Fauna of the Mammoth Steppe*. The University of Chicago Press, Chicago.

Hailey, T.L. (1979) *A handbook for pronghorn antelope in Texas. Federal Aid Report Series 20*. Texas Parks and Wildlife Department, Austin.

Hailey, T.L. & DeArment, D. (1972) Droughts and fences restrict pronghorns. *Antelope States Workshop Proceedings*, **5**, 22–24.

Hansen, A.J. & Brown, D.G. (2005) Land-use change in rural America: rates, drivers, and consequences. *Ecological Applications*, **15**, 1849–1850.

Harrington, J.L. & Conover, M. (2006) Characteristics of ungulate behavior and mortality associated with wire fences. *Wildlife Society Bulletin*, **34**, 1295–1305.

Holzenbein, S. & Marchinton, L.R. (1992) Emigration and mortality in orphaned male white-tailed deer damage in cornfields. *Journal of Wildlife Management*, **56**, 147–153.

Hoskinson, R.L. & Tester, J.R. (1980) Migration behavior of pronghorn in southeastern Idaho. *Journal of Wildlife Management*, **44**, 132–144.

Johnson, D. (1988) When antelope don't roam free. *New York Times*, November **18**, A16.

Jones, P.F., Penniket, R., Fent, F., Nicholson, J. & Adams, B. (2005) Silver sagebrush community associations in southeastern Alberta, Canada. *Rangeland Ecology and Management*, **58**, 400–405.

Kie, J.G., Bleich, V.C., Medina, A.L., Yoakum, J.D. & Thomas, J.W. (1994) Managing rangelands for wildlife. *Research and Management Techniques for Wildlife and Habitats.* (ed T.A. Bookhout) pp. 663–668. The Wildlife Society, Bethesda.

Kolar, J. (2009) *Pronghorn Migration and Resource Selection in Southwestern North Dakota*. M. Sc. Thesis, University of Missouri-Columbia, Columbia.

Leftwich, T. & Simpson, D. (1978) The impact of domestic livestock and faming on Texas pronghorn. *Proceedings of the Eighth Biennial Pronghorn Antelope Workshop.* May 2–4, 1978, Jasper Park Lodge, Jasper.

Mapston. R. (1970) Casper antelope pass studies. Pages 116–128 in Proceeding of the Fourth Antelope States Workshop. January 27–28, Scottsbluff.

Martinka, C.J. (1967) Mortality of northern Montana pronghorn in a severe winter. *Journal Wildlife Management*, **31**, 159–164.

MacCallum, A.D. (1957) Barbed wile in Texas. *The Southwestern Historical Quarterly*, **59**, 207–219.

McKenzie, J.V. (1970) Two "killer winters" 1964–1965 and 1968–1969 in North Dakota. *Antelope States Proceedings,* **4**, 36–40.

Mitchell, G.J. (1980) *The Pronghorn in Alberta*. Alberta Department Lands and Forests, Fish and Wildlife Division and University Regina, Regina.

Mitchell, G.J. & Smoliak, S. (1971) Pronghorn antelope range characteristics and food habits in Alberta. *Journal of Wildlife Management,* **35**, 238–250.

Oakley, C. (1973) The effects of livestock fencing on antelope. *Wyoming Wildlife*, **37**, 26–29.

Ockenfels, R.A., Alexander, A., Ticer, C.L. & Carrel, W.K. (1994) Home ranges, movement patterns, and habitat selection of pronghorn in central Arizona. *Technical report 13*. Arizona Game and Fish Department, Phoenix.

Ockenfels, R.A., Carrel, W.K. & van Riper, C. (1997) Home ranges and movements of pronghorn in Northern Arizona. *Arizona Game and Fish Department Wildlife Bulletin* **3**, 45–61.

O'Gara, B.W. & McCabe, R. (2004) From exploitation to conservation. *Pronghorn Ecology and Management* (eds. B.W. O'Gara & J.D. Yoakum), pp. 42–73. University Press of Colorado, Boulder.

O'Gara, B.W. & Yoakum, J.D. (eds) (1992) Pronghorn management guidelines. *Pronghorn Antelope Workshop* **15**, 1–101.

O'Gara, B.W. & Yoakum, J.D. (2004) *Pronghorn: Ecology and Management*. University Press of Colorado, Boulder.

Opdam, P., Verboom, J. & Pouwels, R. (2003) Landscape cohesion: an index for the conservation potential of landscapes for biodiversity. *Landscape Ecology*, **18**, 113–126.

Palmer W.L., Payne, J.M., Wingard, R.G. & George, J.L. (1985) A practical fence to reduce deer damage. *Wildlife Society Bulletin*, **13**, 240–245.

Pruett, C., Patten, M.A. & Wolfe, D.H. (2009) Avoidance behavior by prairie grouse: Implications for development of wind energy. *Conservation Biology,* **23**, 1253–1259.

Raper, E., Christiansen, T. & Petch, B. (1989) Sublette antelope study: final report. *Annual Big Game Herd Unit Report*, pp.124–169. Wyoming Game and Fish Department, Cheyenne.

Reed, D.F., Pojar, T.M. & Woodward, T.N. (1974) Use of one-way gates by mule deer. *Journal of Wildlife Management*, **38**, 9–15.

Riddle, P. & Oakley, C. (1973) The impact of a severe winter and fences on antelope mortality in southcentral Wyoming. *Proceedings of the Western Association of State Game and Fish Commissioners*, **53**, 174–188.

Rouse, C.H. (1954) *Antelope and sheep fences*. Preliminary Report, U.S. Fish and Wildlife Service, Washington, D.C.

Sawyer, H. & Lindzey, F. (2000) *Jackson Hole Pronghorn Study*. Wyoming Cooperative Fish and Wildlife Research Unit, Laramie.

Sawyer, H. & Rudd, B. (2005) *Pronghorn roadway crossings: A review of available information and potential options*. Federal Highway Administration; Wyoming Department of Transportation; Wyoming Game and Fish Department, Cheyenne.

Sawyer, H., Lindzey, F. & McWirter, D. (2005) Mule deer and pronghorn migration in Western Wyoming. *Wildlife Society Bulletin*, **33**, 1266–1273.

Scott, M.D. (1992) Buck-and-pole fence crossings by four ungulate species. *Wildlife Society Bulletin*, **20**, 204–210.

Sheldon, D.P. (2005) *Pronghorn Movement and Distribution Patterns in Relation to Roads and Fences in Southwestern Wyoming*. Thesis, University of Wyoming, Laramie.

Sheriff, K. (2006) *Modeling Temporal and Spatial Variation in Pronghorn Antelope Population Dynamics in Southern Alberta in Relation to Environmental Gradients*. Master of Environment Design Thesis, University of Calgary, Calgary.

Spillett, J.J. (1965) *The effects of livestock fences on pronghorn movements*. Thesis, Utah State University, Logan.

Spillett, J.J., Low, J.B. & Sill, D. (1967) Livestock fences—how they influence pronghorn antelope movements. *Agricultural Experiment Station Bulletin* **470**, Utah State University, Logan.

Telfer, E. & Kelsall, J. (1984) Adaptation of some North American mammals for survival in snow. *Ecology,* **65**, 1828–1834.

Ticer, C.L., Boe, S.R., Ockenfels, R.A. & deVos, Jr., J.C. (1999) Factors affecting home ranges and movements of pronghorn on a shortgrass prairie of northern Arizona. *Pronghorn Antelope Workshop Proceedings*, **18**, 84–90.

Trombulak, S.C. & Frissell, C.A. (2000) Review of Ecological effects of roads on terrestrial and aquatic communities. *Conservation Biology*, **14**, 1523–1739.

van Riper, C. & Ockenfels, R. (1998) The influence of transportation corridors on the movement of a pronghorn antelope over a fragmented landscape in northern Arizona. *International Conference on Wildlife Ecology & Transportation,* **1998**, 241–244.

Ward, A.L., Cupal, J.J., Goodwin, G.A. & Morris, H.D. (1976) *Effects of Highway Construction and Use on Big Game Populations*. Rept. FHWA-RD-76–174. U.S. Federal Highway Administration, Washington, D.C.

West, D.R. (1970) Effects of prolonged deep snow and cold winters on pronghorn mortality and reproduction in South Dakota. *Proceedings of the Biennial Pronghorn Antelope Workshop*, **4**, 41–49.

White, P.J., Bruggeman, J.E. & Garrott, R.A. (2007) Irruptive population dynamics in Yellowstone pronghorn. *Ecological Applications*, **17**, 1598–1607.

Wishart, W.D. (1970) A brief historic review of the pronghorn antelope in Alberta. *Proceedings of the Biennial Pronghorn Antelope Workshop*, **4**, 128–130.

Yoakum, J.D. (1978) Managing rangelands for pronghorn antelope. *Proceedings Pronghorn Antelope Workshop*, **8**, 321–326.

Yoakum, J.D. (2004) Habitat Conservation. *Pronghorn Ecology and Management* (eds. B.W. O'Gara & J.D. Yoakum), pp. 571–630. University Press of Colorado, Boulder.

Chapter 16
Use of Electric Fencing and Associated Measures as Deterrents to Jaguar Predation on Cattle in the Pantanal of Brazil

Sandra M.C. Cavalcanti, Peter G. Crawshaw, and Fernando R. Tortato

Introduction

As top predators, jaguars, *Panthera onca*, constitute an important component of the megafauna of the Neotropics. Like other large carnivores, they require vast areas of relatively wild habitat and a stable prey base. Therefore, their populations are often restricted to protected areas or inhospitable, remote areas where human densities remain low. The Pantanal, a seasonally inundated plain of over 140,000 km^2 in the centre of South America is one such place, harbouring abundant wildlife and being considered of high importance for the long-term persistence of jaguars (Sanderson et al. 2002). Nevertheless, cattle ranching has been a traditional activity for over 200 years in the Pantanal, and the region harbours the largest beef cattle herd of Brazil (IBGE 2009). As in other areas of the world where large carnivores coexist with domestic animals (Fritts et al. 1992; Mizutani 1993; Srivastav 1997), this proximity between cattle and jaguars results in conflict with ranchers that makes it one of the greatest causes of mortality for the species, throughout its range (Sanderson et al. 2002).

Under certain conditions, jaguars will kill livestock (Schaller and Crawshaw 1980; Schaller 1983; Crawshaw and Quigley 2002; Crawshaw 2003; Azevedo and Murray 2007; Cavalcanti and Gese 2010). Therefore, despite other cultural factors (Amâncio et al. 2007; Cavalcanti et al. 2010), jaguar predation on livestock may pose a real economic predicament to ranching operations and significantly contribute to the conflict with ranchers and ranch hands, leading to retaliatory killing

S.M.C. Cavalcanti (✉)
Instituto Pró-Carnívoros, Av Horácio Neto, 1030, Atibaia, SP 12945-010, Brazil
e-mail: scavalcanti@procarnivoros.org.br

P.G. Crawshaw
CENAP/ICMBio, Av. dos Bandeirantes, s/n°, Balneário Municipal, Atibaia, Brazil

F.R. Tortato
Panthera, R. Barão de Melgaço, 2754, 8° andar, sala 803, Cuiabá, MS, Brazil

M.J. Somers and M.W. Hayward (eds.), *Fencing for Conservation: Restriction of Evolutionary Potential or a Riposte to Threatening Processes?*,
DOI 10.1007/978-1-4614-0902-1_16,

(Cavalcanti and Gese 2010). For this reason, livestock depredation is an important issue to be resolved in jaguar conservation.

Throughout the world, there has been an increased interest in the use of non-lethal methods of depredation control (Shivik 2000; Shivik et al. 2003). These methods range from the traditional management practices, use of guard animals and exclusion methods to the more recent use of chemical repellents and aversive agents, to the rapidly growing technology of visual and acoustical stimuli (Koehler et al. 1990; Andelt and Hopper 2000; Shivik 2006). Examples of these later methods include devices with movement detection sensors, strobe lights, sirens, bells, recorded distress calls, scarecrows, loud music or noise from radios, propane exploders, fladry and electronic training collars (Koehler et al. 1990; Linnell et al. 1996; Shivik 2006). Such practices are viewed as humane, practical and with a potential to minimise problems without necessarily removing the predator. Despite these new advances in the field of predation management, information regarding the use of such devices in Brazil is scarce and to protect their livestock, ranchers still resort to the use of tools they have used for many generations.

The use of electric fences to deter terrestrial predators was first reported by McAtee (1939). Their use has been tested as deterrents for several species of predators such as coyotes (*Canis latrans*), wolves (*Canis lupus*), bears (*Ursus arctos*), and lynx (*Lynx rufus*; Acorn and Dorrance 1994; Levin 2002; Mertens et al. 2002). In general terms, fencing may be more viable for protecting smaller herds (or flocks) of livestock in smaller pastures, as opposed to large open ranges, particularly arid ranges, where they typically spread out in search of food and water. However, it would be desirable if fences provided a reliable, economical way to decrease predation on cattle in field conditions in the Pantanal. In this chapter, we describe the use of an electric fence in a large cattle ranch in the Pantanal as a method to deter predation from jaguar. In addition, we describe the trials we conducted to assess the use of electric fences as a deterrent for jaguars and pumas, *Puma concolor,* in captive situations.

Study Area

Evaluation of Electric Fences as Jaguar Deterrent in Field Conditions

The use of electric fences in field conditions was implemented in the Santa Tereza ranch (18° 18′ 38″ S; 57° 30′ 10″ W), located in the municipality of Corumbá, on the westernmost region of the Pantanal, Mato Grosso do Sul state, Brazil. The ranch, 63,000 ha in size, encompassed a high diversity of habitats, from the open water of the Baía Vermelha on the border with Bolivia, to cerrado (open shrublands) and flooded areas, to mountain tops at about 900 m, on the southern extension of the Serra do Amolar mountain range (Fig. 16.1).

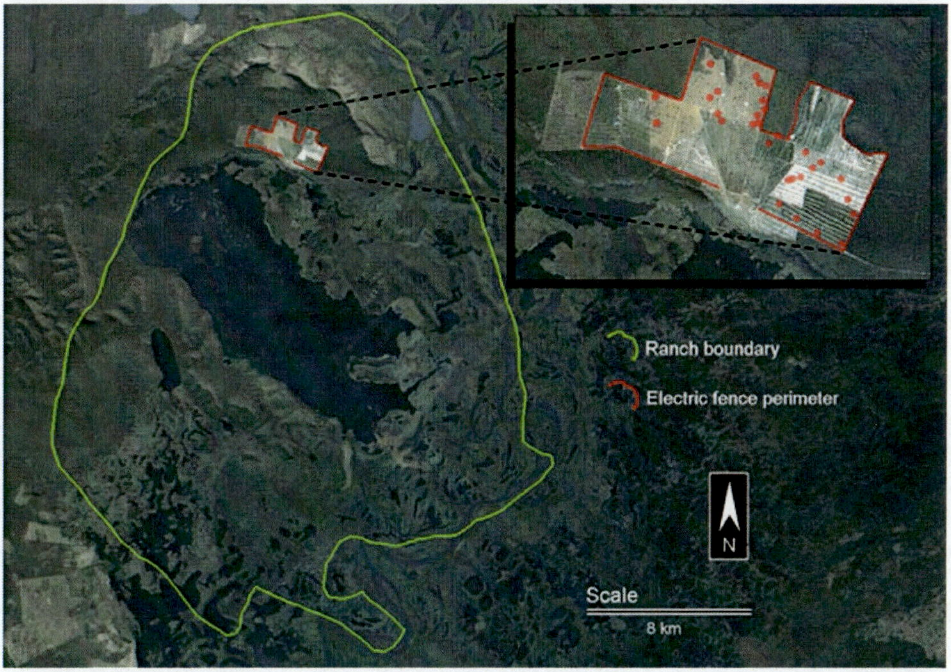

Fig. 16.1 Santa Teresa ranch, Corumbá, MS, Brazil. Most of the area is characterised by open range, with the big Baía Vermelha (lake) in the central portion and the Serra do Amolar mountain chain encircling the area from the west to the northeast. The area in the detail is 900 ha of introduced grasses and surrounded by electric fence. The red dots are locations of jaguar kills within the enclosed area

The ranch was purchased by the current owners in early 2006 and included a herd of 1,300 cattle in a semi-feral state, which grazed the open range for most of the year and spent the wet period in the shrublands above flood level. During the first year, the owners rebuilt/restored fences and recuperated about 900 ha of pasture land that had been opened and planted by the original owner but was kept without the regular grazing of cattle in recent years (R. Jank, pers. comm.). These 900 ha of overgrown pasture mostly comprise brachiaria grass, *Brachiaria humidicola,* and islands of colonião grass, *Panicum maximum*. Characterised by dry ground above flood level, this area was divided into 12 pastures of approximately 75 ha each, to manage the different age/classes of a cattle herd of 2,000 head (Fig. 16.1).

The cattle business was characterised by a beef cattle fattening operation, with the vast majority of the herd comprising heifers of up to 350 kg (78%). Cattle management was divided into two distinct phases according to the drought and flooding cycles, characteristic of the Pantanal. During the floods (mid-February through August), the cattle herd was confined in these 900 ha of high ground pastures. In the drought period (September to mid-February), about 80% of the herd was released into the lower areas of the ranch, where they found large extensions of native grasses.

After an initial bout of recurrent predation, where nice head of cattle were killed in 11 days in February 2007, the owners of the ranch contacted the authors to request assistance with the problem. In contrast with the predominant attitude among ranchers in general, they expressed concern regarding jaguar conservation and sought to reduce the conflict and economic losses to an acceptable level.

Responses of Jaguars and Pumas to Electric Fences in Captivity

As a complement to the field data presented in this chapter, we conducted trials to assess the efficacy of electric fencing as a deterrent for jaguars and pumas, in two zoological parks in the state of São Paulo. The zoos were the Parque Zoológico Quinzinho de Barros (23°30′23.28″ S, 47°26′11.68″ O), in Sorocaba, where we assessed the behaviour of two jaguars (one male and one female) and three pumas (one male and two females), and the Parque Zoológico de Guarulhos (23°26′33.77″ S, 46°33′11.90″ O), in Guarulhos, where we assessed the behaviour of two jaguars (one male and one female) and one puma (male). In the Sorocaba Zoo, animals were kept in a circular enclosure divided into four segments, which encompassed an area of approximately 156 m^2 each (front: 23.85 m, back: 7.95 m, sides: 10.60 m each, and height: 2 m.). In the Guarulhos Zoo, enclosures measured $13.7 \times 4.7 \times 4$ m, encompassing an area of approximately 65 m^2. In both zoos, with the exception of the male puma in the Guarulhos zoo, all the other study animals lived together with conspecifics within the enclosures and were fed once a day.

Methods

Evaluation of Electric Fences as Jaguar Deterrent in Field Conditions

In February 2008, existing conventional livestock fences on the perimeter of the pasture area (13,745 m) were retrofitted with two electrified wires at heights of 25 and 50 cm from the ground. These additional wires, powered by solar panels (Duboi, Campo Grande, MS, Brazil) installed near one of the extremities of the fence, were external to the conventional fences, which had five non-barbed wires at heights of 25, 50, 75, 100 and 125 cm. The electrified wires did not complete the entire perimeter of the external fence. About 630 m of the fence, near the ranch headquarters were not electrified (Fig. 16.1). The voltage used in the fence ranged from 5,000 to 7,000 V. The fence was checked by the ranch foreman on a weekly basis. Whenever the voltage was detected to be low, the entire perimeter of the fence was verified to identify and clean branches or other material that could be isolating the system.

The use of electric fences was accompanied by additional preventive measures. Night-time surveillance of the different pastures was achieved through the use of a tractor and a strong light beam. A ranch hand visited different cattle bed sites throughout their surveillance period, and whenever a group of cattle was observed to be uneasy (i.e. the presence of a cat nearby was suspected), fire crackers were lit up and fired in the direction of the closest forest fragment.

We accompanied ranch hands in their daily activities on horseback, searching for carcasses of cattle killed by jaguars or killed by other causes, monitoring jaguars through sightings and other indirect methods (i.e. tracks and scats). Coordinates of these events were recorded using a global positioning system (GPS; Garmin 12 XL, Garmin International, Inc., Olathe, KS). Sightings of mammalian native species were also noted, with date, time and coordinates recorded.

In addition to accompanying the ranch hands in their daily activities, additional daytime surveillance of the cattle was achieved with the use of a motorcycle. During these outings, we opportunistically recorded jaguar tracks, carcasses of their native and domestic prey and wildlife sightings with the use of GPS.

Responses of Jaguars and Pumas to Electric Fences in Captivity

Electric fences were built with five electrified non-barbed wires at heights of 20, 40, 60, 80 and 110 cm from the ground, cutting off a section of the enclosure where food (raw meat, poultry and beef chunks) was placed. In addition to the food, to increase motivation, the fence cut off a section of the enclosure that included a favourite rest area, or access to it, where animals chose to spend most of their time (Fig. 16.2). The number of posts depended on the design required to close off the selected section of the enclosure. In the Sorocaba zoo, we used a configuration with five posts, for both enclosures (jaguar and puma). In Guarulhos, we used four posts in the jaguar enclosure, and three posts in the puma's. The voltage used in the fences ranged from 7,800 to 9,900 V. There was little variation between the voltage attained in the different wire strands of the fence. All equipment used was provided by Farmtech S.A. Produtos Veterinários (Porto Alegre, RS, Brazil), representatives in Brazil of Speedrite® and Tru-test®, from New Zealand. In the Sorocaba zoo, construction of the fence was supervised by a technician (A. S. Balbino) provided by Farmtech S. A. In Guarulhos, we built the fence ourselves, but followed the same guidelines and general design for the fences.

In both zoos, the trials were organised as 3 daytime and one night-time in 30-min sessions for each individual of each species where we recorded behaviour in relation to the fence. The behaviour of the animals was filmed during the day and photographed at night. In addition, a description of their behaviour was recorded with a digital voice-recorder.

As an additional motivation, study animals were not fed for the duration of the trials. In Sorocaba, the first trial with the jaguars started on a Friday, and the second only continued on the following Monday. Although the animals had been fed the morning of the first trial, they were kept without food, from their habitual feeding Friday morning to the end of the trials on the following Wednesday.

Fig. 16.2 Electric fence set up at the jaguar enclosure at Sorocaba Zoological Park, Brazil, blocking access to favourite resting spot

Results and Discussion

During the flood season of 2007, the cattle herd was brought to the higher ground fenced pastures but it was not until the following year that the electric fences were installed. Data on cattle mortality from 2006 to 2007 were reported by ranch hands and were not verified by research personnel because there was nobody stationed fulltime at the ranch to do so. According to their records, in 2006 jaguars were responsible for 24% of all losses ($n=11$ of 46 total losses). During 2007, the percentage of losses attributed to jaguar kills escalated to 86% ($n=24$ of 28 total losses). In 2008, when we had a researcher spending 20 days/month in the field to monitor jaguars and cattle depredation, total cattle losses attributed to jaguar predation represented 10% of the total losses ($n=50$ of 504 total losses). In that year, total cattle loss escalated to over 500 head, but the vast majority of them were recorded as missing cattle (82%, $n=413$). There was a difference observed in the percentage of losses attributed to jaguars as recorded by research personnel and by ranch hands (10 and 13%, respectively), although this difference was not as high as the ones observed in other studies (Crawshaw and Quigley 1984, 2002; Cavalcanti 2006, 2009).

Jaguars killed mostly heifers (73% of all kills), followed by steers and adult cows (10% each) and calves (7%). Given that heifers comprised 78% of the herd, jaguars did not seem to select for any specific age/class category but rather selected the prey according to their availability.

When we were first contacted by the ranch owners, we recommended that they install electric fences in relatively small enclosures where better control of conditions could be exerted. Electric fences can be effective in decreasing predation losses, particularly in areas where predation is moderate to severe, and particularly if used to enclose the most vulnerable age classes and/or used during the most critical periods. In Venezuela, Scognamillo et al. (2002) tested an initial design with three electrified wire strands (30, 60 and 90 cm) with voltage ranging between 2,500 and 3,000 V, which encompassed an enclosure of 18 ha (with a perimeter of 1,697 m). However, the authors recorded eight attacks by two jaguars and one puma in the first 3 weeks of the experiment. They later modified the existing fence to include an additional wire (20, 40, 60 and 85 cm) and increased the voltage to 4,500–5,000, after which the attacks ceased.

In our study area, we found that the use of the electric fence, as implemented, had limited effect in reducing the probability of livestock depredation by jaguar. Although losses attributed to jaguars in 2008 decreased considerably in relation to the previous year, they can still be considered high. The design adopted by the ranch, to fence only the perimeter of a 900-ha management area (13,745 m), left too many opportunities for jaguars to enter the area and take cattle. Potential for predator entry, as well as electrical malfunctions, washouts and physical damage, is likely to increase as fenced areas increase in size.

On different occasions ($n=7$), we observed sites where a jaguar appeared to have tried to enter the fence without success, as evidenced by several scrapes on the outside of the electric wires (Fig. 16.3). In addition, we followed the tracks of jaguars walking along the fence for distances of over 400 m, until they found where the electric wires were discontinued such as at gates left for the entrance of the cattle. These gates were also subsequently equipped with electric wires to isolate the cattle management areas and thus prevent the entrance of predators. Therefore, we surmised that the predator entrance was made more difficult but not precluded.

Night-time surveillance of the different pastures was initiated in early 2007, as per our recommendation, after we were first contacted by the ranch owners. However, given that we did not have research personnel in the field during 2007, an unbiased evaluation of this measure could not be made. According to the ranch foreman, at one period when night surveillance and the use of firecrackers was conducted by one well-prepared, conscientious employee, it achieved the proposed objectives, and had a marked effect on reducing attacks (F. Machado, pers. comm.). However, the change in personnel and insufficient time schedules through the night left open opportunities that resulted in increased predation. Employees often used firecrackers at will with no relation to the presence or threat of cats to demonstrate that the person was "working". This may likely have hampered the positive effect of the method, making the cats habituated to the explosions, not linked with their proximity to the cattle.

Night-time surveillance is best conducted by an employee specifically appointed and trained for this function. In this case, overload of duties during the day made it difficult to complete the required number of hours during the night. Attempted night-shifts starting at 24:00 h and finishing at 05:00 h or from 21:00 to 03:00 h were shown

Fig. 16.3 Photograph showing where a jaguar tried to enter the fenced enclosure and was impeded by the electric fence (photo F. Tortato)

not to cover the crepuscular hours when jaguars have been shown to be usually active (Crawshaw and Quigley 1991; Cavalcanti and Gese 2009). In four instances, jaguars were seen close to the herd between 18:00 and 21:00 h. Therefore, optimum surveillance timing would include from dusk to dawn, or 18:00–06:00 h, requiring one full-time employee or two employees to divide the task.

We also recommended the maintenance of the pasture and other vegetation at short heights in the cattle management modules, but this was not implemented. However, it was shown to be important, since 51% of all predation events happened at clumps of tall grass or very close to them (within 50 m). Jaguars do not pursue their prey for long distances and usually rely on ambush and a short distance chase. Therefore, they depend on cover to launch their attacks. With the elimination of these patches within the fenced management modules, the opportunities for these attacks would likely decrease.

Cultural factors play an important role in the conservation of jaguars in cattle ranches in the Pantanal and make the interface between ranch personnel and researchers a difficult issue. From our experience in this and other projects, it is clear that most ranch hands do not believe in the efficacy of anti-predation measures and think that jaguar conservation is actually incompatible with cattle operations (Crawshaw & Cavalcanti, pers. obs.). Several studies have shown that most ranches over-estimate the amount of jaguar predation (Schaller and Crawshaw 1980; Crawshaw and Quigley 1984; Azevedo and Murray 2007; Cavalcanti 2009). Ranch hands often perceive that if jaguars are protected, their numbers will rise exponentially and predation will increase accordingly. It is difficult for them to accept that factors other than just the availability of cattle control their population size, such as social behaviour, and that jaguars still rely on native prey, such as peccaries and caiman (Cavalcanti and Gese 2010). It is important to educate ranch employees and their families regarding jaguar ecology to build an understanding of the integral part large cats play in healthy ecosystems. This may allow them to recognise the importance of jaguar conservation and therefore, motivate them to fulfil their important mission in reducing the conflict and achieving coexistence.

During our evaluations in captivity, we observed that each one of the jaguars sustained at least one shock from the fence. Mean interval between first introduction to the fence and first shock was 1 min. The male jaguar in Sorocaba took two shocks in the first session, with the second shock 2 min after the first one (Fig. 16.4). The female also immediately approached the fence, sustaining a shock 1 min after being released into the enclosure. In the next 2 daytime sessions, she spent most of the time at the front of the enclosure, apparently indifferent to the fence. However, during the night session, she appeared to be bolder, approaching the fence at least 5 times, and receiving another shock (Fig. 16.4).

The male and the female jaguar in the Guarulhos zoo also received one shock each in the first session. The male received two shocks in the second session; the first, 7 min after entering the enclosure, and the second, 11 min later. In this setting, a narrow corridor of approximately 1 m was left between one of the sides of the enclosure and the electric fence. In two occasions, the male rushed through this corridor, pressing himself close to the outside wall and as away from the electric fence as possible. During the third session, the male approached the fence twice, with the obvious intention of getting to the meat, but was deterred, most likely due to the previous experience with the shocks. Most of the time during the trials, both the male and female retreated to the shady, covered section of the enclosure.

Results for the pumas were all similar, considering all individuals in both zoos. Each of the animals received a total of two shocks, one each in the first session and one in the night session. The only difference was regarding the greater time interval for the male puma in the Guarulhos zoo, with the shock sustained 13 min after being introduced in the enclosure (Fig. 16.4).

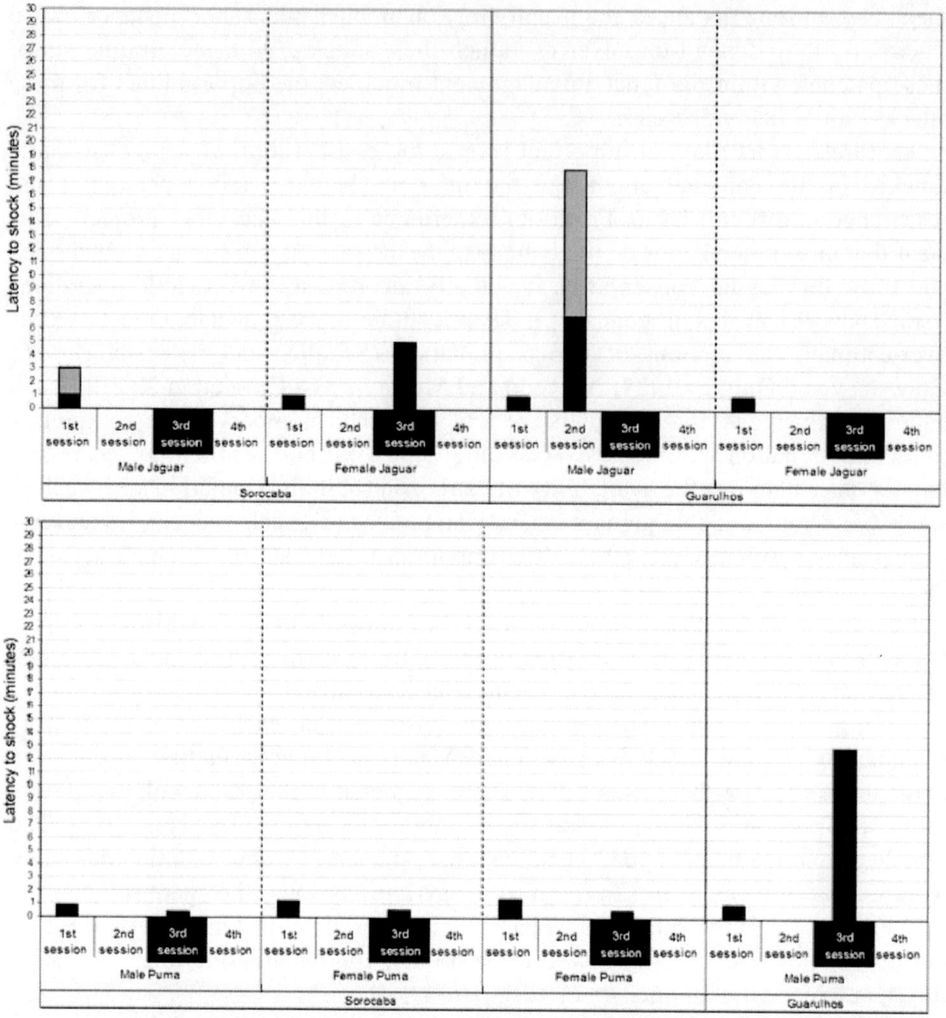

Fig. 16.4 Latency to shock sustained by four jaguars and four pumas, recorded in Sorocaba and Guarulhos zoological parks in the state of São Paulo, Brazil. The *black bar* is the first shock and the *grey bar* is the second shock sustained by individual animals. The third session, in *black*, was conducted at night

The pumas in the Sorocaba zoo seemed to be extremely sensitive to disturbances to their surroundings. After the first session, their behaviour, from openly exploratory, went to extremely wary, looking very alert and skittish. In the last session, the male showed patches of bare skin, which according to the zoo vets (A.V. Nunes, pers. comm.), was due to stress. Even though the male at Guarulhos did not show any physical signs of distress, his behaviour became very reserved in that he spent most of the time after the shocks, lying on a platform at one of the corners of the enclosure. This was the most evident difference in the behaviour of the two species, regarding the electric fence. While pumas became skittish and alert, jaguars reacted as if mostly aloof, most of the time disregarding the fence.

Fig. 16.5 Female puma upon receiving a discharge from the electric fence at the Sorocaba Zoological Park, Sorocaba, SP, Brazil

For the two species, the first contact with the electric fence was determined mainly by exploratory behaviour when animals were investigating novel stimuli in their surroundings. The first shock for all animals was within one and a half min after releasing the animals back into the enclosures. Although our samples do not allow for statistical analyses, they do demonstrate that after one or two shocks, the animals learned to avoid the fence, spending most of the time away from it. However, we noted an apparent increase in exploratory behaviour in the night sessions, especially for the pumas, in that both males and females received shocks in the night sessions (Fig. 16.5), even though they were apparently avoiding the fence during the day. The same was the case for the female jaguar in the Sorocaba zoo.

The effect of the fence was evident also by the contrast in the behaviour exhibited by all animals, of both species, as soon as the fence was removed: immediately, animals proceeded to eagerly explore all that was previously fenced out. On several occasions, jaguars of both zoos displayed marking behaviour, at specific points of the enclosure. However, since we have no information on previous marking rates, we cannot link observed markings to the presence of the fence.

On several occasions, animals approached the fence cautiously and appeared to sense the electricity with the long whiskers (vibrissae), which function as tactile sensors (Kitchener 1991), backing away before receiving a shock.

After the initial shocks, all animals remained as distant as possible from the fence, most of the time apparently indifferent to it. The male and female jaguars from Sorocaba also exhibited what could be interpreted as displacement activities

(Ewer 1968; Kitchener 1991). On a few occasions, after intently looking in the direction of the fence (and/or the food), they would go directly to the tree trunk and sharpen their claws for several minutes. In another instance, they rolled on their backs, playfully.

Conclusion

Total exclusion of predators may be not practical because a predators' response to an electric fence is influenced by various factors, such as the animal's motivation and previous experience with fences. In our study area, cattle grazed spread out in the open range for quite some time before being confined into a smaller pasture area, which considerably increased their density. Optimal foraging theory predicts that predators should choose the most profitable prey (MacArthur and Pianka 1966). However, profitability may be influenced by a combination of search time, encounter rates and energetic costs of capture. During the wet season, when herbaceous fields and drainage vegetation habitats become extremely dense and difficult to move through, high search time and low encounter rates may make hunting in these habitats less profitable for jaguars. The option of a readily available prey item confined into areas of more profitable hunting grounds may be a strong enough motivation for jaguars to overcome the electric fence.

In our observations within the captive settings, we did not record a scraping behaviour (shown by wild jaguars) on the outside of the electric fence at points where the animal was apparently kept from entering the enclosure (Fig. 16.3). It is possible that conditions in captivity, especially easy food availability, mask (or dampen) what could be interpreted as motivation to overcome the fence. However, it is probably safe to surmise that the ease, with which animals in captivity learn to avoid electric fences after one or two electrical discharges, is also applicable to wild individuals. What is likely to differ, from our captive observations, is the drive to overcome the fence. This possibility underscores the need to consider the adaptive capabilities of the predators when planning the use of electric fencing.

According to the owners of the ranch (R. Jank, pers. comm.), the measures decreased in efficacy in the following order: (1) predation was reduced significantly when the herds were moved to open, short native grasses during the drought season, which provided ample visibility preventing the undetected approach of predators. With the flooding season and the cattle being confined in taller vegetation, predation increased considerably; (2) night patrolling with a tractor in a central corridor in between the fenced pastures holding the cattle at night was also effective in reducing predation; (3) the electric fence, as used by the ranch, encompassing a very long perimeter, likely deterred some individuals, but others managed to overcome the fences and succeeded in preying on the cattle; and (4) fireworks and lighting (as employed by the ranch), after habituation by the jaguars, were the least effective measures.

The opinion of the ranch owners, as the results of our study, highlights some of the problems associated with inadequate use of electric fences and other management alternatives in decreasing livestock depredation by large cats in the Pantanal. Nonetheless, we believe that the combined information from the wild and captive conditions is useful in developing solutions to human-jaguar conflict as it relates to livestock depredation.

Despite the costs incurred in fencing for conservation (Hayward and Kerley 2009), we view the use of electric fencing as a possible aid to reduce the conflict between large cats and ranchers. However, it is important to understand that there is seldom a single "right" preventive or control measure that will resolve the issue of livestock depredation. The efficacy of any method will depend on a combination of factors that range from the biology of the predator and its motivation, to the environmental characteristics of the area, to the size and species of the herd to be protected, to cultural factors of the local community, and to the motivation of the ranch owner and his/her willingness to deal with the problem. Thus, it is important that both researchers and ranch personnel work together to test the new methods that should comprise their toolbox of prevention and control methods.

In the Pantanal, for example, the inherent characteristics of the area make it difficult to effectively control the access of predators to cattle, and as long as this is an issue, it is likely that predation will occur, to a certain extent. Ranchers should focus on increasing their production potential, curtailing losses due to rudimentary herd management and poor husbandry practices, which can be more significant than jaguar depredation in many cases (Hoogesteijn et al. 1993). In addition, Cavalcanti and Gese (2010) illustrated the possible role native prey abundance could have on jaguar predation of cattle. Maintaining native prey populations in the Pantanal may not only help alleviate losses incurred from depredation, but may contribute in the efforts to conserve jaguars in the long term.

Acknowledgements We thank Teresa Bracher and Roberto Jank Jr. for the opportunity to work at the fazenda Santa Tereza. Isabel Villa-Lobos helped in many ways during the study. José Augusto Ferraz de Lima, director of the Pantanal National Park, allowed PC the time necessary to conduct the work at the ranch. Ronaldo Morato, director of CENAP/ICMBio, also provided extra flexibility in PC's office hours, at the writing stage. We thank Adauto Velloso Nunes, director of the Parque Zoológico Quinzinho de Barros, in Sorocaba, and Claudia Igayara, from the Parque Zoológico Municipal de Guarulhos, for authorisation to conduct the experiment with their animals. The personnel from both zoos provided invaluable help and were a pleasure to work with. Marcelo Cirne Lima, from Farmtech S.A. Produtos Veterinários (Porto Alegre, RS, Brazil) provided the equipment used in the zoo experiments and the work of one of their supervisors to help with the construction of the fence at the Sorocaba zoo. We thank A. Balbino, from Farmtech, for his competence and disposition to work with us.

References

Acorn, R.C. & Dorrance, M.J. (1994) An evaluation of anti-coyote electric fences. *Proceedings of the 16th Vertebrate Pest Conference* (eds. W.S. Halverson & A.C. Crabbe). (pp. 45–50). Published at University of California, Davis.

Amâncio, C.G., Crawshaw Jr., P.G., Tomás, W.M., Rodrigues, R.B. & Vergara da Silva, M. (2007) *Compreendendo populações locais para facilitar a conservação da onça-pintada no Pantanal, Brasil.* Anais do V Congresso Brasileiro de Unidades de Conservação, Foz do Iguaçu, PR.

Andelt, W.F. & Hopper, S.N. (2000) Livestock guard dogs reduce predation on domestic sheep in Colorado. *Journal of Range Management*, **53**, 259–267.
Azevedo, F.C.C. & Murray, D.L. (2007) Spatial organization and food habits of jaguars (*Panthera onca*) in a floodplain forest. *Biological Conservation*, **137**, 391–402.
Cavalcanti, S.M.C. (2009) *Pantanal Jaguar Project – Northern Corridor*. Activity Report, November 2008 to January 2009. Panthera, USA.
Cavalcanti S.M.C. & Gese, E.M. (2009) Spatial ecology and social interactions of jaguars (*Panthera onca*) in the southern Pantanal, Brazil. *Journal of Mammalogy*, **90**, 935–945.
Cavalcanti, S.M.C. & Gese, E.M. (2010) Kill rates and predation patterns of jaguars (*Panthera onca*) preying on livestock and native prey in the southern Pantanal, Brazil. *Journal of Mammalogy*, **91**, 722–736.
Cavalcanti, S.M,. Marchini,C.S., Zimmermann, A., Gese, E.M. & Macdonald, .D.W. (2010) Jaguars, livestock, and people in Brazil: realities and perceptions behind the conflict. *Biology and Conservation of Wild Felids*. (eds. Macdonald, D. W. & Loveridge, A.J.) Oxford University Press, Oxford.
Cavalcanti, S.M.C. (2006) Aspects of livestock depredation by jaguars in the southern Pantanal, Brazil. *Manejo e Conservação de Carnívoros Neotropicais*. (eds. Morato, R.G., Rodrigues, F.H.G., Eizirik, E., Mangini, P.R., Azevedo, F.C.C. & Marinho-Filho, J.) IBAMA/CENAP, São Paulo.
Crawshaw Jr., P.G. 2003. Uma perspectiva sobre a depredação de animais domésticos por grandes felinos no Brasil. *Natureza & Conservação*, **1**, 13–15.
Crawshaw Jr., P.G. & Quigley, H.B. (1984) Ecologia do jaguar ou onça pintada no Pantanal. Final Report to the Brazilian Institute for Forest Development-IBDF, Brasília, 110pp.
Crawshaw Jr., P.G. & Quigley, H.B. (1991) Jaguar spacing, activity, and habitat use in a seasonally flooded environment in Brazil. *Journal of Zoology*, **223**, 357–370.
Crawshaw Jr., P.G. & Quigley. (2002) Hábitos alimentícios del jaguar y el puma en el Pantanal, Brasil, con implicaciones para su manejo y conservación. *Jaguars en el Nuevo Milenio*. (eds. Medellín, R.A., Equihua, C., Chetkiewicz, C.L.B., Crawshaw Jr., P.G., Rabinowitz, A., Redford, K.H., Robinson, J.G., Sanderson, E.W.& Taber, A.B.) Universidad Nacional Autónoma de México/Wildlife Conservation Society, México DF.
Ewer, R.F. (1968) *Ethology of mammals*. Logo Press Limited, London.
Fritts, S.H., Paul, W.J., Mech, L.D. & Scott, D.P. (1992) *Trends and management of wolf-livestock conflicts in Minnesota*. U.S. Fish and Wildlife Service, Resource Publication 181, Washington, D.C.
Hayward, M.W. & Kerley, G.I.H. (2009) Fencing for conservation: Restriction of evolutionary potential or a riposte to threatening processes? *Biological Conservation*, **142**, 1–13.
Hoogesteijn, R., Hoogesteijn, A. & Mondolfi, E. (1993) Jaguar predation and conservation: cattle mortality caused by felines on three ranches in the Venezuelan Llanos. *Symposium of the Zoological Society of London*, **65**, 391–407.
IBGE. (2009) *Produção da Pecuária Municipal*. Volume 36, 2008, Brasil. Instituto Brasileiro de Geografia e Estatística, Rio de Janeiro.
Kitchener, A. (1991) *The Natural History of the Wild Cats*. Cornell University Press, Ithaca, New York.
Koehler, A.E., Marsh, R.E. & Salmon, T.P. (1990) Frightening methods and devices/stimuli to prevent mammal damage – a review. Proceedings of the 14th Vertebrate Pest Conference. (eds. Davis, L.R.& Marsh, R.E.). University of California, Davis.
Levin, M. (2002) How to prevent damage from large predators with electric fences. *Carnivore Damage Prevention News*, **5**, 5–8.
Linnell, J.D., Smith, M.E., Odden, J., Kaczensky, P. & Swenson, J.E. (1996) Strategies for the reduction of carnivore - livestock conflicts : a review. NINA Oppdragsmelding **443**, 1–188.
MacArthur, R.H. & Pianka, E.R. (1966) On optimal use of a patchy environment. *The American Naturalist*, **100**, 603–609.
McAtee, W.L. (1939) The electric fence in wildlife management. *Journal of Wildlife Management*, **3**, 1–13.

Mertens, A., Promberger, C. & Gheorge, C. (2002) Testing and implementing the use of electric fences for night corrals in Romania. *Carnivore Damage Prevention News*, **5**, 2–5.

Mizutani, F. (1993) Home range of leopards and their impact on livestock on Kenyan ranches. *Symposium of the Zoological Society of London*, **65**, 425–439.

Sanderson, E.W., Redford, K.H., Chetkiewicz, C.L.B., Medellin, R.A., Rabinowitz, A., Robinson, J.G. & Taber, A. (2002) Planning to save a species: the jaguar as a model. *Conservation Biology*, **16**, 58–72.

Schaller, G.B. & Crawshaw, P.G. Jr. (1980) Movement patterns of jaguar. *Biotropica*, **12**, 161–168.

Schaller, G.B. (1983) Mammals and their biomass on a Brazilian ranch. *Arquivos de Zoologia, Museu de Zoologia da Universidade de São Paulo*, **31**, 1–36.

Scognamillo, D., Maxit, I., Sunquist, M. & Farrell, L. (2002) Ecología del jaguar y el problema de la depredación de ganado en un hato de Los Llanos Venezolanos. *El Jaguar en el Nuevo Milenio*. (eds. Medellin, R.A., Equihua, C., Chetkiewicz, C.L.B., Crawshaw, P.G. Jr., Rabinowitz, A., Redford, K.H., Robinson, J.G., Sanderson, E.W.& Taber, A.B.) Universidad Nacional Autónoma de México/Wildlife Conservation Society, México DF.

Shivik, J.A. (2000) Improving methods for resolving conflict: aversive and disruptive stimulus applications for managing predation. *Proceedings and Agenda for Defenders of Wildlife's Carnivores*. November 2000, Denver.

Shivik, J.A. (2006) Tools for the edge: what's new for conserving carnivores. *BioScience*, **56**, 253–259.

Shivik, J.A., Treves, A. & Callahan, P. (2003) Nonlethal techniques for managing predation: primary and secondary repellents. *Conservation Biology*, **17**, 1531–1537.

Srivastav, A. (1997) Livestock predation by Gir lions and ecodevelopment. *Tigerpaper*, **24**, 1–5.

Index

M.J. Somers and M.W. Hayward (eds.), *Fencing for Conservation: Restriction of Evolutionary Potential or a Riposte to Threatening Processes?*,
DOI 10.1007/978-1-4614-0902-1, © Springer Science+Business Media, LLC 2012

Y

Z